Human Factors for Engineers

Human Factors for Engineers

Edited by
Carl Sandom and Roger S. Harvey

The Institution of Electrical Engineers

Published by: The Institution of Electrical Engineers, London,
United Kingdom

The Institution of Electrical Engineers,
Michael Faraday House,
Six Hills Way, Stevenage,
Herts., SG1 2AY, United Kingdom

British Library Cataloguing in Publication Data

Human factors for engineers
1.Human engineering
I.Sandom, Carl II.Harvey, Roger
620.8'2

ISBN 0 86341 329 3

Typeset in India by Newgen Imaging Systems (P) Ltd., Chennai
Printed in the UK by MPG Books Limited, Bodmin, Cornwall

Contents

Foreword

I am very pleased to have been asked to contribute the foreword of this book, especially to make a plea for the importance of integrating ergonomics into the systems development process and communicating the potential added value to systems engineers and project managers. Just for the record, therefore, I would like to emphasise that as far as the Ergonomics Society and the International Ergonomics Association are concerned, Ergonomics and Human Factors are synonymous terms for the same subject, with ergonomics being the UK term and human factors the US one.

Ergonomics (or HF) is the scientific discipline concerned with the fundamental understanding of interactions among humans and other elements of a system, and the application of appropriate methods, theory and data to improve human well-being and overall system performance. Ergonomics is therefore a systems-oriented discipline and practising ergonomists must have a broad understanding of its full scope. That is, ergonomics espouses a holistic approach in which considerations of physical, cognitive, social, organisational, environmental and other relevant factors are taken into account.

I believe that ergonomists and engineers need to work together from the earliest stages of projects, and co-operate to produce the best possible solutions and avoid some of the horrendous failures we seem to be experiencing at the moment, when the human and the system seem to be incompatible. I hope that this book can make its mark in helping ergonomists to understand the language and role of engineers and vice versa.

And finally…. The great playwright George Bernard Shaw said that 'the reasonable man adapts himself to the world; the unreasonable one persists in trying to adapt the world to himself. Therefore, all progress depends upon the unreasonable man.' Ergonomists must be 'unreasonable' in the sense that they will strive vigorously and unceasingly to support a meaningful and productive relationship between systems and humans so we beg forgiveness if sometimes we appear unreasonable; we are only seeking progress.

Meg Galley
President of The Ergonomics Society

Preface

I am not a human factors expert. Nor can I even claim to be an able practitioner. I am an engineer who, somewhat belatedly in my career, has developed a purposeful interest in human factors.

This drive did not come from a desire for broader knowledge, though human factors as a subject is surely compelling enough to merit attention by the enquiring mind. It came from a progressive recognition that, without an appreciation of human factors, I fall short of being the rounded professional engineer I have always sought to be.

I started my career as a microwave engineer. Swept along by the time and tide of an ever-changing technical and business scene, I evolved into a software engineer. But as more time passed, and as knowledge, experience and skills grew, my identity as an engineer began to attract a new descriptor. With no intent, I had metamorphosed into a systems engineer.

There was no definitive rite of passage to reach this state. It was an accumulation of conditioning project experiences that turned theory into practice, and knowledge into wisdom. Throughout, I cloaked myself in a misguided security that my engineering world was largely confined to the deterministic, to functions that could be objectively enshrined in equations, to laws of physics that described with rigour the entities I had to deal with. To me, it was a world largely of the inanimate.

Much of this experience was against a backdrop of technical miracles, buoyant trade and strong supplier influence; where unfettered technological ingenuity drove the market, and users were expected to adapt to meet the advances that science and engineering threw at them. It was an era destined not to last.

Of course, from my world of microwave systems, of software systems and of maturing systems engineering, I had heard of human factors. Yet it seemed a world apart, scarcely an engineering-related discipline. People, teams and social systems were an unfortunate, if unavoidable, dimension of the environment my technical creations operated in.

Nor was I alone in these views. Recently asked by the IEE to write a personal view on systems and engineering, I light-heartedly, but truthfully, explained that I have been paid to drink in bars across the world with some of the best systems engineers around. This international debating arena is where I honed a cutting edge to my systems engineering. Above all others, I was struck by one thing: an engineering legacy of thinking about systems as being entirely composed of the inanimate.

Most engineers still view humans as an adjunct to equipment, blind even to their saving compensation for residual design shortcomings. As an engineering point of view for analysing or synthesising solutions, this equipment-centred view has validity – but only once humans have been eliminated as contributing elements within some defined boundary of creativity. It is a view that can pre-empt a trade-off in which human characteristics could have contributed to a more effective solution; where a person or team, rather than inanimate elements, would on balance be a better, alternative contributor to overall system properties.

As elements within a boundary of creativity, operators bring intelligence to systems; the potential for real-time response to specific circumstance; adaptation to changing or unforeseen need; a natural optimisation of services delivered; and, when society requires people to integrally exercise control of a system, they legitimise system functionality. In the fields of transportation, medicine, defence and finance the evidence abounds.

Nevertheless, humans are seen to exhibit a distinctive and perplexing range of 'implementation constraints'. We all have first hand familiarity with the intrinsic limitations of humans. Indeed, the public might be forgiven for assuming that system failure is synonymous with human weaknesses – with driver or pilot error, with disregard for procedure, even with corporate mendacity.

But where truly does the accountability for such failure lie: with the fallible operator; the misjudgements in allocation of functional responsibility in equipment-centred designs; the failed analysis of emergent complexity in human–equipment interaction? Rightly, maturing public awareness and legal enquiry focuses evermore on these last two causes. At their heart lies a crucial relationship – a mutual understanding – between engineers and human factors specialists.

Both of these groups of professionals must therefore be open to, and capable of performing, trade-off between the functionality and behaviour of multiple, candidate engineering implementations and humans.

This trade-off is not of simple alternatives, functional like for functional like. The respective complexities, characteristics and behaviours of the animate and inanimate do not offer such symmetry. It requires a crafted, intimate blending of humans with engineered artefacts – human-technology co-operation rather than human–technology interaction; a proactive and enriching fusion rather than a reactive accommodation of the dissimilar.

Looking outward from a system-of-interest there lies an operational environment, composed notionally of an aggregation of systems, each potentially comprising humans with a stake in the services delivered by the system-of-interest. As external system actors, their distinctive and complex characteristics compound and progressively emerge to form group and social phenomena. Amongst their ranks one finds, individually and communally, the system beneficiaries. Their needs dominate the definition of required system services and ultimately they are the arbiters of solution acceptability.

Essentially, the well-integrated system is the product of the well-integrated team, in which all members empathise and contribute to an holistic view throughout the system life cycle. The text that follows may thus be seen as a catalyst for multi-disciplinary, integrated teamwork; for co-operating engineers and human factors professionals to develop a mutual understanding and a mutual recognition of their respective contributions.

In this manner, they combine to address two primary concerns. One is an equitable synthesis of overall system properties from palpably dissimilar candidate elements, with attendant concerns for micro-scale usability in the interfaces between operator and inanimate equipment. The other addresses the macro-scale operation of this combination of system elements to form an optimised socio-technical worksystem that delivers agreed services into its environment of use.

The ensuing chapters should dispel any misguided perception in the engineering mind that human factors is a discipline of heuristics. It is a science of structured and disciplined analysis of humans – as individuals, as teams, and as a society. True, the organic complexity of humans encourages a more experimental stance in order to gain understanding, and system life cycle models need accordingly to accommodate a greater degree of empiricism. No bad thing, many would say, as engineers strive to fathom the intricacies of networked architectures, emergent risks of new technology and unprecedented level of complexity.

The identification of pressing need and its timely fulfilment lie at the heart of today's market-led approach to applying technology to business opportunity. Applying these criteria to this book, the need is clear-cut, the timing is ideal, and this delivered solution speaks to the engineering community in a relevant and meaningful way. Its contributors are to be complimented on reaching out to their engineering colleagues in this manner.

So I commend it to each and every engineer who seeks to place his or her particular engineering contribution in a wider, richer and more relevant context. Your engineering leads to systems that are created by humans, for the benefit of humans. Typically they draw on the capabilities of humans, and they will certainly need to respond to the needs and constraints of humans. Without an appreciation of human factors, your endeavours may well fall short of the mark, however good the technical insight and creativity that lie behind them.

Beyond this, I also commend this text to human factors specialists who, by assimilating its presentation of familiar knowledge, may better appreciate how to meaningfully communicate with the diversity of engineering team members with whom they need to interact.

Stuart Arnold
QinetiQ Fellow and Systems Engineering Head of Profession
QinetiQ Ltd

Author biographies

Co-editor and Chapter 14

Carl Sandom BEng MSc PhD CEng MIEE MErgS
Consultant, iSys Integrity, Dorset, UK

Dr Carl Sandom is an independent consultant with iSys Integrity, specialising in safety management as well as safety and human factors assessments for high-integrity systems. Prior to entering consultancy, he spent over 20 years as an engineer and manager in the Aerospace and Defence sectors with the Royal Air Force and within industry. He is currently serving on the Council for the Institute of Electrical Engineers and is Chairman of the Human Factors Engineering Professional Network Executive Committee. He has published and presented numerous human factors and safety papers and he is a Chartered Engineer and a Registered Member of the Ergonomics Society.

Co-editor and Chapter 1

Roger S. Harvey BSc C Psychol AFBPsS MIEE FErgS FRSA
Principal Psychologist, QinetiQ Centre for Human Sciences, Farnborough, UK

Roger S. Harvey is a Chartered Psychologist at QinetiQ Centre for Human Sciences, and has some 30 years Human Factors research and development experience. His main specialities are project support and assessment, human–computer interface (HCI) design and evaluation, and the application of the UK Human Factors Integration/Systems Engineering initiative for equipment acquisition. In addition, he has previously served in the Intelligence Corps, as a NATO-qualified Imagery Analyst. He is a former Chairman of Council of the Ergonomics Society and was the 1999 recipient of the Branton Medal, awarded by the Ergonomics Society for services to ergonomics.

Chapter 2

Michael A. Tainsh BSc PhD (Wales) C Psychol FErgS Eur Erg

Michael Tainsh is a Principal Consultant at the QinetiQ, Centre for Human Sciences with over 30 years experience in psychology and ergonomics. In recent years his main scientific interests have been associated with computer-based systems for the Royal Navy. This has covered requirements, acceptance and investment appraisal. He has published in academic journals and presented at many conferences. He has been a member of the Council for the Ergonomics Society, and chaired the Working Group for Ergonomics in Schools (see www.ergonomics4schools.com).

Chapter 3

Jan Noyes BSc PhD Cert Ed FErgS MIEE AFBPsS
University of Bristol, UK

Dr Jan Noyes is a Reader in Psychology at the University of Bristol. She is a Fellow of the Ergonomics Society and Member of the IEE. In 1999 she was awarded the Otto Edholm medal for her contribution to ergonomics. She has authored over 120 publications including six books, and was awarded the IEE Informatics Premium Award in 1998. She was also Chair of the 1999 and 2001 IEE People In Control conferences, and is on the Editorial Board of the journal *Ergonomics*.

Kate Garland BSc MSc PhD

Dr Kate Garland completed her BSc (Hons) in Psychology through the Open University, followed by an MSc in Research Methods in Psychology and a PhD (Cognitive Psychology) from the University of Bristol. Research work encompasses two main areas: cognitive psychology (particular interests in long-term memory, semantic memory organisation, metacognition and cognitive styles); and applied psychology (differences in cognitive processes between hard copy and computer material). After posts at the Universities of Bristol and Plymouth, she is now a lecturer at the School of Psychology, University of Leicester.

Daniel Bruneau BSc GRADRAeS

Daniel Bruneau completed his undergraduate degree in Psychology at Goldsmiths, University of London, before taking up a one-year research position at University College London (UCL) investigating the use of eye tracking methods within the human–computer interaction domain. He is currently in the first year of his PhD at the University of Bristol (funded by the ESRC) working in the field of Aviation Human Factors. His thesis is examining optimal ways of displaying predictive information on the flight decks of civil aircraft.

Chapter 4

Jan Noyes BSc PhD Cert Ed FErgS MIEE AFBPsS
University of Bristol, UK

See Chapter 3.

Chapter 5

Leslie K. Ainsworth PhD BSc FErgS Eur Erg
Director and Principal Consultant Synergy Consultants Ltd, Yewbarrow, Grange-over-Sands, Cumbria, UK

Les Ainsworth has been involved in ergonomics for over 25 years and has wide experience in the practical application of ergonomics and psychology to industrial inspection and the assessment of mental and perceptual workload. He jointly authored the definitive text on task analysis. He established Synergy as a specialist ergonomics consultancy in 1988 to provide services to a wide range of safety-critical industries. He now provides advice on the integration of ergonomics in Safety Cases, in design and in human factors assessments.

Chapter 6

Erik Hollnagel MSc PhD
Full Professor of Human–Machine Interaction, Linköping University (Sweden)

Erik Hollnagel is an internationally recognised specialist in the fields of system safety, human reliability analysis, cognitive systems engineering, and intelligent man–machine systems. He has worked in a number of industries and research institutes before joining Linköping University in 1999. He is joint Editor-in-Chief of the *International Journal of Cognition, Technology & Work* as well as the author of more than 300 publications including eight books, articles from recognised journals, conference papers, and reports.

Chapter 7

Sidney Dekker PhD

Dr Dekker is Associate Professor at the Linköping Institute of Technology, Sweden, and Director of Studies Centre for Human Factors in Aviation. He received an MA in organisational psychology from the University of Nijmegen and an MA in experimental psychology from Leiden University, both in the Netherlands. He gained his PhD in Cognitive Systems Engineering from the Ohio State University, USA.

He has worked for the Public Transport Cooperation in Melbourne, Australia; the Massey University School of Aviation, New Zealand; and British Aerospace, UK. His specialities and research interests are human error, accident investigations, field studies, representation design and automation. He is a pilot, type-trained on the DC-9 and Airbus A340. His latest book is 'The Field Guide to Human Error Investigations' (Cranfield University Press, 2002).

Chapter 8

Dr David Embrey

Dr David Embrey is the Managing Director of Human Reliability. He originally trained as a physicist and holds degrees in physics, ergonomics and applied psychology from the University of Birmingham and the University of Aston in England. After receiving his PhD in human factors, he has subsequently worked for 25 years in the area of human reliability in systems, where he has published more than a hundred papers and reports and co-authored two books. His main technical interest has been the assessment and improvement of human performance in systems. He has worked extensively in the petrochemical, nuclear power, rail transport, aerospace systems and marine sectors in Europe, the Far East, USA and South America. His consultancy experience has included work in the USA, Japan and Singapore as well as several European countries. He has worked for a number of large multinationals in the petrochemical sector, and has written a definitive text 'Guidelines for Preventing Human Error in Process Safety'. In the area of training, he has developed many courses in human factors and human error prevention which have been given to companies in Japan, South America, the Middle East and Europe.

Chapter 9

John Wood BTech MDes(RCA) FErgS Eur Erg

John Wood is Managing Director of CCD Design & Ergonomics Ltd, a leading UK ergonomics consultancy. The design of control rooms has been one of John's specialities. He is Chairman of the ISO Working Group preparing an International Standard on control room ergonomics and also chairs the BSI Panel shadowing this work. He is a Fellow of the Ergonomics Society and has received the Otto Edholm and William Floyd Awards from the Society for his contributions to ergonomics.

Chapter 10

Professor Robert D. Macredie

Robert Macredie has over 10 years of research experience, and has worked with a range of organisations, ranging from large, blue-chip companies, through small businesses, to government agencies. Professor Macredie's key interest lies in the way in which people and organisations use technology. Educated at the University of Hull (BSc and PhD), he is now Professor of Interactive Systems and Head of the Department of Information Systems and Computing, Brunel University. He has done work on a range of issues associated with people, technology and organisations and has published over 150 research contributions in these areas.

Jane Coughlan BA MRes PhD
Brunel University

Dr Jane Coughlan is a Research Fellow at the Department of Information Systems and Computing at Brunel University. She holds a BA in Psychology, MRes in Informatics (both from Manchester University) and a PhD in Information Systems (Brunel University). Her research interests include the facilitation of communication and the exchange of stakeholder requirements within various contexts, ranging from the interpersonal with users and designers, to organisational and company wide communication scenarios.

Chapter 11

Martin Maguire PhD

Martin has a background in both computer studies and ergonomics. Following computer graphics work at Leicester University, he moved to Loughborough University as a member of staff at HUSAT, now part of the Ergonomics and Safety Research Institute (ESRI). His main interests are in the design of interactive systems, pervasive computing, and consumer products to be usable by the general public, including older and disabled people. He has developed a framework for user requirements analysis as part of the EU Respect project. His research and consultancy work has covered usability aspects of kiosk systems, advanced services for bank machines, new digital TV and web services, household products, mobile phones and environmental systems.

Chapter 12

Iain S. MacLeod BA MSc C Psychol C Eng FErgS AFBPsS MBCS Eur Erg
Principal Consultant, Atkins, Bristol

Iain has been involved in the Aerospace industry for over 40 years. At various times he has worked as an aviator, engineer, and Human Factors consultant. As a military aviator, Iain accrued almost 8000 flying hours worldwide and is a graduate of the Royal Air Force GD Aerosystems Course. Since leaving the military he has worked in industry on many aerospace/defence projects and bodies of research. Projects include Airborne STandOff Radar (ASTOR), Nimrod MRA4, Merlin Mk1 Helicopter, Canberra PR9, Apache WAH-64 Ground Support System, and European

Space Agency Columbus Research. Iain has published over 60 papers in the fields of human factors, systems engineering, cognitive ergonomics, safety, and engineering psychology.

Chapter 13

Ed Marshall BA FErgS C Psychol AFBPsS Eur Erg
Director and Principal Consultant Synergy Consultants Ltd, Yewbarrow, Grange-over-Sands, Cumbria, UK

Ed Marshall is an ergonomist with nearly 30 years experience of applying ergonomics to the control of complex human–machine systems. He has specialised in the study of human issues in the nuclear industry where he has undertaken simulator studies, provided formal human factors support to safety cases and assessed system interfaces. Previously he worked for the CEGB and the OECD Halden Reactor Project. He is a past Chairman of Council of the Ergonomics Society and is visiting lecturer in Ergonomics at University College London.

Introduction

The inspiration for this textbook was provided by many of the delegates attending the first IEE residential course 'Human Factors for Engineers'. The initial success of this course has led to it becoming an established event and it is now held biennially, alternating with the IEE Conference Series 'People in Control'. Both of these events are organised by the Executive Team of the IEE Human Factors Engineering Professional Network, for which the editors of this textbook are Chair and Co-Chair respectively. The delegates at the residential course provided feedback that suggested an introductory Human Factors textbook for systems engineers would be a valuable adjunct to the course; this book is the result of those suggestions and is planned to be amongst the supplementary reference texts for the course.

As the title suggests, this book has been written primarily, but not exclusively, for engineers of all backgrounds and disciplines. It provides a practical introduction to Human Factors targeted at a range of specialisations, including practising systems engineers, engineering managers, project managers, safety engineers and also engineering students from undergraduate to postgraduate level. Following the publication, towards the end of 2002, of the International Standard on Systems Engineering (BS ISO 15288) we believe that this book will provide engineers and other equipment acquisition stakeholders with practical advice and examples for the successful integration of human factors issues into the Systems Engineering life cycle.

We have endeavoured to provide an underpinning of theoretical knowledge together with some pragmatic design guidance, and each of our authors is an acknowledged national or international expert in their field. Thus the coverage of the chapters in the book includes information about the over-arching Human Factors Integration methodology, a review of human skills, capabilities and limitations, and of key human factors methods and tools. Other important topics include task analysis, automation, human error and human reliability. Detailed design considerations are dealt with in chapters covering control room design, interface design and usability. The book concludes with chapters covering verification and validation, simulation and human factors aspects of safety assessment. After much thought we have decided to exclude coverage of the detailed consideration of anthropometry and anthropometric data within this book, principally because a number of other texts cover these specialist topics in detail. In each chapter there is broad coverage of the appropriate subject matter and, wherever possible, worked examples together with exercises or questions and suggestions for further reading.

Most importantly, we believe that this textbook comprehensively demolishes some of the myths about Human Factors; the myths are that it is about common sense, that it is expensive to include in project plans, and that it is time consuming. None of these could be further from reality, and the true situation is the antithesis of each of these myths. If there is a common thread throughout much of this book, and so eloquently argued in the Preface, it is the debunking of these myths.

We would like to thank the President of the Ergonomics Society for providing the Foreword; Stuart Arnold, the Editor of the Systems Engineering International Standard (BS ISO 15288), for providing our Preface; and Dr Robin Mellors-Bourne and Sarah Kramer of IEE Publishing, who have each been so enthusiastic about this project from the start. Finally, and most importantly, our largest measure of gratitude must go to our authors, each of whom has taken our frequent e-mail and telephone messages with great humour and understanding.

Dr Carl Sandom
Roger S. Harvey

Chapter 1

Human factors and cost benefits

Roger S. Harvey

1.1 Introduction

This chapter can be taken to be the overture to an orchestral piece. The chapters that follow are the movements of the work; each movement or chapter has a particular theme or subject that is a part of the overall piece known as 'human factors'. The first group of chapters deals with important techniques, including Task Analysis, a review of HF Methods and Tools, Automation, Human Error and Human Reliability. This is followed by a trio of topics dealing with important design areas, Control Room Design, Interface Design and Usability. Finally the third group deals with Verification and Validation, Simulation and Safety Assessment.

This opening chapter sets the scene for the reader by examining the subject matter coverage of human factors and its close cousin, ergonomics. It then introduces and defines the concept of 'user-centred design' and, most importantly, places human factors within a far broader context. This context is systems engineering and the goal of truly integrated systems, in which users, equipment and the operating environment are appropriately matched for optimal, safe and effective use and performance.

Perhaps not surprisingly, examples of systems in which this goal of true integration has been met tend, almost by definition, to go relatively unnoticed by users and others. This frequently gives rise to the completely erroneous belief that 'good human factors' is only the application of common sense. On the other hand, there are systems that are poorly integrated, or complicated to use, or just badly designed, or have complex instructions or that demand excessively lengthy training regimes, or are combinations of all these. Sometimes users (and maintainers) may literally have to grapple with the system in order to make it work. Unfortunately it is these sorts of systems that tend to be remembered, for all the wrong reasons, by users, designers and other stakeholders.

Finally, this chapter examines the issue of the relative costs and pay-offs of human factors. As can be seen above, some of the problems that poorly integrated systems force upon users and others can lead to problems such as lengthier training, excessively complex operating procedures including frequent 'work-arounds', more demanding maintenance schedules, or error-prone performance by users and maintainers alike. Additionally during the system design and development phases themselves it may appear tempting to reduce HF budgets in various ways, perhaps by shifting scarce funding or staff resources to other activities. The issue then becomes one of Balance of Investment. Of course, even for those well-integrated systems that we tend to take for granted, the human factors design and development activities themselves, that are scheduled within the system acquisition programme, come with their own costs and pay-offs. In each of these differing circumstances it is extremely instructive to examine the manner in which system acquisition can be impacted in both qualitative and quantitative terms by these costs and pay-offs. This introductory chapter will deal predominantly with the quantitative issues, including a review of the findings of a research study commissioned to explore the impact of differential Human Factors Balance of Investment upon system acquisition, whilst some of the qualitative issues will be touched upon in Chapter 2.

1.2 Human factors or ergonomics?

It is helpful to review some of the history associated with the use of the term 'human factors', and set this alongside the somewhat analogous term 'ergonomics'. Each has been influenced by a slightly different set of circumstances of usage and context. In terms of the origin of the word 'ergonomics' it is particularly interesting to note that for many years it had been assumed to have been derived first by K. F. H. (Hywel) Murrell (1908–1984) in December 1949. The prompting for his derivation can be traced back to an Admiralty meeting, convened in central London during July 1949 in order to consider the formation of a 'Human Performance Group'. This Group was intended to act as a discussion forum to bring together Government and University scientists and others engaged in the general areas of operational research, work science and applied psychology. Before 1949 had ended it had reconvened to become the Ergonomics Research Society, the first such society in the world, having been given the name that Murrell had doodled on a piece of paper in December of that year. The original sheet of paper with those doodles shows that he formed the word from the Ancient Greek words *ergo* (work) and *nomos* (laws), but it also shows that he pondered on two rather more unlikely alternative words, anthroponomics and ergatonomics. An image of the sheet has recently been published in a newsletter of the International Ergonomics Association [1]. The sheet itself is preserved in the archives of the Society, which some years later dropped the term 'Research' to become known as the Ergonomics Society. However there is now evidence of a far earlier citation of the word, in 1857 in a Polish newspaper, although this initial use does not seem to have survived until Murrell independently derived the word 92 years later.

Broadly speaking, the usage of these terms, human factors and ergonomics, developed along a transatlantic divide with human factors being used more widely in North America, whilst ergonomics took root in a predominantly European context.

In the USA, the term broadened out with human factors engineering and engineering psychology entering the language. But whilst human factors and ergonomics may have developed on opposite sides of the Atlantic they did have one very important element in common in that each was given a considerable impetus by the Second World War. With the outbreak of war it soon became clear that there needed to be significant investment of financial, intellectual and engineering resources in order to improve many diverse areas contributing to the war effort, including pilot training, personnel selection and target tracking performance. It had become immediately obvious that serious performance shortfalls existed with some manned equipment and both human factors and ergonomics were given a great boost when efforts to address these and many other problems brought together a host of professional specialisms, including psychologists, physiologists, and engineers. This impetus continued on both sides of the Atlantic in the post-war years with the formation of a number of defence-related research laboratories devoted to human factors, or ergonomics, research and development, and several others which were established within labour-intensive industries such as steel and coal-mining. Some more of this history is recounted in Chapter 2.

It is also interesting to note that whilst the USA equivalent of the Ergonomics (Research) Society was formed in 1957 as the Human Factors Society, it too subsequently underwent a name change when in recent years it became known as the Human Factors and Ergonomics Society. By this means it attempted to rationalise and broaden its subject matter and coverage within North America. The gradual eradication of this dichotomy has been encouraged by the IEA with their adoption, in 2000, of the following definition of the discipline of ergonomics.

> The Discipline of Ergonomics.
>
> Ergonomics (or human factors) is the scientific discipline concerned with the understanding of interaction amongst human and other elements of a system, and the profession that applies theory, principles, data and methods to design in order to optimise human well-being and overall system performance. Ergonomists contribute to the design and evaluation of tasks, jobs, products, environments and systems in order to make them compatible with the needs, abilities and limitations of people. [2]

There are two important points to note in the IEA agreed definition. First, note that ergonomics and human factors are defined to be synonymous, even though some residual national differences of usage will undoubtedly continue for some years yet. Second, and most importantly, it can be seen that the definition includes a statement that humans are amongst the several elements that make up a system. This point is expanded upon in the next section, and in Chapter 2.

1.3 Human-centred design and systems engineering

The discipline of ergonomics or human factors places humans as the centre of attention in system development. Systems cannot be considered to be truly integrated without the appropriate matching of users (not forgetting maintainers), the technology or equipment they will use, and the environment within which the equipment will be operated. Taking people into account in this way, by means of human- or user-centred design, has long been a cardinal principle for human factors and this is now enshrined within the national and international Standard BS EN ISO 13407 'Human-centred design processes for interactive systems' [3].

This Standard provides guidance on the human-centred design activities that take place through the life cycle of interactive systems. Although the Standard was originally written with special reference to computer-based systems, it is readily applied to any other type of system. In Clause 4 it stresses that making interactive systems human-centred brings substantial benefits by increasing usability and such highly usable systems:

- are easier to understand and use, thus reducing training and support costs;
- improve user satisfaction and reduce discomfort and stress;
- improve user productivity and operational efficiency of organisations;
- improve product quality, and provide a competitive advantage.

Arising from this, in Clause 5 the Standard lays down four key principles for the human-centred approach to design:

- encourage the active involvement of users in design, and clearly understand the user and task requirements;
- establish the appropriate allocation of functions between users and technology;
- iterate design solutions;
- adopt a multi-disciplinary approach to system design.

These principles, more than any others, underpin the foundations of human factors and ergonomics. They represent the key tenets of the science, and human factors tools and techniques such as task analysis (see also Chapters 5 and 8), workload assessment, and job and team design can each be linked to one or more of those tenets.

However 'twas not ever thus! As will be seen in Chapter 2 there was a time, some 40 or so years ago, when the opportunities and prospects for technology began to offer much, especially in the defence sector. But the benefits that could be delivered to users seemed to fall short of these expectations. It soon became clear that there was an ineffective match between the technology, the users and their operating environment, and the methodology known as MANpower and PeRsonnel INTegration (MANPRINT) was introduced within the US Department of Defense as a means to assist in achieving the most effective match. Subsequently adopted within the UK Ministry of Defence, and then revised and re-named Human Factors Integration (HFI) for broader adoption in the defence and civil sectors of industry, these two programmes have done much to encourage the adoption of the human-centred approach within complex equipment design.

The Standard BS EN ISO 13407, is derived from, and links to, what can be termed a most important 'umbrella' Standard. This is the recently published Systems Engineering Standard BS ISO/IEC 15288 [4]. This Standard provides a common series of processes and a unifying framework covering the life cycle of all systems from their conception through to retirement and disposal. Most importantly the framework can be 'tuned' appropriately to suit any organisation's particular needs in terms of its overall purposes and the outcomes that it is aiming to achieve. Within BS IS EN 15288 the varied life cycle processes are described in terms of four types of process groups and it is from these that processes can be tailored to suit the individual needs of organisations, their structures, functions and stakeholders. Many organisations have derived their own life-cycle stages which themselves derive from the generic life-cycle stages tabulated in 15288.

Together with 15288 and 13407 there are two additional documents and collectively these form an important 'quartet' of related standards:

- BS ISO/IEC 15288
- ISO PAS 18152
- BS EN ISO 13407
- BS ISO/IEC 15504

As noted above, 15288 provides the total spectrum of processes covering the life cycle of systems. From this framework 13407 has been derived so as to provide guidance on the human-centred design activities that take place throughout the life cycle of interactive systems. The third document, ISO PAS (Publicly Available Specification) 18152 [5], is intended for use in process risk assessments (also known as Capability Maturity Modelling) and it has been carefully structured to show the links between 13407 and the fourth document BS ISO/IEC 15504 [6]. In this way we have *Human Factors* processes (as contained within the PAS) expressed in a manner completely analogous with *Software* processes (15504) and all are derived from the over-arching, common, framework of 15288.

It is important to note that within the wider context of the engineering framework of life-cycle processes HFI can be considered as only one of several related, methodologies contributing to system acquisition; each is related insofar as they contribute to the processes of the system life cycle. Thus many engineers will be familiar with the through-life management programmes known as Integrated Logistics Support (ILS), intended to provide in-service support at the optimum whole life cost, and Availability, Reliability and Maintainability (AR + M). These various methodologies of HFI, ILS and AR + M inter-relate strongly and each has a complementary role to play within system development. Scheduled and resourced appropriately they vividly demonstrate the seamlessness of operability, maintainability and availability throughout the system life cycle. Thus, for example, operability does not end when maintainability and/or availability starts or vice versa because each contributes in various ways to the evolution of system design. A good example of the manner in which they seamlessly contribute to one another can be found with the ILS Use Case Study. Amongst other things this may contain user role and activity descriptions that can also be used within the HFI methodology to assist with the evolution and development of Target Audience Descriptions (TADs). The TAD is the means by which a picture of the skills and capabilities, and the associated training regimes, can be put together for design and other engineers.

1.4 Human factors costs and pay-offs

What is meant by costs? What is meant by pay-offs? As a starting point readers are directed to the excellent review of cost/benefit analyses by Sage [7]. It is also useful to consider the implications of various human factors techniques themselves. Thus, for example, the design and resourcing activities producing a specific human–computer interface will come with costs, but may lead to greater benefits to the system such as reduced operator workload, and thence reduced system manpower requirements or reduced risk of injury or a less intensive training regime. In these circumstances the resultant benefits would, it is hoped, outweigh the initial cost or investment. That is to say the initial (small) investment has substantial 'leverage'. But even lengthier chains of relationships can result. An initial task analysis may suggest that a larger team size than had been anticipated may be necessary. This team might therefore need more workstations, and thus a larger building to house them, but the result might be considerably foreshortened task times, and possibly more accurate and less error-prone task performance, all of which would eventually lead to lower costs. For example, the enhanced accuracy in performance in a weapon system might lead to substantial savings in ammunition expenditure.

Strictly speaking some of what might at first appear to be 'cost savings' are actually 'cost avoidances' but these can nevertheless be very substantial indeed. Thus the initial investment of design effort may have an extremely significant 'leverage' effect. As pointed out earlier, the United States defence programme known as MANPRINT was introduced partly as a result of the apparent failure of technology to deliver the necessary benefits in performance that it might initially have promised. What had been highlighted was that extremely large investments in technology appeared to be leading to *lower* performance. High initial costs were leading to extremely small apparent benefits and thus the 'leverage' was very small indeed. It became clear that the issue was predominantly a failure to appropriately match the physical and cognitive skills and capabilities of users and maintainers with their operating environment and the technology provided for them. In some instances the shortfalls were extreme and were leading to costly training programmes, or had outcomes such that weapon systems failed to reach their designed performance parameters [8].

Following the introduction of the MANPRINT programme there were some very dramatic examples of large cost avoidances. One example frequently cited is that of the US T-800 aero-engine. The predecessor system needed a grand total of 134 different tools for effective maintenance leading to costly equipment inventories and manuals and lengthy training courses. A detailed analysis of the maintenance schedules and required tools for the proposed T-800 was undertaken with a view to drastically simplifying maintenance complexity and reducing the number of tools required. The result was a new engine with far lower through-life costs because it only needed a maximum of six tools for any maintenance task and had a greatly simplified series of maintenance schedules and diagnostic test regimes. Other cost avoidances cited by Booher [8] include:

- A small series of studies examining human error during the use of an Anti-Aircraft Missile System resulted in a multi-million dollar cost avoidance because hit probabilities were increased to the extent that considerably fewer missiles were needed during training.

- A multi-million dollar investment in a rollover protection cage for a wheeled vehicle fleet led to a 98 per cent reduction in non-fatal injuries, and in excess of 90 per cent reduction in vehicle damage costs.

Considerably more detail about MANPRINT and its United Kingdom analogue (Human Factors Integration) is given in Chapter 2. In particular this chapter also provides information on a UK MoD research study of some of the *qualitative* benefits of HFI.

But how should we consider the benefits of Human Factors activities themselves? Can these be quantified? More importantly, can we examine the effects of differential investment in various Human Factors activities and their impact upon the outcome of an equipment development programme? Work being carried out under the auspices of the US Federal Aviation Administration (FAA) has examined Human Factors Assessments as carried out during the Investment Analysis process itself [9]. This work has examined the human-system performance contribution to overall programme benefits and, most importantly, the estimated costs associated with mitigating human factors risks and with conducting the overall engineering programme support. In this way the human factors components related to benefits, risks and costs can all be examined in an integrated manner within the investment analysis process as a whole.

Similar work has been carried out within the UK. The introduction of the HFI methodology within the MoD equipment acquisition system has undoubtedly contributed greatly to improvements in military operational effectiveness and reductions in whole-life and performance risks. However, until fairly recently there has only been limited quantitative data to allow comparative judgements to be made about the success of differential Balance of Investment (BOI) strategies within equipment procurement. A research study carried out in the period 1999–2001 examined Human Factors BOI strategies within equipment procurement, and developed an Influence Model of HFI activities to describe the interactive influences of HFI upon the outcome of a system procurement [10].

This study examined and quantified, for the first time, the linkages between HFI and the costs and benefits associated with a military system, and addressed ways of modelling the link between investment in HFI and the outcome, in terms of equipment acquisition consequences, of changes in these investment strategies. With this approach the interactive HFI ‘Influence Diagrams’ for a generic Light Anti-Armour Weapon (LAW) system were successfully elicited in a ‘workshop-style forum’ with appropriate Operational Analysis (OA) and HF subject matter experts. These Influence Diagrams were used to build a System Dynamics (SD) model in order to examine the dynamic interactions between Balance of Investment options and the resultant military capability. This was developed using the commercially available modelling tool POWERSIM to produce a stock-flow diagram. In the stock-flow diagram the measurable quantities (including HF issues such as hazard concerns, safety issues, as well as financial budgets, sample charge-out rates, etc.) were clearly delineated as separate, but linked, flows. Additional factors needed to allow computation of rates of change were also included.

The results illustrated the significant potential of a dynamic interaction between HFI investment options (BOI strategies) and the impact of this upon the resulting 'military capability'. In particular, influence diagrams were constructed for the generic LAW system so as to represent the continuous and interactive application of investment in HF specialist activity for each of the six HFI domains (Manpower, Personnel, Training, Human Factors Engineering, System Safety and Health Hazard Assessment). See Chapter 2 for a more detailed exposition of these 'domains'.

The impact upon military capability was expressed by means of a shortfall in the number of systems operating following the procurement programme. No specific meaning was attached to the results for the 'exemplar' system (a LAW system) beyond showing that the model produced an appropriate response when input data are changed and the data were considered to be indicative only. Extracts from the results are shown below.

The *baseline condition was considered to be the optimum funding of HFI effort* whereby the outcome was to resolve all the known HF issues of concern for each domain. The details and the outcome of this were as follows:

Length of development programme = 65 months
Number of LAW systems required for development and fielding = 200.

With *reduction in funding of effort within Hazard and Safety domains by 90 per cent* the details and outcome were as follows:

Length of development programme = 75 months (i.e. an extra 10 months)
Number of LAW systems required for development and fielding = 228 (i.e. an extra 28 systems).

With *reduction in funding of effort within the Training Domain by 70 per cent* the details and outcome were as follows:

Length of development programme = 100 months (i.e. an extra 35 months above the baseline condition)
Number of LAW systems required for development and fielding = 200.

It is clear that variations from the baseline vividly demonstrated that insufficient funding for the HF activities could lead to a requirement for greatly increased numbers of fielded systems, and incur very considerable delays before these systems could be fully supported by appropriate personnel. All these could of course be translated into considerable increases in overall acquisition costs. Expressing this in terms of investment strategies, it is obvious that altering the differential BOI strategies for HFI had a marked impact or 'leverage' upon project costs. Apparent savings taken early on (i.e. reductions in funding for HF activities) resulted in dramatically increased costs brought about by the increased numbers of fielded systems required or the increased programme lengths or both. This was a dramatic demonstration of the quantitative impact of an HFI programme upon equipment acquisition and project costs. Further information about HFI is contained within Chapter 2.

1.5 References

1 International Ergonomics Association, 'The origin of "Ergonomics"' *Ergonomics International*, 75, Feb 2003 (http://www.iea.cc/newsletter/feb2003.cfm)

2 International Ergonomics Association, 2000

3 BS EN ISO 13407:1999. 'Human-centred design processes for interactive systems'

4 BS ISO/IEC 15288:2002. 'Systems engineering – system life cycle processes'

5 ISO/PAS 18152:2003. 'Ergonomics of human-system interaction – specification for assessment of human-system issues'

6 BS ISO/IEC 15504-1:1998. 'Software process assessment. Concepts and introductory guide'

7 SAGE, A. P.: 'Systems management' (J. Wiley, New York, 1995)

8 BOOHER, H. R. (Ed.): 'MANPRINT: An approach to systems integration' (Van Nostrand Reinhold, New York, 1990)

9 HEWITT, G.: 'Human factors assessments in investment analysis: definition and process summary for cost, risk and benefit'. FAA Paper v1.0, 28 Jan 2003

10 HARVEY, R. S., WICKSMAN, R., BARRADALE, D., MILK, K. and MEGSON, N.: 'Human factors investment strategies within equipment procurement' in McCABE, P. (Ed.): 'Contemporary ergonomics', (Taylor and Francis, a paper presented at the Ergonomics Society Annual Conference 2002)

Chapter 2
Human factors integration

Michael A. Tainsh

2.1 Background to HFI prior to the mid 1980s

The application of the human sciences to the design of jobs and equipment, along with the associated issues such as training, has a long history within the UK.[1] The defence needs of the nation during the Second World War, and later, did much to provide the impetus for this application. Perhaps one of the best and earliest examples of reports from this tradition is Sir Frederick Bartlett's Ferrier lecture of 1943 in which he presented conceptual models and empirical information on a range of gunnery and other tasks. Such work was developed in the 1950s at the Medical Research Council, Applied Psychology Unit in Cambridge with the pioneering project on the 'Cambridge Cockpit'. By the 1960s, the Royal Air Force Institute of Aviation Medicine (RAF IAM), within the UK Ministry of Defence, had wide-ranging programmes in the areas of physiology and ergonomics. This was followed shortly after by similar programmes within the Army Personnel Research Establishment (APRE) for land forces, and the Applied Psychology Unit (APU), Teddington, for the Royal Navy. Within industry Prof. Brian Shackel established a laboratory in EMI, during the 1950s, and this contributed greatly to many design projects – including notable work on the combat systems of the Royal Navy.

All of the above were groundbreaking projects and programmes and some of them had outstanding success. Collaboration between Rolls Royce and Associates, the MoD and APU Teddington on nuclear reactor control developed a novel design for the control panels within the nuclear area of submarines with the result of making substantial manpower reductions, and improving training requirements while maintaining safety levels. This design was subsequently used in UK 'T' class submarines

[1] Work in Germany, Italy and USA has a similar history during this period.

and provided a very well documented case of human factors benefits through the application of systematic equipment design processes. On the other hand there were acknowledged shortcomings that were well documented for the UK's Type 42 Destroyers. These led to important studies that under-pinned the need to address human factors issues in the procurement of future naval Command Systems [1].

Work in the 1970s also began on the provision of appropriate human factors standards within military equipment procurement projects. This work included the 'Human Factors for the Designers of Naval Equipment' [2] which was published in landscape format, intended for use within drawing offices, and later the tri-service 'DEFSTAN 00-25' [3], which is still in use. The use of demonstrator systems was critical at this time to enable stakeholders to understand the consequences and benefits of applying human factors within the context of design and training. Substantial resources were put into such demonstrator programmes [4].

At this time, work addressing human factors issues tended to start from the concept of the Target Audience Description (TAD). The emphasis was on fitting the job to the known characteristics of the user, e.g. by using DEFSTAN 00-25 with its early emphasis on anthropometry and with its data on body size, strength and stamina. Training also was a major concern. However, work in the 1970s showed this was inadequate by itself: it did not cover sufficient of the aspects of system requirements to ensure an effective system performance. In particular, such solutions were seen to fail to take full account of operational requirements.

In the early 1980s, it was decided by the RN procurement authorities within the MoD that as a result of success in the area of computer based systems, it would be useful to document principles, method and techniques. This would ensure that the lessons learned from research and practice could be used by as wide as possible a set of procurement projects. These documents were published with additional support from the Department of Trade and Industry, and became known as the MoD/DTI Guidelines [5].

However, for all this scientific success and the development of standards and procedures, the application of human factors/ergonomics within the UK was limited. Its take-up depended on personal contact, established practices and individual preference.

2.2 Post mid 1980s

Additional books were to appear in the area and the most notable was that edited by Hal Booher [6]. The US MANPRINT programme was the first to propose a broad approach to handling the whole of human factors (rather than just the engineering and training aspects) in the context of equipment acquisition. Its simple premise was that of integration of the domains of manpower, personnel, training, human factors engineering, health hazard assessment and system safety [7]. (The name MANPRINT originated as an acronym from its overall emphasis on MANPower and PeRsonnel INTegration.) This integrated approach was intended to make it possible to manage the human factors contribution to acquisition projects by including human factors in a manner that was consistent with what were the then current project management practices. Most importantly, it would enable high-level assessments to be made of the relative value of investments in the various human factors aspects of a system, between human factors and other contributions and the means of handling risk throughout the life of the project (see also Chapter 1 of this book).

The standards and supporting documentation appeared rapidly over the 1990s both in the areas of the International Standards Organisation and the MoD (mainly in the maritime area: Sea System Publications (SSPs) 10, 11 and 12 [8, 9, 10]). However, two series of events have had a major impact:

(*a*) The introduction of new financial controls and the requirement on UK Defence Acquisition to take account of Whole Life Costs (WLCs) – see MoD's Acquisition Management System (AMS) [11];

(*b*) The adoption of far-reaching health and safety legislation which brought the UK MoD in line with civil standards, and required risks to personnel to follow the ALARP principle (As Low As Reasonably Practicable) – for example, see JSP 430 [12] and Chapter 14 of this book.

The need for controlling and reducing costs in defence acquisition programmes is nothing new, but the current initiative has brought together the personnel management and budget with the equipment management and budget in order to jointly address the factors associated with manpower and personnel issues in equipment acquisition, and the costs associated with them. This has increased the priority of the human factors work in these areas substantially.

Similarly, the improved perception of the need to address health and safety issues has been associated with the disadvantages and costs of not addressing them.

As a result of work since mid 1980, the traditional human factors areas of training and equipment design have been reconsidered in line with the MANPRINT approach and its concept of domains. Furthermore these domains have been placed within the context of the recent MoD procurement reorganisation 'Smart Acquisition', as a general requirement for work to be carried out within the Defence Procurement Agency and Defence Logistics Organisation of the MoD for all equipment acquisitions and through life support. These human factors domains now form the scientific foundation for all work within the UK successor programme to MANPRINT, simply known as Human Factors Integration (HFI).

2.3 Scope of application of HFI

2.3.1 Application areas

While the origins of HFI lay mainly within the defence sector prior to the 1980s, there have been major efforts to approach the issues and application in an entirely generic way. Work in the 1980s, particularly for the UK in the 'The Alvey Programme', which was jointly funded by the MoD, DTI, SERC, industry and academia, ensured that the issues addressed have been generic – useful to as wide an audience as possible. This approach to the work has been continued through the

European Union Framework programmes where human factors has always been well represented.

There have also been notable centres within civil industry including the railways, iron and steel industry, coal, process industries including glass making, automobiles, office design and practice, aerospace including air traffic control and various research associations. Further, there are products that are commonly acknowledged to be 'difficult to use' – the television adverts for video-recorders claiming that they were so easy to use that a cat could manage, can be interpreted as an acknowledgement, by the suppliers, of these difficulties. The handling of such difficulties has varied from industry to industry and application to application depending on many factors.

2.3.2 Customer and supplier

There are two vital components to any HFI programme: customer and supplier. In this chapter the approach that will be taken is mainly from the supplier side, i.e. how to satisfy the HFI requirements. Quite simply, the supplier side is the more difficult, which is why an emphasis upon it is provided here. Conceptually, if you understand the supplier side then the customer side is relatively straightforward.

2.4 Life cycle management and risk

2.4.1 Smart acquisition within the MoD

Prior to the introduction of 'smart acquisition', with its emphasis upon faster, cheaper, better equipment procurement, within the MoD in 1999, Human Factors expenditure was always vulnerable to cost cutting because it tended to be expressed in terms of manpower savings (long term aspects), or additional work for enhancements on trainability or operability. Hence, Human Factors work tended to be viewed as 'gold-plating', and known failures as 'unfortunate lapses', even though the evidence of the accruing benefits was inescapable to the Human Factors community and to many of the equipment users themselves.

The policy of smart acquisition changed the priority of HFI work within equipment acquisition with immediate effect because of its emphasis on Whole Life Costs (WLCs) managed via a Through-Life Management Plan (TLMP). This aims to deliver projects within performance, time and cost parameters and to execute the project processes within a framework of managing risk.

The affordability of many MoD acquisitions can now be understood to rest heavily on HFI issues because manpower and training costs can contribute up to approximately 50 per cent to the WLCs. The major portion of the HFI costs will be associated with costs of manpower. At current values, it is reasonable to assume £1 million (as a very rough average estimate) per person over the life of a system in salary alone. Hence for any large system involving significant numbers of personnel that is open to examination and savings, financial costs will rapidly become an area of high priority – there will be a renewed set of imperatives to manage the risks associated with HFI associated costs.

2.4.2 Acquisition and design outside the MoD

Clearly, the Whole Life aspects described above would also apply to an industrial manufacturing plant or production line which has, or might have, high levels of manpower associated with it. However, whilst HFI is applied to acquisitions within the Defence Procurement Agency (DPA) and Defence Logistics Organisation (DLO) of the MoD, it can equally be applied to acquisitions external to the MoD ranging from smaller scale products such as hand tools or other artefacts to major plant. The general approach and methodology is independent of scale or application – the only important aspect is the involvement of people.

2.4.3 Design drivers

The priorities that support the specification of any engineering solution may vary. In the military setting, in most cases, the highest priorities are most likely to be identified with goals associated with operational effectiveness. However, allied to operational goals will be HFI goals. The priorities associated with a specification will in turn influence the priority of work within the HFI domains. In the commercial work the highest priority may be improved sales, while in the industrial application the highest priority may be production levels and costs. These high level priorities are referred to as 'design drivers' – these are the variables that will need to be controlled and optimised [13]. It is essential that they are fully understood in order to ensure the effectiveness of the HFI programme. For example, there is no point in worrying about manpower reduction if there is a surplus of low-cost manpower likely to continue for the life of the equipment.

The design drivers should be identified within the requirements statements, and need to be fully understood so that acceptance processes can be developed that ensure an effective product across its planned life cycle (LC).

2.4.4 Managing risk

Currently within the MoD the central concept determining the approach to the TLMP is the management of risk. Strategies for mitigating risk are used as the main foundation for project planning. The risk involved here may apply to effectiveness, UPC, WLC, schedule or whatever project or project characteristic has sufficient priority.

Risk has two components: likelihood of the risk occurring and the importance of the risk. Both of these components are defined as variables which are themselves generally defined in terms of categories such as high, medium and low. The precise definitions of these categories may vary on a case by case basis. It may be a matter of judgement based on prior history, as to exactly how these may be defined. The risks that project teams or individuals may be willing to consider and the means for mitigation, may vary dependent on the circumstances of the HFI programme, the industry and application.

2.4.5 The applicability of the approach

There are many organisational, training or design applications where WLCs may not be important but other design drivers such as availability of space, aesthetic appeal or safety will be more relevant. The design drivers and the problems to be solved will vary from industry to industry and from application to application, but in each case the approach would, in principle, be the same. The problem needs to be well understood, and information on the lessons learned from past work needs to be gathered and analysed for relevance.

The techniques to support work on lessons learned and their introduction at an early stage of the LC are well known to the general scientific/technical and HFI communities. Some of the most popular techniques include consumer/customer surveys, workshops, gathering of expert opinion, examination of historic records, past safety records, engineering designs, legal considerations, government/company policy and so on. It is important to appreciate that at an early stage HFI will work like any other branch of concept development and ideas may be shared and discussed with any stakeholder or potential stakeholder. The approach is therefore entirely generic and although HFI techniques have been developed from within the defence sector there is much to be gained from their universal application.

2.5 Starting an HFI programme

2.5.1 The stages of the life cycle

Although the HFI domains have now been identified and examples of successful work provide indicators on how to progress, this is still not sufficient to provide a basis for a successful programme of work. Further, the HFI programme must be integrated with the contributions of other members of a project team throughout the LC. There must be a notion of where to start, how to progress and where to stop. The answers to these questions are provided with reference to the LC. Prior to the development of life-cycle concepts, human factors work tended to be developed along the lines of an academic/scientific investigation with an emphasis on hypotheses testing. More recent approaches bring HFI practice in line with project requirements.

The first problem then is how to start the HFI process. The domains of HFI provide a set of concepts at a high level to guide thinking but they do not always lead to topics and issues that are immediately comprehensible to managers, administrators, users, engineers and other stakeholders in a project.

The question then is:

> how can HFI be introduced into an acquisition process to enable its full benefits to be realised and achieve the goal of fitting the job to the user and the user to the job?

The key to this is fitting in with the project LC and Life Cycle Management (LCM) as described in ISO 15288 [14].

ISO 15288 describes four process groups:

- Agreement processes
- Enterprise processes
- Project processes
- Technical processes

The system LC is part of the description of the enterprise and helps ensure that the goals and policies of the enterprise are met. It provides a framework within which HFI specialists and others can work and understand each other's contribution. Perhaps some of the most important technical issues (from the point of view of this chapter) involve the formulation of stakeholder requirements and system requirements along with their verification processes. Their importance lies in their technical contribution to the contractual aspects of the technical work – they are the major constraints within which the technical work takes place.

LCM is the name of a life-cycle concept that is taken from ISO 15288. It recognises the need to mitigate technical risk and control WLCs. HFI is brought within this cycle when there is a need to take account of the six HFI domains:

- manpower
- personnel
- training
- human factors engineering
- health hazard assessment
- system safety

The stages within the life cycle are given in Table 2.1, along with their purposes and decision gates.

Table 2.1 Stages, their purposes and decision gates

Life-cycle stages	Purpose	Decision gates
Concept	Identify stakeholder's needs Explore concepts Propose feasible solutions	Decision options: • execute next stage • continue this stage
Development	Refine system requirements Create solution description Build system Verify and validate system	• go to previous stage • hold project activity • terminate project.
Production	Mass produce system Inspect and test system	
Utilisation	Operate system to satisfy user's needs	
Support	Provide sustained system capability	
Retirement	Store archive or dispose of system	

The HFI work will contribute to the reports provided to support decision making on the options but this will not be covered here as these are seen as mainly addressing financial control issues rather than the core of the technical content.

The issues to be addressed here may include UPCs and WLCs including affordability and value for money. Again, these are seen as generic issues where priorities may vary from application to application.

2.5.2 Technical processes of ISO 15288

This chapter concentrates attention on the technical processes as defined with ISO 15288. The processes include requirements definition and analysis, architectural design, implementation, integration, verification, transition, validation, operation, maintenance and disposal. No attempt will be made to treat these comprehensively but the main attention will be paid in the remainder of this chapter to the requirements processes, verification and validation: these topics lie at the heart of HFI concepts.

2.5.3 Concept

2.5.3.1 Identification of issues and risks

Table 2.2 maps the Technical Areas (TAs) of interest within an exemplar naval project on to the HFI domains. The project was a large computer-based system with high levels of manpower. The TAs are general topics that are well understood by the stakeholders within the project. The set of TAs describes the full set of HFI issues for the project in the language that is common amongst stakeholders. It is highly preferable to work in language[2] that is understood generally by stakeholders (rather than in academic terms which may be poorly understood), and at a level of detail that is appropriate, rather than force technical terms upon the project (especially if there is no full-time HFI specialist within the project team). The matrix provides an initial statement of TAs that need to be covered and helps ensure that all TAs are covered by cross-checking to see what is covered under each domain.

The first stage is simply to agree with the stakeholders a set of TAs and a matrix such as shown in Table 2.2. This is likely to form the basis for a statement of requirements.

Each TA may be seen as defining a set of Technical Issues (TIs), which are open to expression as a requirement and risk analysis. The level of detail associated with the statement of issues will depend entirely upon the application and its history, the stakeholders' understanding of the application, perceived risks, resources available and schedule. The issues will be important in informing requirements statements and in turn verification procedures.

The matrix that is laid out in Table 2.2 was constructed for a large-scale control system, and is presented here for illustrative purposes only (the original had

[2] Experience with writing HFI manuals suggests that if the users cannot handle HFI concepts within the terms of their everyday language then they will reject them.

Table 2.2 Matrix to show the mappings of the HFI domains (vertical axis) onto Technical Areas (TAs)(horizontal axis) – the number of stars (up to five) indicates the degree of contribution from the domains to the TAs. In some cases the TAs may be subdivided as shown. (For further information on TAs, see Section 2.5.3.1)

Application areas	Operational scenario	Team organisation		User characteristics	Human–computer interaction		Equipment	Maintenance and support		Training	Environmental conditions	System safety	Health hazards
	Mission and task, context	Watch keeping	Team integration	Manpower	Personnel	Mission critical tasks, and operability	Layout and spatial arrangements	Maintenance	Records and support	Training	Environmental conditions	System safety	Health hazards
Manpower	★★	★		★★★★★	★★★★			★★★	★	★★★★			
Personnel		★		★★★★	★★★★★			★★★		★★★★			
Human engineering	★★★★	★★★★★	★★★★★	★★★	★	★★★★★	★★★★★	★★★★★	★★★	★★	★★★★★	★★★	★★★
Training			★★★★	★★★		★★		★★★	★★	★★★★★			
System safety		★				★	★	★			★	★★★★★	
Health hazards						★	★	★			★		★★★★★

over 600 items). It is taken from an application where individuals, teams and equipment are expensive with substantial implications for both system effectiveness and WLCs. The original was used at the concept stage of a major international project in the 1990s. It has been used widely since that time in a variety of computer-based control and communications applications.

2.5.3.2 Early human factors analysis

The need to address a wide range of HFI issues early in the life of a project has led to the development and use of Early Human Factors Analysis to ensure the fullest consideration of all relevant TAs and TIs. This technique is so important that it is described more fully below. It will lead to the identification and prioritisation of TAs along with their associated risks, strategies to mitigate them and resources required (including personnel, equipment, schedule and costs). In turn this may provide valuable source material for the specification of an HFI programme and its contribution to the TLMP.

2.5.4 Development

While requirements definition and analysis is associated with concept, development is associated with the architectural design process: the synthesising of solutions.

The TAs and associated risks will have been identified in concept at a high level. The list identified in the case of Table 2.2 is expanded and developed in Table 2.3. This is the most critical stage for HFI – if risks are not addressed here they will become increasingly expensive to mitigate and if not handled effectively the risk will be transferred to the utilisation and support stages – the operators, maintainers and support staff will shoulder the consequences.

Each of the TIs and risks will need to be addressed or given a reason for no action. This will form the basis of the updated TLMP which will be carried through the project for the remainder of its life cycle.

Exemplar strategies to be considered for the clarification of the issues (in the case of the system described in Table 2.2) and implemented in the development stage include:

- Operational analysis is required to provide the context against which all HFI work will be judged. Clearly the HFI element will not be the lead here (this will probably be led by experts in the likely use of the equipment) but the work will form part of the HFI programme, and may influence important aspects of equipment design, manpower and safety.
- Watchkeeping schedules and team organisation. The allocation of tasks between individuals within a team and the allocation of individuals to schedules may be entirely known for an established system but equally may by itself be subject to investigating as new demands are encountered. The selection of watchkeeping schedules may have important implications for manpower levels and safety, and hence require investigation.

Table 2.3 A brief description of possible TAs and TIs. (Such a list of TAs and TIs could easily be used on the customer side to develop a list of requirements, and hence acceptance tests [12])

Technical areas	Brief description of the technical issues
Operational scenario	The complete set of tasks and goals to be achieved in all scenarios, by the whole system needs to be described. The set of tasks and goals to be achieved needs to be analysed into their components so that the contribution of the equipment and the contribution of the users can be fully described. The operational rules to be followed in all scenarios shall be fully described so as to ensure a good match with job and task design.
Team organisation and integration	The organisation of the various teams, and their integration within the total system in which the users operate, maintain or support the system shall be considered.
Manpower	The total number of users, maintainers and support personnel shall be specified along with a statement of their roles and responsibilities. The manpower requirements shall be considered to take account of through-life issues.
User characteristics	The equipment design needs to match the likely quality of the personnel available at the time of operation. This shall take account of: • employment legislation • anthropometric characteristics • educational and training background • operational experience • mental capabilities.
Mission critical tasks, human–computer interaction and operability	A set of mission critical tasks should be identified for assessment and acceptance purposes. These should be agreed with the appropriate authorities. Performance requirements shall be met for mission critical tasks. These may refer to both individual and team achievement levels. High levels of operability and interoperability should be achieved through the use of standardised design features. The design of human–computer interaction should aim for ease of operation and training. The layout of the displays and controls should take full account of the characteristics of the potential users.
Equipment layout	The layout of the workstation shall take account of the need for users to work in a seated position while others may be positioned around it. Full account shall be taken of reach and visibility requirements.

Table 2.3 Continued

Technical areas	Brief description of the technical issues
Equipment layout	The design of the room accommodating the workstation shall make full allowance for ingress and egress, especially in emergency conditions. A chair shall be provided that shall accommodate a specified set of the likely user population and adequately protect the seated person from hazards as defined by the operational scenario.
Maintenance	The need for maintenance needs to be fully considered so that access can be provided. The need for full time maintainers should be identified. Equipment should be designed for ease of maintenance to avoid the need for highly skilled manpower.
Training	The training needs arising from the design of the equipment and likely TAD shall be fully identified. The possibilities for training aids to be built into the equipment should be considered. The training facilities should take account of the needs for individual and team performance assessment and debriefing.
Environmental conditions	The habitability standards shall be met to ensure the comfort of the users. Full account shall be taken of the noise, heating, ventilation and lighting levels when assessing the users' performance on mission critical tasks. The design of the displays shall take account of the requirements for any special lighting conditions, e.g. high levels of ambient lighting or the use of red lighting.
System safety	System safety will be investigated and the equipment designed to ensure that there is as low as possible a risk from accidental damage to the user or the equipment.
Health hazard assessment	Health hazards will be investigated and a hazard analysis performed to ensure that there is as low as possible a risk of injury to the user.

- User characteristics. The techniques for estimating the manpower requirements will have to take into account at least two classes of tasks: those specifically related to the equipment and events under the users' responsibility and those that belong to the responsibility of the wider community within which the equipment is located. The manpower requirement will, in the first instance, be determined by functional analysis but will then take into account such issues as health and safety levels and employment policy. The risks here include the system's effectiveness and WLCs. They may also include availability for utilisation due to poor availability of trained personnel.

- The size, shape, strength and stamina of the personnel will require definition, along with their training background, educational attainment and background experience shall be specified. The derived document is conveniently known as a Target Audience Description (TAD). Increasingly the use of highly detailed TADs is becoming the norm in many spheres of equipment acquisition and engineering. Failure to comply with appropriate norms may result in very costly re-engineering in later stages to overcome problems that can be handled routinely during earlier development.
- Manpower/personnel analysis will be required to ensure that requirements of design drivers are met.
- Human–Computer Interaction (HCI) – mission critical tasks and operability. For many computer-based systems associated with control, this may be seen as a critical area. Its importance is entirely dependent upon the operational requirements, Safety Integrity Levels (SILs) and automation. Derisking strategies may involve the use of simulations, technical demonstrators and mock-ups of many varieties. For systems such as those found in avionics it may involve expensive work with high fidelity equipment, but equally a cardboard model may suffice for others. The consequences of failure here may include system effectiveness, low levels of operability (including high levels of errors) and reworking at a later stage at substantial expense. Further information on HCI and usability is provided in Chapter 11 of this book.
- Equipment layout. Stories of failures to understand layouts effectively are legion and failure may be expensive depending on the situation. Poor layout may not only mean inconvenience but can also mean increased manpower levels or safety risks. A thorough knowledge of the TAD and users' tasks, is essential in this stage to ensure that characteristics such as reach, visibility and communication are correctly estimated. This information is likely to be important not only to HFI specialists but others associated with the broader aspects of system engineering. The strategy here will be to seek to optimise the use of space.
- Maintenance. Increasingly with the concept of WLC, the major personnel costs associated with a system lie with the manpower levels and as a consequence maintenance staff may make up a major proportion of the personnel assigned to a system. This aspect carries with it elements of human engineering to ensure maintainability, and manpower to carry out the work and safety levels. This work will feed into the manpower specifications, training and human engineering. Allied to maintenance is support, and together these may be major manpower drivers. Maintenance and support strategy may well need to take full account of any Integrated Logistic Support (ILS) concepts.
- Training. Costs associated with training and the means of training delivery have changed substantially over the last decade. Training Needs Analysis and Training Media Analysis are required early to ensure that training targets can be met to budget and schedule. Given that it frequently takes two years for a system to reach peak effectiveness with effective training, a failure in this area may have substantial effects of performance. Of course, the training and career paths of the system personnel may well take place over a number of years. The training

strategy needs to take full account of the equipment development strategy, or there may be penalties of costs and schedules.

- Environmental conditions. Failure to meet standards in this case may vary greatly dependent on whether these are part of a life support system or merely supporting the comfort of the users. The risks here may vary substantially depending on the location of the system and hazards that might be expected. A full specification will be required early to ensure that any consequences for the broader aspects of system engineering may be taken into account.
- System safety. It is now standard practice to develop a safety case associated with any new procurement/design work. The risks of failure here are multitudinous depending on the application. Within the MoD the ALARP principle has been adopted to provide guidance on how to proceed. The safety case will involve a multidisciplinary approach but the role of the human is likely to be central. Strategy here may need to be highly integrated with any overall system safety work (see also Chapter 14).
- Health hazard assessment. Work here is required to meet standards such as the Montreal Protocols. The risk here may include the possibility of standards changing throughout the life of the system as a result of legislation. It is most likely that this line of work will feed through directly into the safety case.

2.5.5 Production

2.5.5.1 General

This stage will involve final definition and agreement of HFI issues prior to the start of the production process and progressive acceptance, once the production process has started to yield deliverables.

The production process itself may involve little HFI activity – although there may be human factors aspects of production itself (for reasons of space these aspects are not considered within this coverage). The area of greatest HFI activity will be associated with verification and validation: the definitions used here are in line with those proposed in ISO 15288. (See also Chapter 12 which covers verification and validation in detail.)

2.5.5.2 Verification

ISO 15288 states that: ‘the purpose of verification is to demonstrate that the characteristics and behaviour of a product comply with specified design requirements resulting from the Architectural Design Process’.

The aim is to verify the design of the product and ensure that the role of the user is appropriate to the operational requirement. It is recognised that some aspects of verification may start within the development stage but in major projects it is much more likely that verification can only start once a product is available for inspection.

The process of verification may involve at least four major phases:

- Assessment of the facilities needed to support the functional requirements of the user’s tasks.

- The assessment of the design characteristics to ensure that they meet acceptable operability criteria.
- The assessment of task performance to ensure that they meet appropriate operational criteria.
- The provision of facilities, their operability, task characteristics and the surrounding environmental and other conditions, to ensure that they meet all the criteria, standards, and policy statements as contained in the baseline and requirements documentation.

The process of verification has been the subject of considerable scrutiny as it is generally expensive both in resources and time. It is likely that one of the objectives of designing the verification procedure will be to optimise costs by concentrating on the items of highest priority. However the cost of failure to test may also be high. The cost of reworking during the utilisation phase may be disproportionately high in relation to the costs at concept.

One advantage of dividing up verification into a number of phases is that each phase can be carried out in a location and with personnel appropriate to the phase and this may served as a cost reduction measure. Some tests may be carried out in easily available environments using easily available participants, while others may have to be carried out in the field with populations of specific potential users.

2.5.5.3 Validation

ISO 15288 states that: 'The purpose of validation is to provide evidence that the services provided by the system or any system element when in use comply with the needs and requirements of their respective stakeholders'. Typically this is the final stage in a process of progressive acceptance when the whole system may be put in a working situation to check features, performance, etc.

The importance of progressive acceptance is clear when placed in a context with validation trials. It is to ensure that there are 'no surprises' at this final stage, of a category that could have been handled at an earlier and probably less costly stage. Trials can be expensive as they may involve large numbers of people and instrumentation; in these circumstances any failure due to issues that could have been handled earlier is evidence of poor risk management.

2.5.6 Utilisation

For some years, work in the area of HFI, both in the MoD and more generally in industry, has tended to focus on the early stages in the life cycle when the risks are most open to mitigation, at relatively low cost. However, it is equally true that while equipment or a product may not change through its life, the requirement for it may change substantially and the users' perception of the equipment may change. Its actual use may not accord with the intended use.

Therefore, there is an important need to monitor all activities and characteristics relevant to operations within the TAs, to ensure that the original goals are being met. This class of monitoring work is likely to be part of the TLMP. The activities here might involve:

- customer surveys
- safety analysis including incident and accident reporting
- confidential reporting systems
- training performance analysis
- plans for continuous improvement and development.

Well-established customer and user organisations have the means to report back to the suppliers on the satisfaction they have with the product or service.

2.5.7 Support

As in Section 2.5.6 with operations, there is an important and continuing need to monitor all activities and characteristics relevant to maintenance and support within the TAs, to ensure that the original goals are being met.

The activities here might include a similar list to the one above. However, the costs of maintenance and support, and the manpower component, might make this the subject of a programme of continuing improvement to ensure that standards (including safety) are maintained while costs are optimised.

2.5.8 Retirement

There may be human factors issues at this stage especially if there is a need to dispose of equipment, but it may be easiest to treat this as a discrete stage and run a programme that is somewhat dislocated from the previous stages. However, if that is the case, it is essential that nothing is done in the earlier stages that would jeopardise this stage and that full records are made available – especially in respect of any potential health and safety risks.

2.6 Early human factors analyses

MANPRINT invented the set of human factors domains together with an emphasis on the issues that could be derived from the domains. It also enabled a concept of Technical Areas (TAs) that was understood by the users and engineers and placed HF in a framework that could be understood comprehensively. The domains and TAs could be related as in Table 2.2 and it is important to note that the TAs can be varied in accordance with the application, and the stakeholders' interests, including preferred vocabulary.

Early Human Factors Analysis [15] starts from a baseline of topics, i.e. TAs and issues contained within an Issues Register. This is likely to be part of any overall such register run by the project. It will be generated from project documentation and accepted documents such as standards. In many ways this can be the most critical stage of the whole process for if important documentation is not available or is omitted at this stage then it may become increasingly unlikely for it to be included later. Hence the aim here is to be broad and shallow: to spread the net widely with the intention of making it possible to deepen the investigation later. The issues at this step of the analysis will read like a list of contents for a report.

The next step in the assessment process is to express the issues in a form appropriate to the project. This will ensure that their impact and risk to the project can be understood and handled with specifically created strategies to mitigate important risks, while noting that others will be handled by the usual management practices. Each issue needs to be expressed in a form that shows its relationship to a requirement, i.e. whether the requirement will be satisfied or not. For example, the baseline statement may show that training issues should be addressed but at this second step the issue should be expressed in terms of the training characteristic meeting the requirement or not.

Both impact and likelihood are handled as category scales. At this stage it may be sufficient to use a simple three-way category scale. However, a more detailed analysis may be preferred expressing impact in terms of effectiveness, schedule of delivery or costs – the degree of detail involved will depend on the information available, the purpose of the analysis and the way that the results will be used in the future.

The importance of each of the issues is estimated in relative terms. Expert judgement is applied to each issue to decide its category. Table 2.4 gives the description for each category. Figure 2.1 is used to provide a combined score for a three-category analysis and then Table 2.5 is constructed. Table 2.5 provides a starting point for handling the HFI issues from the point of view of the risk that is presented by them in the acquisition project.

The approach described here is now used widely within the projects covered by the Ministry of Defence Acquisition Management System and has gained widespread support from both the human factors community and non-human factors specialists. It is in line with approaches from other disciplines, e.g. safety [12]. It has the great advantage of enabling judgements of importance to be made against clearly defined criteria that have widespread currency, and arriving at a comprehensible

Table 2.4 Categories of impact and likelihood of issue occurring

Likelihood of requirement not being met	Confidence in information on which judgement is made	Description of impact
High	Performance goals are ambitious or ill defined.	Failure to meet specified performance.
Medium	Performance goals are achievable and defined.	Failure to meet specified requirements but within manageable bounds.
Low	The approach has already been demonstrated.	Failure to reach requirement or lower acceptable bounds.

Impact Likelihood	High	Medium	Low
High	6	5	2
Medium	5	4	1
Low	4	3	1

Figure 2.1 Issue score matrix

Table 2.5 Risk record

Technical issue	Impact	Likelihood	Issue score	Mitigation strategy
TI1				
TI2				

table (Table 2.5) that can be assessed by experts to ensure that the results comply with stakeholder concepts and criteria.

As yet there is no agreed method of taking the issues from Table 2.5, for example, and deciding whether one approach is better than another – a risk may be seen worth taking if there is a sound mitigation strategy available and the prize is high enough.

2.7 The future

2.7.1 Assessment of progress within the MoD on new projects

This final section describes recent work within the MoD to examine the implementation of HFI procedures and methodologies. It is included in this chapter as a means of providing a form of case study for readers.

The introduction of HFI within the MoD started in 1989 and it has now influenced a substantial number of major projects that are coming to fruition with the possibility of establishing the consequences and way ahead. A recent project funded under the MoD's Corporate Research Programme set out to investigate how work to support the implementation of HFI principles had influenced acquisition programmes within the 170 Integrated Project Teams (IPTs) of the Defence Procurement Agency (DPA) and the Defence Logistics Organisation (DLO) [16].

A substantial number of representatives from IPTs were visited. Projects were selected by (financial) size (using the four categories taken from the Acquisition Management System (AMS)): above £400 million, between £100 and £400 million, between £20 and £100 million, and under £20 million, and across the four defence operational capability areas (manoeuvre, strategic deployment, strike and information superiority). This selection was made to ensure a true representation of projects in initial phases within the DPA and DLO. Those IPTs which could not be visited were sent a questionnaire by post addressed to the leader. This helped to ensure that the results were truly representative. All projects were guaranteed anonymity to avoid disclosure of sensitive issues. All approached were most willing to help. Some of the most important general results are presented here.

Figure 2.2 shows the amount of funds that projects declared they were willing to allocate to HFI work in their stage of the project. There were four categories 0–5 per cent, 5–10 per cent, 10–20 per cent and 'do not know'. It is clear from the figure that many projects were preparing to allocate substantial proportions of their resources to HFI. Furthermore they believed that they had a sound understanding of the potential benefits.

Figure 2.3 shows part of the technical basis for the work. The projects declared their understanding of 'lessons learned' that guided their work. Not surprisingly their main guidance was taken from human engineering considerations. However, HFI and Project Management and Systems Engineering (PM/SE) issues were clearly important. These are closely followed by manpower, safety and Integrated Logistics Support (ILS).

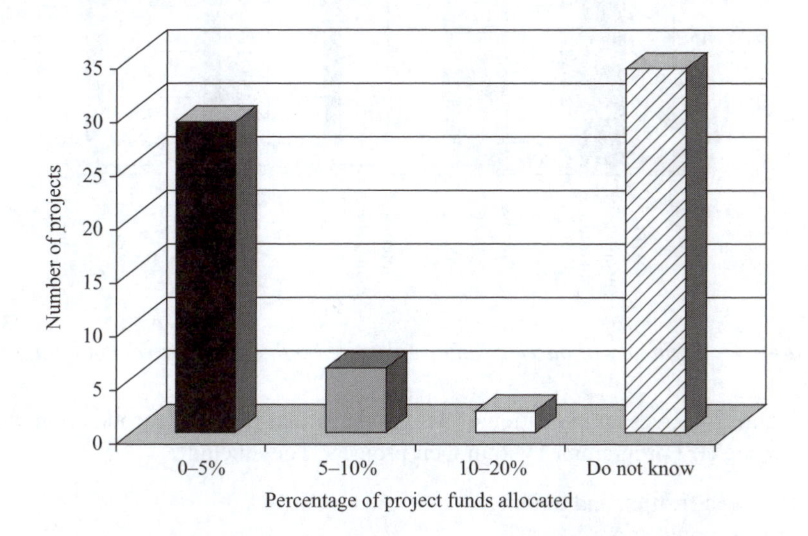

Figure 2.2 Funds allocated to HFI in the current planning stages

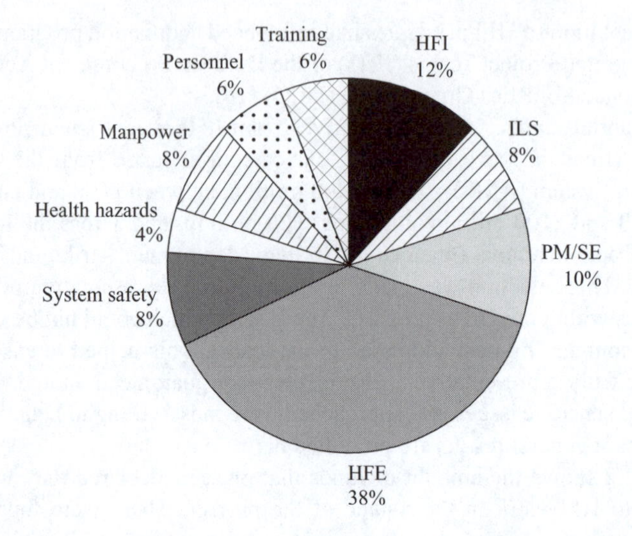

Figure 2.3 The 'lessons learned' that influence current HFI planning

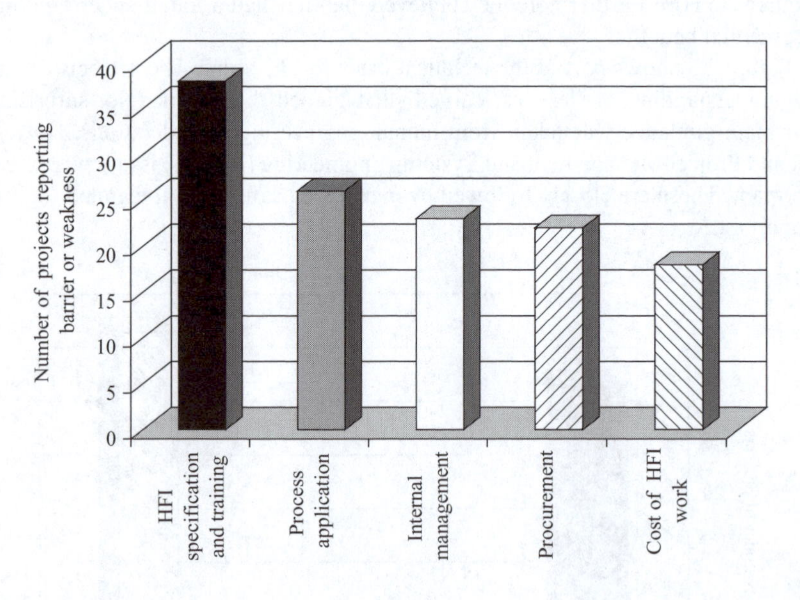

Figure 2.4 Summary of barriers and weaknesses influencing current planning

Figure 2.4 gives an indication of the problems that IPTs are encountering when developing HFI programmes within their projects. They include:

- HFI specification and training;
- process application;
- internal management;

- procurement constraints;
- cost of HFI work.

The first three of these barriers address issues that are internal to the MoD and are being addressed. The procurement constraints are more difficult. These refer to issues arising from acquisitions such as Commercial Off The Shelf (COTS) products where items come directly from a supplier. Perhaps the most difficult to understand is the perceived cost of HFI. There is little doubt that the discipline is seen as expensive even though it is a fraction of the cost of many other such specialisms and the allocation of funds seems to be well underway.

Perhaps more than anything else Figure 2.4 reminds the HFI community that HFI methods and techniques, costs and benefits need to be made more comprehensible to customers and stakeholders within a project. However, a number of major projects were prepared to claim that they had already experienced benefits.

2.7.2 Projects in 'utilisation'

The survey also asked those IPTs that were responsible for equipment currently in service about their HFI work and the issues they were addressing.

Table 2.6 provides a summary of the most important issues under each TA that was reported as being addressed. It also identifies the percentage of the issues that it is believed are being addressed.

There are a number of ways of interpreting Table 2.6. The first might be to say 'well done, past errors are already being corrected'. Another might be to say 'why did these failures that require fixing arise in the first place?' A third might be to ask 'what is the cost of these fixes and what is the price of getting it right in the first place?' It is clear from a number of the confidential reports that some issues being

Table 2.6 Topics and issues being addressed with DLO

Technical area	Most important technical issue	% of technical issues addressed so far
Operational scenario	Role change	62
Team organisation and manpower	Communications capability	68
User characteristics	Personnel availability	66
Equipment operability	Operation of controls	58
Equipment layout	Workspace	22
Maintenance	Maintenance access	27
Training	Trainer representation/fidelity	25
Environmental conditions	Temperature	28
System safety	Unsafe loads	86
Health hazard	High risk conditions	68
Additional topics	Survivability	40
Additional issues	Operational effectiveness	38

addressed were both expensive and preventable. Once again, projects declared that they believed that the HFI approach was delivering benefits.

The report on the work on the DPA and DLO has been widely accepted and it is generally believed that the results obtained are comparable to those that would have been obtained in similar industries elsewhere.

It would therefore appear that much HFI-related work is taking place, and in particular taking place early in the cycle and during in-service activities. However, it would appear that there are plenty of items to address on current equipments so the early work should be in great demand to enable the application of lessons learned.

2.8 Conclusions

The UK approach to HFI has had a long and sometimes distinguished history. Based initially upon the approaches of academic psychology, it has come to be an important part of project management and in many circumstances may be seen as a very high priority item.

However, the methodology and techniques associated with much of this work are still unclear to some, and the tools available are rarely rugged enough for industrial application. The tools available and listed in handbooks betray their academic backgrounds in all but a few instances. However, it is important that there is widespread acceptance of the approach, it is entirely generic and it is often seen to work. Equally, its absence can lead to undue costs and failures. The UK MoD has been a strong supporter of HFI and its successes are now becoming apparent. The basis for all work to be controlled with the TLMP is currently seen as EHFA along with the development of appropriate mitigation strategies. The current international standards such as ISO 15288 are seen as the source of much useful information on the process aspects and enable the links to the systems engineering community to be strengthened.

The way ahead, as seen from the work within the UK MoD, is to expect a much greater integration of HFI practice within engineering and acquisition teams – it may be expected to become general practice in line with ISO 15288.

2.9 References

1 'BR-93002 Future command system feasibility study – NST7868 – high level design for a command system', 1982
2 'Human factors for designers of naval equipment' (Royal Naval Personnel Committee, 1969)
3 'Defence Standard 00-25: Human factors for the designers of equipment' (Ministry of Defence, UK, available as a download from www.dstan.mod.uk), 1998–1997 (currently under revision)
4 TAINSH, M. A.: 'On man-computer dialogues with alpha-numeric status displays for naval command systems', *Ergonomics*, 1982, **28** (3), pp. 683–703
5 MoD/DTI: 'Human factors guidelines for the design of computer based systems', 1988

6 BOOHER, H.: 'MANPRINT – An approach to system integration' (Van Nostrand Reinhold, New York, 1990)
7 WOLVERSON, R. C., and WHEATLEY, E. H. I.: 'Implementation of a MANPRINT programme in the British Army', 1989
8 SSP 10: 'Naval equipment procurement human factors integration management guide', 1999
9 SSP 11: 'Naval equipment procurement human factors integration technical guide', 1999
10 SSP 12: 'Guidance for the design of human computer interfaces within Royal Navy systems', 1999
11 'The MOD acquisition handbook' (2004, 5th edn.)
12 JSP 430: 'Ship safety. Management system handbook', vol. 1, issue 1, 1996
13 TAINSH, M. A.: 'Human factors contribution to the acceptance of computer-supported systems', *Ergonomics*, 1995, **38** (3)
14 ISO 15288: Information Technology – Life Cycle Management – System Life Cycle Processes
15 'Early human factors analysis: user guide'. Produced under TG5 (Human Factors and Effectiveness studies) of the MoD Corporate Research Programme
16 TAINSH, M. A., and WILLIAMS, R.: 'An audit of the effectiveness of HFI within Defence Acquisition: Final Report'. QINETIQ/CHS/CAP/CR020005/1.0, 2002

Chapter 3

Humans: skills, capabilities and limitations

Jan Noyes, Kate Garland and Daniel Bruneau

3.1 Introduction

When designing for humans there is a need to know about and understand their cognitive skills and limitations. Humans, for example, are excellent at pattern-matching activities (and have been described as 'furious pattern-matchers'), but are poor at monitoring tasks, especially when the target is a low-frequency one. It is quite likely when monitoring say a radar screen for the appearance of an infrequent target that the human operator will miss it completely when it finally appears. These two activities are part of an array of information processing activities, which humans carry out throughout their waking hours. Pattern-matching is a perceptual skill while monitoring requires focused attention skills sustained over a period of time. One approach to considering human skills and design issues has been to think of the human as an information processor. This provides the basis for the current chapter, the aim of which is to look at the cognitive activities associated with human information processing in order to locate our skills, capabilities and limitations. Examples will also be provided to show how our knowledge of human information processing relates to the design of objects in everyday life and advanced technologies.

3.2 Humans as information processors

Humans are constantly bombarded by information. We continually receive information via our senses from our environment, which we either act upon or ignore. Information that we act upon will usually lead to some type of observable response. The information can take a number of forms from light, which falls on our retinas, to complicated instructions given to us by a colleague. One particular approach, often used by psychologists, is to view the human as an information processor.

Some may argue that viewing humans in terms of information flow between various components, e.g. information stores and transformational processes, misses some of the essence of being a human. However, it provides a useful starting point for organising information about the human system, especially when considering Human Factors and design issues [1]. In the information processing approach, people are viewed as autonomous 'interactors'. They live within the world as free entities interacting with their surroundings as and when necessary. Although this may be the case for most people, there will be some who do not have the opportunity and/or the ability to live autonomously.

A further issue concerns the nature of the information. We are subject to a huge range of information from electromagnetic energy to auditory, tactile, vestibular and kinaesthetic material. Some might argue that it is contrived to include everything under the label of 'information'. The situation is further complicated when considering how we react to simple, primeval information such as an aggressive dog snarling at us, compared to complex information such as that needed to fly an aircraft in adverse weather conditions. The reaction to the former might be considered primarily a physiological response as the body prepares for attack while the latter demands one involving the highest mental functions. By their very nature, these higher level thought processes do not lend themselves to be considered easily in terms of information processing. However, in order to study the mind, it is often thought of as a 'general purpose, symbol-processing system'. The study of psychology aims to identify the symbolic processes and representations, which make up our higher order functioning.

The basic mechanisms by which people perceive, think about and remember information are collectively known as cognition. A common approach is to consider cognition in terms of three or four stages. Kantowitz [2], for example, suggested three stages: perceptual, in which information is collected via the senses; cognitive, where the central processing of this information takes place; and action, as the brain elicits a response. This usually takes the form of a motor action. These stages have been formalised over the decades into information processing models.

3.3 Models of information processing

Many models of information processing have been developed, and they all follow a similar form. In its simplest form, the human could be considered as a 'black box' with information being input to it, and output emanating from it. The activities that take place within the black box are not observable or directly quantifiable. The only way we can work out what is happening is by considering input and output, and their possible relationships, in order to make inferences. It could be argued that this difficulty of not being able to observe what is happening is endemic to the study of cognition. Despite this, considerable progress has been made in psychology and related subjects in understanding the human mind.

Like the black box example above, models of information processing tend to adopt a linear approach. Human information processing is usually considered as a sequential activity from input (sensations) to output (motor activity/actions). Information comes in at one end, is processed by a number of cognitive systems, and is then output in a different form at the other end.

In summary, sensory information is received by various receptor cells around the body for sight, hearing, touch, smell, taste and knowing where body parts are in space. Although these senses are distributed across the body, together they form the sensory register [3]. This sensory information is then centrally processed. However, before this can happen, the information has to have been perceived by the individual, which will only occur if they are paying attention to it. Central processing, therefore, involves perception and attention before the higher order processing can take place. The higher order processing could be summed up as the 'thinking' which the individual carries out, e.g. the decision-making, problem solving, reflection and judgement-type activities. Thinking includes a wide range of mental functions from the more concrete such as decision-making through to the more esoteric such as creativity. Underpinning all human cognitive activity is memory. If we cannot remember, these higher order functions will break down. There will also be some effects with regard to our attention and perceptual processes. They will become less efficient as we fail to remember what to do with information to which we are attending and perceiving. Assuming central processing has successfully taken place, this will often lead to some sort of motor response. The response will need to have been selected and then executed. In terms of information processing, this is the final stage of the model.

Although information processing models provide a framework and a useful starting point for considering human information processing, they have a number of limitations. Like a lot of models, they are gross over-simplifications. Attempting to represent and capture the complex and intricate cognitive functioning that takes place by drawing a number of boxes on a piece of paper is understandably inadequate. Memory, for example, is important to all human activities, and therefore needs to appear at all points in the model. Further, information processing involves many feedback loops as we draw on information previously attended or search for an analogous problem solving situation in memory. We also continually filter incoming information as well as that being centrally processed. Given the amount of sensory signals stimulating our senses at any one time, it is vital that we filter out information to which we do not need to pay attention. As I write this, parts of my body are aware of clothing being worn, whether there is discomfort, how hot/cold it is, etc. Most of this information is not relevant to my current primary task of generating this chapter, and hence, I need sometimes to be aware of it, but most of the time to ignore it.

The simplicity of the models means that they do not readily accommodate fundamental cognitive approaches. For example, they tend to focus on the individual. They do not take into account what would happen to processing when the individual is part of a group of people. When we are part of a group with the resultant social influences and interactions, our information processing is likely to be very different from when we are acting alone. The models are also presumptive. They assume the individual is always in a state of readiness, and waiting for information to process. Hence, they do not easily allow for unexpected information. Finally, they do not accommodate individual variability. The models suggest we process information by moving through these sequential steps in an organised and efficient manner.

In reality, our processing is likely to be much more individualised and iterative as we adopt various strategies and experiment with novel approaches. Thus, the richness and diversity of human information processing are lost in the models.

3.4 Sensations

Our knowledge of the world in which we exist is dependent on our senses. At its most basic, sensory systems are commonly divided into five sensory modalities: vision, audition, olfaction (smell), gustation (taste) and somesthesis (touch). The unique reciprocal exchange of information that occurs between human and machine begins with information being fed into sensory receptors. Sensory receptors can be classified into two groups: exteroceptors, which receive information from the environment through one of the five sensory modalities, and interoceptors, which alert the individual to internal states such as hunger. Collectively, these are referred to as the sensory register.

The unique abilities of the human sensory system are often overlooked in that we tend to accept without question our sensory modalities such as hearing, seeing, smell and touch. However, closer examination of how our different sensory systems interact with the environment reveal that they are not as straightforward as one may first imagine. For example, the auditory system is able to isolate or discriminate between different sound sources from a large area. This is often described as the 'cocktail party' effect where we are able to focus attention on one sound stream, i.e. in this case a conversation from a barrage of simultaneous conversations, but are still able to hear meaningful information from outside of our immediate focus. The 'cocktail party' example that has been extensively tested is the ability of an individual deep in conversation to hear their name when spoken in a totally separate conversation on the other side of a crowded room. A similar effect has been noted with parents (usually mothers) of new babies, who have the ability to hear their baby crying in noisy surroundings, where other people will have no recollection of a baby's cries. The meaningfulness of the auditory information is thus of vital importance.

Our ability to use auditory localisation to isolate sound sources is considered to be the result of the difference in intensity in the frequency and broadband sounds received by both ears [4]. However, there are certain limitations to auditory localisation. In particular, in Reference 4, Young considered the small environment of an aircraft cockpit to illustrate limitations of our ability to isolate sound streams. A flight deck environment produces high levels of ambient noise, and sound reflections become more pronounced due to the size of the flight deck and severe cut-off from external noise. These all act as mediators in the pilot's somewhat poor ability to localise and isolate sound streams in this type of environment.

Within the broad field of human factors and design, the information we receive through vision is perhaps the most dominant of all the sensory modalities. Indeed, the way in which sensory systems, e.g. the visual system, encode information has profound implications for the way in which display systems are designed. For example, presenting information in central vision will make that information more legible than information presented in the periphery [5]. The use of eye tracking technology is extremely useful in attempting to understand the ability of the human visual system to scan a computer interface such as a web page. It has been found, for example, that in a product search using web pages, web sites with high information content (termed 'information clutter') will result in less efficient scanning of the web page. These are represented by longer fixations in ambiguous areas of page. This is in contrast to finding required products in a web page with less information clutter, where fixations are centred on the search and navigation areas of the web page [6].

Confusion often exists in trying to understand the distinction between sensation and perception. It is therefore useful to consider sensation as any sensory modality that receives information from the outside world, and transmits it along nerves impulses to the brain. Perception, on the other hand, is the interpretable process or transformation of these nerve impulses into a recognisable pattern of sight, sound, etc. [7].

3.5 Perception

Perception is one of the 'clever' parts of information processing. Let us take the visual channel, for example, whereby light merely falls on the retina. It is our perception of these light patterns that interprets this information into the visual scene at which we are looking. Hence, perception can be defined as the way in which we interpret sensory inputs, and how we process and organise information to form a coherent representation. Approximately 80 per cent of incoming information is visual [8] and this is reflected by a predominance of visual research within the cognitive field.

Visual perception involves the eyes and the balancing mechanism of the ear. It is prospective and predictive, in so far as it is influenced by our expectations based on our experience of the properties of the object. There are three main theories of visual perception. The Gestalt theory of the 1930s reflects a holistic approach, where we search for the 'best fit' between what we see and our stored knowledge of objects. This fit is achieved rapidly, even if the stimuli bear little relation to an individual's stored information, due to a natural tendency for objects to contain organised patterns. Gestalt theory identified a number of principles reflecting pattern organisation. The 'law of proximity' states that stimuli in close proximity are identified as a group even if they differ in appearance. This follows from the tendency for stimuli that are similar to be visually grouped together (the 'law of similarity'). Simple and smooth stimuli are preferred to more complex and sharp patterns (the 'law of good continuation'), while fragmented patterns tend to be processed quicker than those that can be visually closed (the 'law of closure'). A key issue in perception is expectation, and the extent to which we perceive 'reality', i.e. objects as they actually are or the way in which we expect them to be. The Gestalt principles listed here provide some insight into how we perceive objects.

Gibson [9, 10] argued for a theory of 'direct perception', in which stimulus inputs are processed in successive stages from simple to more complex analyses. Information in the visual scene provides sufficient cues to determine what the object is

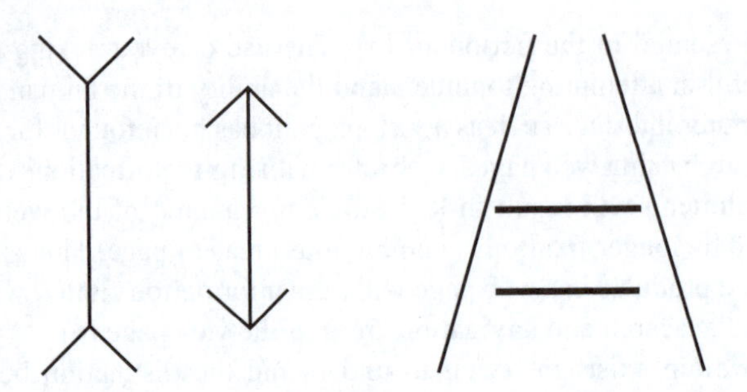

Figure 3.1 Examples of Müller-Lyer and Ponzo illusions, respectively

(i.e. the meaning we attach to it), its relative position and its movement. By contrast, Gregory [11] argued that rather than being data-driven, visual perception is a top-down process, which involves the development of a perceptual hypothesis, based on our experience. The hypotheses that are created when we see something are usually very accurate. However, this is not always the case, and these perceptual errors can result in inappropriate behavioural responses. Evidence for these errors comes mainly from research using visual illusions, and is, according to Gregory, caused by the misapplication of constancy scaling.

Perception failures, commonly known as illusions, can occur either if concentration on one part of a visual task is so strong that we fail to notice another part of the task, or if our concentration lapses due to fatigue or anxiety. Perception can also be prone to error due to our expectations, i.e. we see or hear what we expect to see or hear. Optical illusions for geometric images, e.g. the Müller-Lyer and the Ponzo illusions, demonstrate how perception can be misleading (Figure 3.1). These show how 'central' lines of the same length can appear to vary due to surrounding features.

Others show how distance and/or depth, variations in contrast, or motion, can lead to perceptual errors. The 'impossible' Schuster illusion and the Penrose triangle demonstrate that we can be initially deceived, while the Necker cube shows how depth can be interpreted in two ways, with the circle being situated either on the front or rear plane (Figure 3.2). Whilst as images on a page these illusions may appear to have little bearing on human responses, they can occur quite readily (and with potentially catastrophic effects). For example, crew landing an aircraft at night can experience perceptual problems (illusions) because the runway lights sometimes appear to move. Knowledge of these situations and potential errors that may arise is vital in order to learn to accommodate and reduce the risks.

Depth and distance perception can be caused by stereo kinesis where, for example, an object rotating on a vertical axis may appear to reverse direction, or auto kinesis where stationary objects appear to be moving. A common example of this as given above is the phenomenon of a single or group of stationary lights at night (dark visual background) appearing to move. This is of particular concern in low-level aircraft navigation and landing, but can also lead to difficulties interpreting lights of other aircraft at higher altitude.

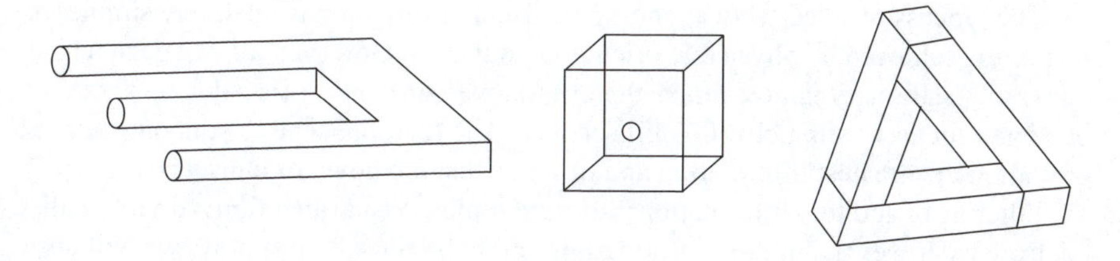

Figure 3.2 The Schuster illusion, Necker cube and Penrose triangle, respectively

Movement of a vehicle next to us or across our visual field can create illusions of movement when we are stationary, or a false impression of current speed. This may lead to inappropriate responses, such as premature breaking or adjustment of speed. The slope of approach terrain to level runways can influence height perception. Where the ground level rises up, the altitude of the aircraft in relation to the runway appears higher than it is, and conversely, ground falling towards the runway creates the illusion of it being at a lower altitude than it is. These effects can be even greater when the surface of the runway is sloping or where the width or texture (both perceptual cues) varies from normal expectations. Whilst visual illusions are the result of automatic processing, the potential effects can be reduced by effective training. Training in the acceptance and identification of illusory phenomena should be combined with additional visual cues designed to augment 'natural' observations.

Arguably, a sufficiently loud auditory stimulus is less easily ignored than visual stimulus; it can be perceived more readily by someone on the move, and single warning tones are less prone to misinterpretation than comparative visual information [12]. However, sound usually comprises a mixture of tones, with individual tones being identified by their pitch, which is determined by the frequency of the sound wave. A mixture of light wavelengths will create an individual, but entire colour, whereas a mixture of tones is usually perceived at the level of the individual tone components. Speech perception is distinct from auditory perception in that the influence of experience is greater. The relatively simple categorisation of an auditory stimulus into, for example, a warning tone, is contrasted with the ability to perceive at the level of phonemes (e.g. 'da', 'ma') and from this discrimination reconstruct the information in terms of whole words with associated semantics.

3.6 Attention

Attention comprises a number of filtering mechanisms that screen the vast input of information being received by our senses at any one time, generally selecting only that which is relevant to the task in hand. Without this process our senses would be overloaded and irrelevant information would dominate, severely limiting our ability to complete any task. Selective attention is the process in which we direct our attention to the stimuli of interest.

The process of selectively attending to visual information is relatively simple, in that this is achieved by physically orientating our eyes. However, a fixed gaze whilst outwardly stationary is not; rather the eyes move constantly (saccades) to fixate at one point for approximately 300 milliseconds. The fixations tend to seek out sources providing the greatest information and/or items that are novel or unusual.

Filtering of auditory information is more complex. Visual attention is directionally controlled. However, auditory information cannot be filtered so simply; we will hear whatever is within our auditory range no matter what the source. In this case, attention is achieved more by mental than physical means, although the direction from which the sound emanates is used as a cue. We also focus on the characteristics of the sound (pitch, intonation) and for speech, the speaker's lip movements. In addition, the meaning (especially relevant for speech) will aid attention, and can be used, although with some difficulty, in the absence of other cues.

Unattended information, both visual and auditory, is minimally processed, although there is sufficient input for us to respond and change our direction of attention, for example, hearing our name or other 'key' words or phrases, seeing movement or colour change at the periphery of our vision. However, as we remember little if anything of the unattended information, this suggests that it is filtered shortly after the perceptual processing stage. The limited impact of unattended information can have disastrous effects. For example, an Eastern Airlines flight crashed into the Florida Everglades due to the crew selectively attending a single malfunction and in so doing, failing to notice altimeter readings warning of descent [13]. As a result, no one was flying the aircraft with the result that it crashed with considerable loss of life.

Research into attention usually falls within two areas, namely that which looks at how effectively we select certain inputs over others (focused attention), and that which studies processing limitation by examining how we perform when attending to two or more stimuli at once (divided attention). Auditory information appears to be selected early in order to limit the demands on processing capacity. Information outside focal attention continues to be processed, but at a reduced rate. A similar low level of processing appears to occur for unattended visual material. Visual attention is not just a matter of focusing on a specific area; it can be more sophisticated, being able to select only certain stimuli within that area, or important parts of the area (e.g. it is possible to focus on one movie image even when a second movie is also being displayed on the same screen over the top of the first).

Many operational tasks require vigilance, namely focused and sustained attention and concentration. Tasks include those where signals must be monitored over long periods of time and specific targets detected. These may be found in industrial inspections, avionics applications and medical monitoring (for clinical and neurological disorders). It is difficult to maintain peak levels of attention, particularly as tasks are typically monotonous and simple, and the targets are often weak, intermittent and unpredictable. There are various vigilance paradigms, including free response where the target events can occur at any time, and inspection where events occur at reasonably regular intervals. In addition, the task may require the detection of changes in auditory or visual intensity, or may be more complex and involve some application of cognitive processing. There are a number of problems associated with vigilance tasks. The ability to distinguish targets from non-targets declines over time, and the observer may become more or less cautious in their reporting of the target presence. The position of the decision criterion for detection can induce a specific pattern of reporting. For example, a high (less stringent) criterion will encourage reporting of fewer targets and fewer false alarms (identification of a target when it is not), while a low criterion will be reflected by higher target reporting with an increase in false alarms.

Explanations for vigilance decrement include 'expectancy theory' [14], 'resource theory' (e.g. [15]), and 'arousal theory' [16]. For example, operators will adjust to the level of targets based on expectations, and if they miss a target the likelihood of further targets going undetected increases. In addition, the level of resources required in vigilance tasks is high, and the high cognitive workload may be difficult to accommodate. Arousal theory is supported by findings showing reduced brain activation is related to reduced detection rates, which suggest that prolonged task performance leads to a lowering in the arousal state of the central nervous system. Vigilance performance will depend on the nature of the task (signal strength, modality, discrimination level, event rate and probability), and can be affected by factors including situational (noise, climate, time-of-day), and personal (operator's age, sex, experience, personality, whether they have taken stimulants or depressants). In order to maximise vigilance, target sensitivity should be improved either directly, by increasing the salience or varying the event rate, or indirectly by adequately training observers and showing examples to reduce memory load.

Any form of focused attention, including vigilance tasks, can be easily disrupted. For example, a control room operator can be distracted by the conversations and actions of others in the vicinity, or by a very busy and changing visual display. Any disruption will, in effect, cause attention to be divided between the primary task and the peripheral activities.

The ability to perform two tasks concurrently will depend on the nature of the tasks. The key is the degree to which the processing involved in one task interferes with that required for the second. For example, it is relatively easy to operate machinery or drive a car (if experienced) whilst holding a conversation, but it is not so easy to converse and compose a letter at the same time, for while both can be achieved it should be noticeable that attention is being constantly switched between the two tasks. The latter occurs because some processing resources (verbal) need to be shared between the tasks resulting in lack of processing capabilities and interference. Minimising the effect of interference between the processes involved in the tasks will increase efficiency in the tasks. Interference is lowest when the two tasks use different modalities, for example, one uses visual processes and the other verbal. Driving a car will involve a combination of visual and procedural (motor skill) processing, while a conversation will engage verbal processing, and hence the limited level of interference. One point to note here is that reading, although a visual act, also uses verbal processing to convert visual symbols into meaningful words; in our brain we effectively say what we read.

There are three main factors that affect dual-task performance, namely task similarity, task difficulty and practice. As can be seen from the driving and conversing example, these two tasks are quite different, using different stimulus modalities. However, similarity extends also to stages of processing (to what degree do the tasks use the same processing at input, internal processing and output), the extent that each rely on the same or related memory codes (e.g. verbal, visual), and the similarity of the response. The more difficult a task, usually the more processing capacity is needed and hence, the reduction in overall efficiency. Practice has the effect of automating some processing and/or responses, thus reducing interference with a second task. A novice driver or operator of equipment is likely to find it much more difficult to converse at the same time than would someone highly practised. Automatic processes tend to be fast, require limited attention, are usually unavoidable (the response occurs whenever the stimulus is present), and are largely unconscious in nature. Training to achieve a high level of automation for a task is advisable, but practice at being able to switch between automatic and 'manual' operation in emergency situations is also important.

It is important that wherever possible, the design of display or auditory systems incorporate features that can help to direct attention when needed. Some systems can give prior warning of a potential target. This cue needs to be timed to permit an appropriate change in attention prior to the target appearance, and it needs to be of a suitable frequency for auditory systems, or position (central or peripheral) for visual systems.

3.7 Memory

If you were asked to remember a nine-digit telephone number (whether presented visually or verbally), some might achieve this (although probably only a few). However, if presented with two such consecutive numbers, it is unlikely that this could be accomplished. By contrast, if you were asked the name of your first school, most would remember, and would respond quickly. This is remarkable given that you may not be primed, i.e. this question may be totally unexpected and not predicted, and it may be 50 years since you attended that school. These two examples demonstrate both the limitations and scope of the human memory, and the distinction usually made between the capabilities of short-term or working memory [17] and long-term memory.

The working memory holds a limited amount of information, typically some five to nine items [18], for a limited amount of time. For example, given a telephone number, it is necessary to retain the information from the time it is heard or seen until dialling is completed. However, as input will be lost within a matter of seconds, it is necessary for us to use a rehearsal process (mental repetition of the item) to retain the telephone number. The revised working memory model [19] proposes a modality-free central executive that carries out general attention-type processing and largely dictates other short-term memory processing, a phonological loop that retains speech-based information, and a sketch pad that holds visually and spatially encoded material. Information is held in working memory for processing and is then either discarded, transferred to long-term memory (after sufficient exposure), or is displaced by other inputs.

Long-term memory is far more resilient and appears to have unlimited capacity. The quality of memory retrieval (and consequently its application) is dependent on a number of factors. The extent of initial encoding and the degree of rehearsal or elaboration are of primary importance. However, also important is the distinctiveness of the material (including its relevance to 'self'). These factors influence the way in which we mentally represent information and how knowledge is organised.

Another main reason for poor recall is interference. Previous learning may interfere with subsequent learning (proactive interference) and vice versa (retroactive interference). Therefore, the nature of both initial and later instruction can be a critical factor in performance outcome. Forgetting can occur over time, but the effect of time is perhaps, overstated. Whilst we have a tendency to remember more recent events, this is largely due to distinctiveness or relevance, and we are very able to remember events and knowledge obtained many years previously providing it is reused (rehearsed) during that intervening period.

Memory failure, from whatever cause, can easily cause error. A 'plane crash was attributed to forgetting the basic procedure to check the state of cargo and identify missing safety caps on oxygen generators' [20]. Likewise, unscheduled instructions can be readily forgotten in emergency situations when cognitive load is at its highest. Understanding of memory limitations can be invaluable in the design of systems that reduce the potential for error. Detailed checklists can be made available for many procedures. However, it is advisable that these be changed occasionally in order to restrict any tendency to shortcut what can become routine, and any influence of expectations. Potential sources of error in tasks that impose a high demand on working memory capacity can be reduced by designing displays that group and space information most efficiently for the observer, and by the provision of various cues and reminders that limit the amount of information that needs to be processed at any one time.

Memory is an important cognitive process not only in its own right, but also as an integral part of both perception and attention. For example, perceptual input can be interpreted based on our expectations, which are determined by our stored experiences; and our ability to divide our attention will be influenced by the extent to which tasks may be automated (procedural memory). Indeed, the working memory model suggests that attention processing is carried out by the central executive, thus indicating a high level of integration with memory processes.

3.8 Higher order processing

Higher order cognition represents the complex network of processes that enhance the understanding and production of information that has been assimilated initially through sensation, followed by perceptual categorisation of a situation, before being attended to via specific attentional mechanisms and then stored in memory. In summary, higher order processing transforms the information that has been assimilated and stored into meaningful concepts that the individual can act upon.

Some examples of higher order processing concepts are illustrated in Figure 3.3, and are broadly examined under the general heading of thinking research.

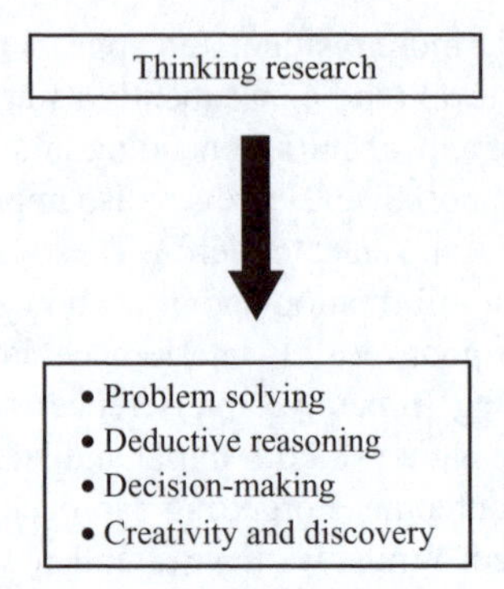

Figure 3.3 Higher order processing concepts

First, it is important to examine the general concept of thinking and thinking research. The term thinking may, at first glance, seem self-explanatory. It is an important component of our everyday lives and we use the term both freely and readily to explain our state of activity as humans. Yet how simple do you believe it would be to explain the concept of thinking without actually using the term 'thinking?' Psychologists and philosophers alike have battled with what thinking entails but perhaps the simplest way to understand thinking is to illustrate its usefulness in our actions. For instance, thinking is important for our ability to make decisions, to choose our own personal goals and to form our beliefs [21]. Because thinking is involved in so many of our mental processes and actions as humans, it is logical to assume that the term 'thinking' naturally becomes a prerequisite, as well as forming a fundamental part of, other high order cognitive processes such as problem solving and decision-making.

Humans are good at a number of different types of thinking. For example, inductive reasoning, and critical thinking, which involve thinking about beliefs, actions and goals, and drawing on experience. This is in contrast to computers, which are adept at deductive reasoning, where a purely logical approach is required. A framework, which links thinking about beliefs, actions and personal goals, is the search-inference framework [21]. This model essentially considers 'thinking space' to be made up of two related factors: search and inference (Figure 3.4).

In an example adapted from Baron [21], the following scenario is an illustration of how the search-inference model operates:

> A software engineer has just been given the task of developing a new database system for a small motor-sport company. The engineer talks with existing software developers about the best programs to use and refers to the manuals for the material, which will provide the content for the database. However, her research reveals that the database she is required to design and build exceeds her expertise and she consults with senior managers about what course of action should be taken. Some suggest designing the database system to the level at which she is competent, while others suggest taking a course to develop her skills further. After consideration, the software development engineer decides to take the course, improve and hone her database skills and develop the database system to its full potential.

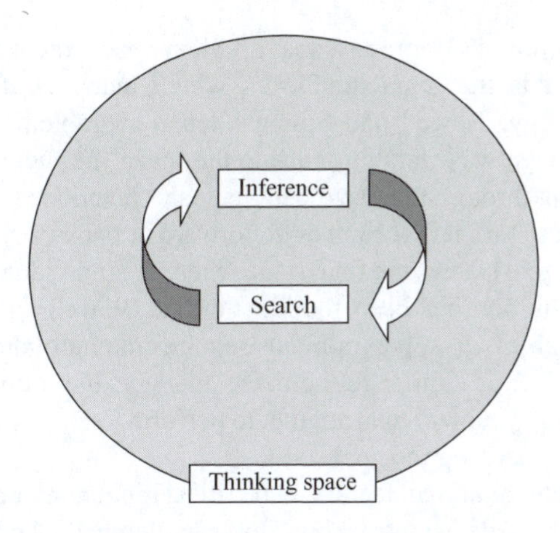

Figure 3.4 An illustration of the search-inference framework

The software engineer is clearly involved in a number of search and inference thinking strategies. According to Baron [21] thinking involves searching for three kinds of objects: possibilities, evidence and goals. Therefore, in the above example, the software engineer sees that there are many possibilities for the task that she has been set and these possibilities also relate to the doubt that she is experiencing. Possible solutions not only come from the engineer herself, but also from other colleagues. In this sense therefore the goals become the benchmark by which the possibilities of action are evaluated. In the scenario above, the engineer's goals consist of undertaking a task to design and build a database. Internally, the engineer has a desire to succeed and complete the task to the best of her abilities. Finally, evidence about the goal of achieving the correct database will lead to either a decision being made to choose either designing the database to the current level of her expertise or undertaking a course to improve her skills in the area. The evidence for each possibility is laid down and the software engineer decides which course of action to take. This is a clear example of how thinking becomes a prerequisite for choosing the correct problem solving strategy and thus also relates closely to decision-making. It is also a clear example of what is commonly referred to as 'inductive thinking' where evidence and past experience are used to form generalisations about what is most likely to be true.

Problem solving provides a good illustration of how humans utilise higher order cognitive process to resolve unfamiliar situations. Wickens and Hollands [22] stated that while planning and multi-tasking are also types of cognitive processing that are able to deal with situations that are not always identical, these methods are arguably different from problem solving approaches. The main reason for the difference lies in the fact that planning and multi-tasking processes use existing cognitive mental models as their starting point. Problem solving has no such model as a starting point and therefore requires the human to develop new cognitive mental methods for the situation.

Newell and Simon [23] proposed a problem-space theory, which is worth mentioning since it is the foundation from which many existing theories stem. Using a maze analogy, Newell and Simon likened a problem to having multiple paths. The initial stage was standing outside the maze (i.e. being presented with a problem). Travelling through the maze with its many junctions where the individual can choose whether to turn left or right or go forward or backward is termed the intermediate stage. The goal state is the final stage where the individual reaches the centre of the maze (i.e. solution has been found). Further, Newell and Simon suggested that the human capability to solve problems requires the individual to pass different knowledge states. This is similar to the maze analogy; the individual begins at an initial knowledge state and will then attempt to perform a search of alternative mental states before finally reaching the goal state.

Within the domain of human factors, two critical higher order cognitive processes are problem solving and decision-making. However, logically it can be suggested that all of the higher order cognitive processes are inter-related in some way. Earlier it was highlighted how decision-making was integrally related to problem solving and thinking in general. However, decision-making can also be considered as a cognitive entity in its own right and therefore deserves further clarification. If we were to assess retrospectively our daily activities both in the actions we perform but more specifically in the higher order cognitive activities that take place, we would see that decision-making would most likely make up the greatest part of our daily cognitive processing activities and actions.

Logic would suggest that decision-making activities involve the selection of one choice from a number of choices, that there is a certain amount of information with respect to each choice and that there is often a certain amount of uncertainty with the choice. Indeed this is the paradigm, which many researchers consider to define decision-making tasks [24].

Theories of decision-making, and therefore the subsequent development of decision-making models, fall into three broad categories:

1. Normative (1950s–)
2. Descriptive (1970s–)
3. Prescriptive (1990s–)

The order of these categories represents the chronological order of dominance of decision-making theories over the last 50 years or so. In other words, normative theories were most dominant or common in the 1950s; descriptive models were most common in the 1970s and so forth, although research today continues to develop notions based on any one of these three broad categories of decision-making.

Normative models (now known as 'classical decision theory') represent early emergent thinking on decision-making. Essentially, normative models specify the cost or benefits associated with different choices. Mathematical models could therefore be applied to these ideas in order to optimise outcomes. Utility Theory is an illustration of a normative model. It works on the premise that humans seek pleasure, and avoid pain. In general decision-making, decisions are made to maximise pleasure (positive utility) and minimise pain (negative utility) [25]. One of the limitations of the normative decision-making models is that they work on the assumption that humans make decisions according to a normative model, i.e. they are programmed to make decisions in an ideal setting taking into account various well-defined parameters. A possible scenario relating to normative models is illustrated below.

> An engineer attempts to make the following decision: whether to stay late after work and complete a set of tasks ahead of schedule in order leave early the next work day or complete the task in the regular work period resulting in not staying late or leaving work early. The engineer will weigh up both positive and negative utilities and may decide that the pleasure gained (positive utility) from leaving early the next day will far outweigh the inconvenience (negative utility) of staying late at work tonight. Therefore, the engineer decides to stay late at work and finish the assigned tasks.

The reality is that humans often do not make decisions that conform to the normative model. They do not weigh up the situation in an organised and methodical way taking into account positive and negative utilities. This lack of reference to the subjective element was one of the reasons leading to the demise of the normative models. They were superseded by the descriptive models, which attempted to 'describe' human decision-making as it actually occurs. In the example given above, a colleague might suggest the engineer joins them for a drink after work, and a spontaneous decision might be made to abandon working late despite arrangements having already been made to leave work early the next day.

Descriptive models are arguably self-explanatory since these models describe how people make decisions and as a result such issues as the irrationality and biases that are so distinct to our human psychological make-up are taken into account [26]. As one example of a descriptive model, Prospect Theory [27] describes the extent to which the probability of various outcomes affects people's decisions. Therefore, people will make a choice based on the highest prospect and not the utility values. For example, an aircraft pilot may perceive the probability of running into detrimental weather conditions to be low (since the pilot may be familiar with the terrain and area), when in reality evidence such as weather reports suggests the probability of bad weather to be high. Like many descriptive theories prospect theory is considered to be vague [28], but nevertheless it highlights the capabilities of humans to use probability estimations as just one of the many strategies in their decision-making process.

Prescriptive models are by far the most accessible models, the reason being that prospective models attempt to study decision making in natural and operational settings [29] as opposed to developing descriptive theories in one context and trying to relate the decision-making process to another area. For instance, it is clear that a decision-making scenario involving deciding whether or not to work late at night or leave at the normal time would not be applicable to a safety-critical situation such as an emergency on the flight deck of an aircraft where there are so many different factors such as time pressure and stress that will ultimately influence the decision-making process and are unique to the flight-deck in that instance. Examining decision-making in this manner is called 'naturalistic decision-making' [30]. It takes into account how people use their experience to make decisions in operational and therefore natural settings [26]. For instance, the capability of humans to learn from mistakes makes experience an invaluable tool in safety-critical domains. It has been documented that human errors, while contributing to detrimental consequences on some occasions, provide a vital source of information for experienced decision-makers. Taking another example from the aircraft flight deck, a study by Rauterberg and Aeppli [31] found that experienced pilots can sometimes be considered 'better' pilots because their expertise results from using previous unsuccessful behaviour (e.g. errors in a flight deck simulator or in a real situation) to make current decisions. In other words, using prior experience resulting from errors can be successfully used to make 'safer' decisions in safety-critical situations on the flight deck.

There are, however, certain human limits in decision-making and an area in which this limitation becomes apparent is in statistical estimation. Wickens [32] suggested using an analogy with the classical procedure of statistical inference. The analogy basically states that in the first instance a statistician or other suitably qualified person such as a psychologist computes a set of descriptive statistics (e.g. mean, median, standard error) of the data; then using the estimated statistics some inference about the sample is made. According to Wickens, humans have limitations in two fundamental areas:

1. ability to accurately perceive and store probabilistic data; and
2. ability to draw accurate inferences on the data presented.

An example of these limitations is seen when people are asked to perceive proportions. It appears that when we try to estimate proportions, let us say for instance that an engineer is trying to estimate the numerical value of a proportion from a sample of data, there is tendency to be conservative in our estimation. In other words, humans tend to underestimate proportions rather than overestimate. Wickens [32] suggested that the reason we tend to be cautious or conservative in our estimation of proportion is due to the fact that any large estimation using extreme values may run a greater risk of producing an estimation that is incorrect. Sokolov [33] offers a more interesting explanation where it is proposed that the salience and frequency of attending to something ultimately influences how conservative we will be in estimating proportion and making a decision on such estimations. In this context salient events are events that are rare and may therefore cause an increase in estimation of its frequency. For example, an aircraft pilot may only attend to the salient characteristics of the flight deck environment instead of characteristics occurring on a more frequent basis. Therefore, the decision-making process is biased towards these salient cues, which means that it could be fundamentally flawed and not lead to the best outcome.

3.9 Action

Once information has been sensed, perceived, stored and then transformed via higher cognitive processes, the final stage according to traditional information processing models is the output. In other words, responses are produced as a result of gathering and interpreting information from the environment. In some situations, a nil response will be the most appropriate action, but in most, some form of motor activity will be elicited: for example, turning a dial to reduce altitude on an aircraft flight deck as a result of instructions from air traffic control. Hence, it is normally accepted that information processing is primarily a cognitive activity that leads to a motor response in some form.

Although cognitive processing can take place extremely rapidly, there is a limit on the response times associated with motor actions. This is primarily because of the size of our bodies and the length of our nervous systems. When we decide that we are going to press a particular key or take some form of action, nerve impulses have to travel from our brain to the outer extremities of our body. The time taken for our nervous system to respond is up against a limit, because of the physical distance involved. Hence, it is impossible for us to react any faster. In most day-to-day activities, this is not an issue. For example, a learner driver practising the emergency stop requires a rapid response to an imperative stimulus [34], but the driving situation usually allows this. The concern is, for example, in a military situation where the capability now exists to fire missiles at such speed that they hit the target before the human even has time to register that they are under attack. Since the human operator is at their limit in terms of reaction time, the solution here is to develop technology that can cope with this situation.

3.10 When information processing goes wrong

The principal feature of the human information processing models is the assumption that information progresses through a series of stages or mental operations that mediate between stimulus input and response execution [35]. There are therefore many points at which information processing can go wrong. When this happens, errors may result. These are discussed in detail in Chapter 7.

The cognitive perspective of human error is based largely on the general information processing theory. Therefore, in this perspective, errors occur when one or more of these mediating operations fail to process information properly. Rasmussen [36] developed a detailed taxonomic algorithm for classifying various types of information processing failures associated with the erroneous actions of operators. Cognitive models are popular because they go beyond the simple classification of 'what' the erroneous action entailed (e.g. the pilot failed to deploy the landing gear) to address the underlying cause of such errors (e.g. was it a decision error or an attention error?). The cognitive approach therefore allows what on first impression would seem to be unrelated errors to be analysed based on similar underlying cognitive features (e.g. lack of attention) which in turn allows the identification of specific trends and, consequently, intervention strategies to be developed. However, information processing models typically do not address contextual or task-related factors (e.g. equipment design) or the physical condition of the operator (e.g. fatigue), nor do they consider supervisory and other organisational factors that often impact performance. The consequences of this may lead to blame being unduly placed on the individual who committed the error rather than on the error's underlying causes, the latter being something over which the individual may has no control.

The ergonomics and system design perspective rarely places the human as the sole cause of an accident [37, 38], but instead emphasises the intimate link and interaction between individuals, their tools, their machines, the general work environment and the organisation. It is clear that the system approach fills in the gaps for which the information processing models fail to account. However, the generality afforded by systems models often comes at a cost of specificity and thus the models lack sophistication when it comes to analysing the human component of the system. Other perspectives include the aeromedical perspective (e.g. physiological mechanism such as fatigue) and pure psychological perspective (regarding flight operations, for example, as a social endeavour).

3.11 Implications for design

The study of human factors allows a better understanding of how human strengths and limitations can promote and therefore influence the design of systems. The overall aim is that they meet the needs of people, rather than people having to adapt to systems which have been built for stand-alone functionality rather than compatibility with human facets. It is logical therefore that, more often than not, an action and its consequence are based on the design of a system. Therefore, a system that is poorly designed will often produce outcomes that are undesirable. This may at one level be merely frustration and irritation experienced by the individual. However, at the other extreme, safety may be compromised. This is especially crucial and true in safety-critical domains such as medicine and aviation, where human life may be at risk. It is this group that will be discussed here.

In recent decades, as aircraft technology has become more sophisticated, the commercial airline pilot's primary role has moved away from flying to a role that could be termed 'manager of systems'. Here, flight deck crew spend the majority of their 'flying' time monitoring automated systems. This change, however, does not mean that pilots are inactive during a flight. Rather, they are more cognitively active as they attempt to monitor and therefore maintain attention with respect to various tasks depending on what flying stage, i.e. take-off, cruising flight mode or landing, is taking place. This has been termed 'cockpit task management' [39]. It is clear therefore that before an action is performed, crew will need to integrate effectively the various tasks and available information. Decisions will need to be made about what actions should be executed at a particular phase in the flight and which tasks should take higher priority during high workload periods, i.e. during take-off and landing [40]. Based on the decisions made from these tasks, the actions performed, whether they are slight adjustment to aircraft systems such as an altitude change for example, should not create a safety-critical situation where the aircraft is put into any danger. The design of the system, therefore, clearly plays a major role in ensuring that the actions performed based on the information presented to the pilots are accurate. However, research into how information should be presented and managed on modern flight decks is still ongoing [41] and there is much still to find out concerning the optimum means for doing this.

While flight decks scenarios clearly demonstrate the skill of pilots in performing actions based on the assessment and interpretation of various sets of information, there are certain limitations, referred to as resource demands [42]. In this instance the ability to perform two tasks well will be highly dependent on whether or not one task poses less of a demand on the crews' attentional resources compared to the other. Two tasks, which by their very nature are highly demanding, will ultimately lead to actions, which may not often be appropriate, and error-stricken. This is particularly applicable to inexperienced personnel [43]. Indeed the demand on cognitive resources allocated to various tasks is termed 'mental workload' and within safety-critical environments such as an aircraft flight deck, automated aids and systems are continually being developed in order to minimise the cognitive demand or mental workload placed upon pilots. Reducing cognitive demands, especially in high workload situations, leads to the execution of actions that are congruent with the safe and optimal operation of a system.

3.12 Conclusions

Information processing models provide a good starting point to consider human skills and limitations for application within the design context. Humans are an interesting paradox in terms of their information processing skills. On the one hand, they have inductive problem-solving capabilities that far outstrip any machine or computer in the foreseeable future. On the other hand, they have severe limitations on the amount of information they can process at any one time. For example, working memory is fragile and limited, and effort is required to retrieve information. This is in contrast to long-term memory, which is often extensive and detailed even after many decades of non-retrieval of that particular information.

In addition, inconsistency, variability and errors are all part of human performance. We cannot depend on humans acting in a consistent manner. They may perform the same task in an identical manner on nine occasions, and then change their actions on the tenth. Take, for example, making a cup of tea – how many of us have forgotten to include the tea or added cold water to the teapot having omitted to boil the kettle? These slips are commonplace and can be explained in terms of automatic processing of frequently carried out activities. Variability within people can be more than between individuals. If we took, for example, braking reaction times in a car, we would find more variability within the individual than between people. Finally, errors are an intrinsic feature of human performance. In the past, there have been movements to design systems by removing the human from their operation in order to address the problem of human error. However, it is hard, if not impossible, to remove humans from system operation, because they design, develop and maintain the systems, and retain overall control, e.g. in the case of needing to abort procedures. Thus, the characteristics of humans are an important consideration not only in the design of systems and products, but also in the allocation of function of tasks between humans and machines.

3.13 References

1 WICKENS, C. D., and CARSWELL, C. M.: 'Information processing', in SALVENDY, G. (Ed.): 'Handbook of human factors and ergonomics' (Wiley, New York, 1997, 2nd edn.)
2 KANTOWITZ, B. H.: 'The role of human information processing models in system development', Proceedings of the 33rd Annual Meeting of the Human Factors Society (Human Factors Society, Santa Monica, CA, 1989) pp. 1059–1063
3 BEST, J. B.: 'Cognitive psychology' (West Publishing, St. Paul, MN, 1995, 4th edn.)
4 YOUNG, L. R.: 'Spatial orientation', in TSANG, P. S., and VIDULICH, M. A. (Eds): 'Principles and practice of aviation psychology' (Lawrence Erlbaum Associates, Mahwah, NJ, 2002)
5 SALVENDY, G. (Ed.): 'Handbook of human factors and ergonomics' (Wiley, New York, 1997, 2nd edn.)
6 BRUNEAU, D. P. J., SASSE, M. A., and McCARTHY, J.: 'The eyes never lie: the use of eye tracking in HCI methodology', in Proceedings of CHI '02 (ACM, New York, 2002)
7 OBORNE, D. J.: 'Ergonomics at work: human factors in design and development' (Wiley, Chichester, 1995, 3rd edn.)
8 ORLADY, H. W., and ORLADY, L. M.: 'Human factors in multi-crew flight operations' (Ashgate, Aldershot, 1999)
9 GIBSON, J. J.: 'The ecological approach to visual perception' (Houghton Mifflin, Boston, MA, 1979)
10 GIBSON, J. J.: 'The ecological approach to visual perception' (Lawrence Erlbaum Associates, Hillsdale, NJ, 1986)
11 GREGORY, R. L.: 'Distortion of visual space as inappropriate constancy scaling', *Nature*, 1963, **199**, pp. 678–680
12 EDWORTHY, J., and HELLIER, E.: 'Auditory warnings in noisy environments', *Noise and Health*, 2000, **6**, pp. 27–39
13 WIENER, E. L.: 'Controlled flight into terrain accidents: system-induced errors', *Human Factors*, 1977, **19**, p. 171
14 BAKER, C. H.: 'Maintaining the level of vigilance by means of knowledge of results about a second vigilance task', *Ergonomics*, 1961, **24**, 81–94
15 PARASURAMAN, R.: 'Memory load and event rate control sensitivity decrements in sustained attention', *Science*, 1979, **205**, pp. 925–927
16 WELFORD, G. L.: 'Fundamentals of skill' (Methuen, London, 1968)
17 BADDELEY, A. D., and HITCH, G. J.: 'Working memory', in Bower, G. H. (Ed.): 'The psychology of learning and motivation', vol. 8 (Academic Press, London, 1974)
18 MILLER, G. A.: 'The magic number seven, plus or minus two: some limits on our capacity for processing information', *Psychological Review*, 1956, **63**, pp. 81–93

19 BADDELEY, A. D.: 'Human memory: theory and practice' (Lawrence Erlbaum Associates, Hove, 1990)

20 LANGEWIESCHE, W.: 'The lessons of ValuJet 592', *Atlantic Monthly*, March 1998, pp. 81–98

21 BARON, J.: 'Thinking and deciding' (Cambridge University Press, Cambridge, 2000, 3rd edn.)

22 WICKENS, C. D., and HOLLANDS, J. G.: 'Engineering psychology and human performance' (Prentice Hall, Upper Saddle River, NJ, 2000, 3rd edn.)

23 NEWELL, A., and SIMON, H. A.: 'Human problem solving' (Prentice Hall, Englewood Cliffs, 1972)

24 WICKENS, C. D., GORDON, S. E., and LUI, Y.: 'An introduction to human factors engineering' (Longman, New York, 1997)

25 STERNBERG, R. J.: 'Cognitive psychology' (Harcourt Brace, Fort Worth, TX, 1999, 2nd edn.)

26 NOYES, J. M.: 'Designing for humans' (Psychology Press, Hove, 2001)

27 KAHNEMAN, D., and TVERSKY, A.: 'Prospect theory: an analysis of decision making under risk', *Ecomometrica*, 1979, **47**, pp. 263–292

28 DONNELLY, D. M., NOYES, J. M., and JOHNSON, D. M.: 'Development of an integrated decision making model for avionics application', in HANSON, M. (Ed.): 'Contemporary ergonomics' (Taylor & Francis, London, 1998) pp. 424–428

29 KLEIN, G.: 'A recognition-primed decision of rapid decision making', in KLEIN, G., ORASANU, J., CALDERWOOD, R., and ZSAMBOK, C. (Eds): 'Decision making in action: models and methods' (Ablex, Norwood, NJ, 1993)

30 ZSAMBOK, C. E.: 'Naturalistic decision making: where are we now?', in ZSAMBOK, C. E., and KLEIN, G. (Eds): 'Naturalistic decision making' (Lawrence Erlbaum Associates, Mahwah, NJ, 1997) pp. 3–16

31 RAUTERBERG, M., and AEPPLI, R.: 'Human errors as an invaluable source for experienced decision making', *Advances in Occupational Ergonomics and Safety*, 1996, **1** (2), pp. 131–134

32 WICKENS, C. D.: 'Engineering psychology and human performance' (Harper-Collins, New York, 1992, 2nd edn.)

33 SOKOLOV, E. N.: 'The modelling properties of the nervous system', in MALTZMAN, I., and COLE, K. (Eds): 'Handbook of contemporary Soviet Psychology' (Basic Books, New York, 1969)

34 MATTHEWS, G., DAVIES, D. R., WESTERMAN, S. J., and STAMMERS, R. B.: 'Human performance: cognition, stress and individual differences' (Psychology Press, Hove, 2000)

35 WICKENS, C. D., and FLACH, J.: 'Information processing', in WIENER, E., and NAGEL, D. (Eds): 'Human factors in aviation' (Academic Press, San Diego, 1988) pp. 111–155

36 RASMUSSEN, J.: 'Human errors. A taxonomy for describing human malfunction in industrial installations', *Journal of Occupational Accidents*, 1982, **4**, pp. 311–333

37 NOYES, J. M., and STANTON, N. A.: 'Engineering psychology: contribution to system safety', *Computing & Control Engineering Journal*, 1998, **8** (3), pp. 107–112
38 WIEGMANN, D. A., and SHAPPELL, S. A.: 'Human error perspectives in aviation', *International Journal of Aviation Psychology*, 2001, **11** (4), pp. 341–357
39 DISMUKES, K.: 'The challenge of managing interruptions, distractions, and deferred tasks', Proceedings of the 11th International Symposium on *Aviation Psychology* (Ohio State University, Columbus, 2001)
40 ADAMS, M. J., TENNEY, Y. J., and PEW, R. W.: 'Situation awareness and the cognitive management of complex systems', *Human Factors*, 1995, **37**, pp. 85–104
41 NOYES, J. M., and STARR, A. F.: 'Civil aircraft warning systems: future directions in information management and presentation', *International Journal of Aviation Psychology*, 2000, **10** (2), pp. 169–188
42 WICKENS, C. D.: 'Pilot actions and tasks', in TSANG, P. S., and VIDULICH, M. A. (Eds): 'Principles and practice of aviation psychology' (Lawrence Erlbaum Associates, Mahwah, NJ, 2002)
43 DAMOS, D. L. (Ed.): 'Multiple-task performance' (Taylor & Francis, London, 1991)

Further Reading

NOYES, J. M.: 'Designing for humans (Psychology at Work)' (Psychology Press, London, 2001)

This book covers the capabilities and limitations of humans within the context of human–machine interaction and the design of work environments. It provides an extension and more detailed discussion of many of the points made in this chapter.

WICKENS, C. D., GORDON, S. E., and LUI, Y.: 'An introduction to human factors engineering' (Longman, New York, 1997)

The authors state in the preface of this book that they 'saw a need for engineers and system designers and other professionals to understand how knowledge of human strengths and limitations can lead to better system design, more effective training of the user, and better assessment of the usability of a system' (p. xvii). This also provides an accurate summary of the book's contents in this extremely detailed and comprehensive introductory text to human factors.

WICKENS, C. D., and HOLLANDS, J. G.: 'Engineering psychology and human performance (3rd edn.)' (Prentice Hall, Upper Saddle River, NJ, 2000)

This is the third edition of the highly successful 1984 and 1992 books. Its aim is to bridge the gap between theoretical research in cognitive and experimental psychology, and human performance. Over the last five decades, our understanding of human information processing and human performance has vastly improved, and this book attempts to show these advances within the context of human–machine interactions.

Chapter 4
The human factors toolkit

Jan Noyes

The reasonable Man adapts himself to the world: the unreasonable one persists in trying to adapt the world to himself. Therefore, all progress depends upon the unreasonable Man.

George Bernard Shaw, 1903

4.1 Introduction

Man has dominated planet Earth for thousands of years, and we have only to look at our environment to see all the things that humans have created. In an ideal world, all the objects used by us should have been designed with this in mind. For example, the tables, the chairs, the lighting in my current surroundings, should all meet my physical and work needs in terms of their design. Not all objects fall into this category, namely, those pieces that have been created to fulfil our aesthetic needs, e.g. objets d'art, paintings, pottery, sculptures, etc. A simple classification, therefore, for objects in the world is that they fall into three categories: those belonging to the natural world about whose design we can do little (although biological advances are now challenging this view), those objects that we create for our aesthetic needs, and those that are developed for us to use. It is the third group that is of interest here. The aim of this chapter is to provide a broad overview of the issues surrounding human factors methods and tools.

In theory, all the objects in the built world should have been carefully designed for the human user. In practice, this is not the case and we are surrounded by poor design. As an example, take domestic products. We have microwave ovens that are difficult to programme, we have cooker hobs with no obvious matching between the controls and the burners, and clock radios and video recorders with banks of identical switches. The end-result is that we make mistakes, become irritated and do not use the range of functionality offered by most domestic appliances. But does this matter? For most of us, it is of minor inconvenience, and often with extra effort or paper aids we can manage to achieve our desired goals. Certainly, in the past in industry, poor design was not considered a priority.

In the heavy industries in the late 19th century, the human was often the expendable 'part' in the system. There was little need to ensure the design of the equipment was suited to the worker. If a person was injured or killed, there was usually another individual to take their place. Since a lot of the work was unskilled, 'replacement' workers quickly learned how to do the manual jobs. However, the situation began to change around the time of the Second World War. The introduction of more advanced technologies, e.g. radar screens, cockpit controls and displays, were found to be problematic for the operators. Many were having difficulties learning to use the equipment and were making mistakes. This might not have mattered in peacetime, but skilled personnel were at a premium during the war. There was little time for training; hence, it was vital to preserve the working population. Consequently, there was a realisation that it was no longer possible to ignore the needs and capabilities of the users when designing these more advanced technologies. In recognition of this, the UK Ergonomics Society and the US Human Factors Society were created in the late 1940s.

Ergonomics derives from the word 'erg' meaning work, and 'nomos' meaning science; hence, it is the study of the science of work. The terms ergonomics and human factors are often taken as synonymous, although there is some evidence that ergonomics has its origins more in the consideration of the physiological aspects of the human, while human factors focuses more on the psychological element. Both societies have now celebrated their 50th anniversaries, and in conjunction with the International Ergonomics Association (IEA) continue to thrive and to grow. Today, in the developed world, there is a growing awareness of the benefits to be accrued from taking the human user into account in the design process. However, designing for humans is not easy, and further, it comes with a cost.

There are a number of reasons why it is not easy to design for humans and some will be briefly considered here. The first issue concerns adaptation – even if the design is poor or inadequate, humans will adapt to it. Take, for example, the QWERTY keyboard designed in the 1860s as part of the Victorian typewriter, and still in very wide use today. (Ironically, it is now appearing in miniaturised form on Personal Digital Assistants [PDAs].) Yet, this design has been recognised by many researchers as not being the optimum keyboard layout (see [1]). Another issue concerns our creativity. Humans are creative and will think of ways to accommodate poor design, e.g. placing beer pump labels on control room panels in order to differentiate specific switches from among the banks of identical ones. A final point is variability. There is a lot of variability in human performance, e.g. reaction time measures can show more intra-variation, i.e. within the person, than inter-variation, i.e. between people. A further compounding factor relates to user expectations; often, prospective users are not sure what they want from a product. Take, for example, white goods. Most have a wide range of functions: washing machines, for example, have perhaps 20 different programmes. However, studies with users indicate that most only use two programmes – a fast and a slow wash. What users say they would like to use is different from what they actually use. As a result, designing for humans is not straightforward.

A further issue relates to the costs of design. When designing a product or a system, there are certain personnel involved in the design process without whom the object could not be created, e.g. software writers, design engineers, hardware and electronic specialists, etc. The design team may view the human factors engineer as a 'bit of a luxury' since the product can be designed without them. (A frequently heard comment from engineers concerns the fact that they are human, so they can use themselves as the user model!) Pheasant ([2], p. 10) summed up this point and others in his five fundamental fallacies of design. These illustrate some of the misconceptions concerning ergonomics and design, and are given in Table 4.1.

Moreover, adding human factors personnel to the design team inflates the costs and the development time. On the surface, there appears little incentive to include human factors. However, there is evidence that there are cost benefits to be attained from the implementation of human factors. Bias and Meyhew [3] in their book on 'cost-justifying usability' provided many examples where incorporating ergonomic methods into the design process resulted in significant savings. Stanton and Young [4] took this a step further by factoring the reliability and validity of various ergonomic methods into the cost-benefit analyses. A detailed discussion on the cost-benefits of human factors is given in Chapter 1 of this book.

In summary, human factors (or ergonomics) is all about design. But there is little point in designing something for humans to use without carrying out some form of assessment that this objective has actually been met. Consideration of how to assess and evaluate design provides the focus of the current chapter, namely, an overview of the methods, grouped together under the umbrella of the human factors toolkit, which can be employed for assessing the design of products and systems.

Table 4.1 The five fundamental fallacies (reprinted with permission from Taylor & Francis)

No. 1	This design is satisfactory for me – it will, therefore, be satisfactory for everybody else. (Fallacy – designing for oneself will be fine for everyone else.)
No. 2	This design is satisfactory for the average person – it will, therefore, be satisfactory for everybody else. (Fallacy – designing for the average will be fine for everyone.)
No. 3	The variability of human beings is so great that it cannot possibly be catered for in any design – but since people are wonderfully adaptable it does not matter anyway. (Fallacy – the population varies greatly and people adapt, so why bother to consider design at all?)
No. 4	Ergonomics is expensive and since products are actually purchased on appearance and styling, ergonomic considerations may conveniently be ignored. (Fallacy – ergonomics is expensive and people are more interested in how things look, so need we bother?)
No. 5	Ergonomics is an excellent idea. I always design things with ergonomics in mind – but I do it intuitively and rely on my common sense so I do not need tables of data or empirical studies. (Fallacy – designing for humans is intuitive and common sense, so why bother with ergonomics?)

4.2 The human factors approach

The human factors approach is one of user-centred design, i.e. the user is placed at the centre and is the focus of the design process. In the design team, the interests of the user fall within the province of the human factors engineer or ergonomist. Often, this person will have a background in psychology. It is suggested that one of the optimum ways of ensuring the user needs are represented in product or system design is for this person to work as part of a multi-disciplinary team. However, it is idealistic to imagine the object can be designed totally from a user-centred design perspective. Often, compromises will have to be made due to engineering, practical and pragmatic considerations. As an example, in the late 1970s as part of my doctoral research, I was designing chord keyboards. Chord keyboards, unlike sequential keyboards such as the standard QWERTY keyboards, are input devices that require patterned pressing of keys. One of the difficulties incurred was locating appropriately designed keys to fit the various models. As there were none available commercially that were totally appropriate for the chord keyboards, a compromise had to made between suitability and availability. In conclusion, the human factors engineer attempts to design always from the perspective of the user, but the reality is that this is not always possible. In order to assess the design of an object, the human factors specialist has an array of techniques that might be used (the so-called 'human factors toolkit'), and these are discussed below.

4.3 Methods for assessment

Methods are often divided into formative, i.e. those that are more appropriate for use during the development of the product, and summative, i.e. those for application with the finished product. (These are discussed in Chapter 12.) The following analogy is sometimes used: 'when the cook tastes the soup, this is formative evaluation, but when the guest tastes the soup, this is summative evaluation'. Many methods, however, are appropriate to both formative and summative evaluation, i.e. all phases of the lifecycle. For example, usability testing is often thought of as a summative method, but there are many benefits to considering usability aspects from the beginning of the design process. The earlier in the lifecycle that human factors is applied, the greater are the benefits in the long term [5].

A further division that is frequently used relates to the objectivity of the methods. Methods can be divided into those that are subjective, i.e. indirect in that the user reports details about their experience with the product, and those that are objective, i.e. direct measures of the user experiencing the product are made. Again, like formative and summative, there is some blurring of the definition as some methods cross the boundaries in providing a means of collecting both subjective and objective data.

In total, there are probably some 25–30 methods in the human factors toolkit (depending on definition and the extent to which they are sub-divided). Some of them are unique to human factors and have been developed specifically by this community for assessment purposes, e.g. heuristic evaluation. Others are methods that are familiar to large parts of the working population, e.g. questionnaires and interviews. It is the intention here to focus on the former group of methods, although key references will be provided for the latter to enable the reader to find out more about these techniques.

A further caveat concerns the nature of the tools and methods covered. The examples given in this chapter relate primarily to the cognitive dimension of human–machine interaction. Take, for example, workload. This can refer to physical workload as measured by physiological measures, e.g. energy expenditure, galvanic skin response, eye blink rate, etc., or mental (cognitive) workload. Mental workload can be measured according to objective task parameters, e.g. number of input modes, input data rate, etc., response parameters, e.g. physiological measures such as galvanic skin response (GSR) or behavioural measures such as reaction time, errors, etc., and subjective ratings. The latter is usually measured by some form of rating scale, e.g. the NASA-TLX (Task Load Index: [6]), the Subjective Workload Assessment Technique (SWAT: [7]), and 'instantaneous self-assessment' [8]. It is intended that this chapter should focus more on the cognitive measures and tools for design. However, it should be borne in mind that there is a whole raft of physiological, psychophysical, biomechanic and anthropometric techniques that may be used by the human factors engineer. For an excellent book on physical aspects of design including anthropometric measures, see the late Stephen Pheasant's book [2].

A final point concerns usability. The assessment of usability in the design of products and systems is key to human factors. However, it is deliberately not being covered extensively in this chapter, because a comprehensive and specific coverage of usability is given in Chapter 11.

4.4 Human factors tools

4.4.1 Subjective methods

Subjective methods focus on asking the user about their experience using the product or system. Thus, they are indirect methods of finding out about the design of the product since the user reports on their expectations and perceptions of using the interface. Some examples include heuristic evaluation, questionnaires, interviews, checklists and focus groups. Questionnaires and interviews (from open-ended through to semi-structured and closed questions) are commonly used tools for data collection, and the reader is referred to Reference 9 for an excellent review.

4.4.1.1 Heuristic evaluation

Heuristic evaluation is a method that has arisen within the human factors community over the last couple of decades. It is essentially a systematic inspection of a user interface in order to check the extent to which it meets recognised design principles, e.g. compatibility with the user's mental models, internal consistency of operations, availability of reversion techniques, etc. Nielsen [10] developed a set of 10 heuristics for use in this context. These included visibility of system status, match between system and the real world, user control and freedom, consistency and standards, error prevention, recognition rather than recall, flexibility and efficiency of use, aesthetic and minimalist design, facilitating users to recognise, diagnose and recover from errors, and help and documentation. In its simplest form, a heuristic evaluation works by having an expert evaluator carry out a number of typical tasks using the product (or a mock-up, if the product is not fully developed) and reporting on how it conforms in terms of the design heuristics, as mentioned above. For a fuller explanation of the methodology for carrying out a heuristic evaluation, see Reference 11. Heuristic evaluations fall into the so-called 'quick and dirty' group of methods. They can be carried out very quickly, but they do need an expert evaluator and it is preferable that this person has a human factors background. A further point to note is that a heuristic evaluation will highlight design problems but not their solutions. However, it is generally thought that once problems have been identified, ways of solving them are often apparent.

4.4.1.2 Checklists

Checklists could be thought of as a sophisticated form of heuristic evaluation in that the questions have already been researched and formed. They provide an 'off-the-shelf' technique because a lot of checklists for evaluating interfaces are available commercially or from the academic sources where they were developed. Their advantages include the fact that they are easy to administer, and can be used by the non-expert to assess the design of an interface. Participants completing them tend to like them because they are fast to fill in. However, like heuristic evaluation, they will highlight problems with the interface, but not provide any solutions. There might also not be a checklist that will fit the particular needs of the interface being evaluated, so this may prove problematic. Some examples include:

SUS (System Usability Scale: [12]);
QUIS (Questionnaire for User Interface Satisfaction: [13]);
CUSI (Computer User Satisfaction Inventory: [14]);
SUMI (Software Usability Measurement Inventory: [15]);
FACE (Fast Audit based on Cognitive Ergonomics: [16]);
MUSiC (Measuring Usability of Software in Context: [17]);
WAMMI (Web-site Analysis and MeasureMent Inventory: [18]);
MUMMS (Measuring the Usability of Multi-Media Systems: Human Factors Research Group at http://www.ucc.ie/hfrg/). This is a new multi-media version of the SUMI (developed by the same group of people).

4.4.1.3 Focus groups

Focus groups have been gaining in popularity recently in a number of different sectors in industry and the research community. They are essentially brainstorming sessions where people meet to consider a problem or situation. (For a good review, see [19].) In the human factors context, they have been used to assess the usefulness of a product in terms of meeting user requirements. They can be quite elaborate with table-top discussions, stooges being present to stimulate the focus group content, and taking place over a number of days (for a detailed example, see [20]). Table-top discussions are group sessions in which a selected group of experts discuss issues based on specific scenarios. Often, these groups are chaired by a human factors specialist [21]. Like heuristic evaluations and checklists, they have been classed as 'quick and dirty'. However, the reliability and validity of the findings emerging from focus groups need to be questioned. Two of the difficulties of working with humans are evident in focus group work. First, what people say they do is actually different from what they do, and second, people will tell you what they think you want to hear, i.e. they attempt to provide the right answers especially when there is no comeback on them. Hence, there are many opportunities for the introduction of bias when running focus groups and compiling the findings.

4.4.2 Objective methods

Whereas subjective methods focus primarily on users' attitudes towards a product, objective methods, as the term suggests, provide an appraisal of the product based on measured data in a more controlled setting. They can be carried out formatively or summatively. The examples to be considered here include observation, task analysis (TA) and human reliability assessment (HRA). It should be noted that the latter two are more extensively covered in Chapters 5 and 8.

4.4.2.1 Observation

Observation of users can provide a rich source of information about what users actually do. My colleague, Chris Baber, quotes the example of an observation study he was conducting of ticket vending machines on the London Underground. He observed that people were folding up notes and trying to push them into the coin slots. It would have been hard, if not impossible, to predict this type of user behaviour. Hence, observation studies have a high level of face validity, that is they will provide reliable information about what people actually do, but a low degree of experimental control. Further, they do not provide information about the use of a system or why people are using it in a certain way. The primary disadvantage with the direct observation technique is the lack of further information to explain why users are behaving in certain ways. This can make redesign difficult. A way around this is to follow observation studies with a debriefing session with the users – this allows reasons why they carried out specific actions to be explored. However, in naturalistic settings, this may not be possible.

Direct observation needs to be accompanied by some means of recording the observations. Common ways include audio and video recording, and event/time logging. Video recording has frequently been used in observation studies, but it is resource-hungry in terms of the time needed to transcribe the video-tapes. Some researchers report that an hour of video-tape takes around 10 hours of transcription for analysis purposes. With event/time logging, the raw observations consist of the events or states along a time-line. Drury [22] provided the example of observing customers queuing in a bank. The raw observation data comprised customer-related events along a time-line. The data collection exercise was event driven and this was transformed into a time/state record for indicating the system states in the queue formation. Drury gave five types of information that can be calculated from this type of observation information. These include: sequence of activities; duration of activities; frequency of activities; time spent in states; and spatial movement. This information can then be represented in the form of Gantt charts, process flow charts and link charts.

Another issue concerns the Hawthorne effect; if users know they are being observed, they will often modify their behaviour accordingly. (The Hawthorne effect arose from the studies carried out at the Hawthorne electrical works in the 1920s. It was found that workers responded whenever the environmental conditions of their workplace were changed by increasing their work output. This happened even when the changes were detrimental. Thus, the experimenters concluded that the workers were responding to the fact that they were being studied, and had improved their work rate accordingly.) The artificial aspects of the experimental setting may result in the Hawthorne effect being a problem when running experiments. Although it can occur in natural, operational environments (as the original Hawthorne studies indicated), it is less likely.

An extension of the observation method is that known as the Wizard of Oz. This method, influenced by the film of that name in which the Professor instructs onlookers to pay no attention to the man behind the curtain, involves having a person 'act' as the product or system by making the responses. The real advantage of this technique is that it allows emerging technologies, e.g. advanced speech recognition that can cope with natural language dialogues, to be simulated. Users can then experience and feed back information on a system that is currently not technically feasible. There is an element of deception when participants undergo experimental studies that employ the Wizard of Oz; hence, the ethics of carrying out this type of study need to be carefully considered.

Indirect observation techniques can also be used to find out how people carry out tasks. Surveys, questionnaires, interviews, focus groups, group discussions, diaries, critical incident reports and checklists can all be administered to individuals to provide details and documentation about their use. In contrast to these self-report techniques, archive material can also be employed to find out about individuals' use of a particular system. These might include analyses of written records, and computer logs of user activities.

4.4.2.2 Task analysis

In contrast to observation, task analysis (TA) is a better-known and established method in the human factors toolkit. The term 'task analysis' essentially covers a range of techniques to describe, and sometimes to evaluate, the human–machine and human–human interactions in systems. Task analysis has been defined as follows: 'methods of collecting, classifying and interpreting data on human performance in work situations' ([23], p. 1529). However, the generality of the term evident in definitions such as this is often not particularly helpful in terms of trying to find out about and execute this technique. Task analysis is a term that includes a number of different techniques. It is particularly useful for large, multiple activity tasks, where techniques such as timeline analysis, link analysis, or critical incident technique would prove inadequate on their own.

One of the best-known forms of TA is hierarchical task analysis (HTA) (see [24] for a comprehensive overview). Here, a task is taken and broken down in terms of goals and sub-goals, and their associated plans. The end-result can be a pictorial representation, often in the form of a flow chart, showing the actions needed to achieve successful completion of the task. On its own, this type of breakdown of the task can be useful in demonstrating where the problems are. Likewise, carrying out a TA for a task being carried out under different conditions, say a paper-based versus a computerised system, can be particularly illuminating. The main advantage of this method is that it provides a systematic breakdown of the tasks and sub-tasks needed to use a product or a system. Often, this in itself is beneficial to finding out about the difficulties people have using the interface. In terms of disadvantages, there is a need to decide the level of granularity required in the TA. Take a simple example, like making a sandwich. When carrying out a TA for this everyday domestic task, decisions have to be made about the level of detail needed. For example, if unsliced bread is used, the availability of a sharpened bread knife needs to be considered, a check on the thickness of the slices, using the crust, etc. Trivial though this example might be, it provides a good demonstration of the type of problems encountered when carrying out a TA.

Other variants on TA include Cognitive Task Analysis (CTA). The cognitive element is an important component of tasks involving automated and complex systems, and CTA methods reflect this. The example given by Shepherd [25] to differentiate physical and cognitive elements is that of button pressing: the action of actually pressing the button is physical whereas deciding which button to press is cognitive. Successful completion of the task will depend on both being carried out, and Shepherd argued that it is more useful to consider a general TA that accommodates all elements of the task rather than focusing on cognitive and non-cognitive TA.

Early CTAs included GOMS (Goals, Operations, Methods and Selection rules) developed by Card, Moran and Newell [26]. This was used in a text processing application and required the identification of rules for selecting methods for allowing the operators to achieve their goals. GOMS was followed by TAG (Task Action Grammar: [27]) and TAKD (Task Analysis for Knowledge Description: [28]). TAG focuses on describing tasks in terms of syntax rules, while TAKD considers task descriptions in the context of rules for knowledge elicitation. A more recent development is Applied Cognitive Task Analysis (ACTA: [29]), which comprises three interview methods to enable the practitioner to extract information concerning the cognitive skills and mental demands of a task. The early TAs have been criticised because they focused exclusively on the role of the user in the design process to the point of excluding other aspects. Diaper, McKearney and Hurne [30] have attempted to compensate for this approach by developing the pentanalysis technique. This technique comprises the following elements: an initial requirements data capture, task and data flow analyses carried out simultaneously, integration of the pentanalysis and data flow analyses, and finally, the development of a final data flow model and pentanalysis.

In conclusion, TA is a term that covers a plethora of techniques to examine the activities carried out by humans in complex systems. Reference 31, for example, has been cited as listing over 100 methods relating to TA. This vast array of different methods does mean that an expert is needed to decide which techniques are appropriate in order to conduct the TA. As a result, a TA is not easy for the non-specialist to carry out. The primary benefit of executing TAs stems from the systematic analysis of human–machine activities that these techniques facilitate. This, as their long history indicates, is one of the main reasons for the continuing interest in developing and using them. Further details on TA techniques are provided in Chapters 5 and 8.

4.4.2.3 Human reliability assessment

Human reliability assessment (HRA) is a generic methodology, which includes a TA, to assess the reliability of a system. A methodology for an HRA was given by Kirwan and Ainsworth [32]. In this methodology, after the problem had been defined, they suggested that a TA should be conducted in order to consider the goals and sub-goals of the users in terms of the tasks and sub-tasks they are carrying out. This would lead to an identification of the particular types of error being made. As a general rule, HRA techniques focus on quantifying the impact of different errors in order to consider means of error prevention, reduction and management. In an HRA, a fault or event tree might be used to model the errors and their recovery paths, as well as specification of the human error probabilities and error recovery probabilities. Thus, a key component of HRA is error identification; there are many techniques available for doing this.

Human error identification (HEI) techniques have been developed over the last 20 years as systems have become more complex and more highly automated. The link between human error and incidents/accidents has fuelled developments in this area, although it is this author's view that errors only play a small part in causing accidents (or the less critical incidents). For more information on errors and their role in accidents, see Reference 33 and Chapter 7 of this book.

Like TA, HEI techniques benefit from facilitating a systematic analysis of the operation of a product or system from the point of view of where the errors are likely to occur. Once the activities where errors are likely to occur have been located, this then allows remedial actions to be taken in order to either prevent or accommodate the operator making them. Many techniques for HEI have been developed: some examples are listed below.

THERP (Technique for Human Error Rate Prediction: [34]);
HEART (Human Error Assessment and Reduction Technique: [35]);
SHERPA (Systematic Human Error Reduction and Prediction Approach: [36]);
PHECA (Potential Human Error Cause Analysis: [37]);
GEMS (Generic Error Modelling System: [38]);
PHEA (Potential Human Error Analysis: [39]);
TAFEI (Task Analysis for Error Identification: [40]).

4.4.3 Empirical methods

The third and final group of methods are the empirical ones. It could be argued that these are similar to the objective methods in that they have a high level of objectivity.

The difference, therefore, is slight and relates primarily to the degree of control: empirical methods are tightly controlled. This is in contrast to the subjective and objective methods that rely primarily on indirect or direct reporting and analyses of user activities. The use of experiments for examining usability, modelling tools and fitting trials and mannequins will be considered here.

4.4.3.1 Experiments

The focus of the experimental approach in human factors work is on usability. This is primarily because usability is a central issue in the design of products and systems. It emanates from the term 'user-friendly', which began to be used in the early 1980s to describe whether or not computer systems had been designed to optimise user interactions. Although the term is still widely used and applied to a range of products today, the human factors community does not particularly like 'user-friendly' because it does not have an agreed definition. Hence, it cannot be measured in any objective way. This is in contrast to the term 'usability', which has been defined by the International Standards Organisation [41] as: 'the usability of a product is the degree to which specific users can achieve specific goals within a particular environment; effectively, efficiently, comfortably, and in an acceptable manner'. Important in this definition is the subjective component as illustrated by the final point concerning acceptability.

The affective element of product/system use has been increasingly recognised in recent years (see [42]). If users find that using a device is unacceptable for some reason, they may decide they do not like it, and will cease to use it. Hence, it has been suggested that the ultimate test of usability is whether or not people use an object. Data logging techniques, e.g. the keystroke level model, may be employed to do this. Although there could be ethical considerations in doing this, i.e. logging an individual's use of a product or system, it is purported to provide an accurate assessment of use. Self-report techniques, where individuals are asked to report their use, are known to be subject to inaccuracy. People, either deliberately in order to create a certain impression, or accidentally, will often provide erroneous reports of their level of use of a product or system.

The first personal computer (PC) was launched in February 1978, and PCs began to penetrate the workplace in the early 1980s. The net result of this was to bring computers and computing to the masses. It is perhaps not surprising therefore that the concept of usability was developed in the 1980s. Emeritus Professor Brian Shackel is credited with this as he was the first to use the term in this context in 1981 [43]. Since then, the original definition has been extended, and many, many publications have been produced on usability. Shackel's early definition in conjunction with his colleague, Ken Eason, focused on the user and the task in terms of learnability, effectiveness, the attitude of the users (the subjective component) and flexibility (nicknamed the LEAF acronym). Over the years, this original work has been modified to include usefulness, acceptability and reusability. Thus, usability metrics are a vital way of assessing the usefulness and appropriateness of a product or system (Table 4.2). These metrics are often investigated in an experimental setting.

Table 4.2 Basic concepts in research experiments (adapted from [44])

Concept	Meaning
Theory	A set of explanatory concepts
Hypothesis	A testable proposition
Variables	Independent variables, which the experimenter manipulates, e.g. keyboard 1 and keyboard 2 representing the two experimental conditions
	Dependent variables, which the experiment measures, e.g. response times and error scores
	Confounding variables, which the experimenter attempts to control, e.g. all males when evaluating the usability of a product intended for use by both sexes
Participants	Individuals who take part in the experiment
	Ideally, participants should mirror the intended user population in terms of demographics, and be randomly selected. Often, this is not possible and opportunistic samples may be used. One problem relates to self-selecting groups. Our work with older adults and computer technology generated a skewed sample in that we could not locate those people who had no interest in computers
Methodology	A general approach to studying research topics
Method	A specific research technique
Quantitative data	Data where the focus is on numbers and statistical analyses
Qualitative data	Data where the focus is on collecting information in a natural setting rather than an artificially constrained one such as an experiment (see [45])

The experiment is a widely used tool especially in the academic world. Experiments require some expertise in terms of design, data analysis and interpretation of results. However, there are a number of useful texts available for carrying out experimental work and writing up the results. These include References 46–48.

4.4.3.2 Modelling tools

There is an array of modelling tools available to the human factors specialist. These range from creating models of a product or system through to computer-aided-design (CAD) techniques and specific simulations. In their simplest form, it is often possible to build paper-based prototypes of a product/system. These might range from part to full prototypes of the complete system. These are particularly appropriate when the product or system is at a conceptual stage. For example, a slideshow presentation might be used to demonstrate the sequence of events when using a system. In contrast, a full prototype might include a physical mock-up of the system. This may involve a fairly faithful representation of the product. (I once took part in an experiment at the former Institute for Consumer Ergonomics, which involved evacuation from a double decker bus. In order for the experimenters to investigate this, a bus had been 'built' in their research laboratory!) Sometimes, bolsa wood or even cardboard cut outs can be used. I recall a talk given by an ergonomist who discussed the benefits of simulating the layout of an office using cardboard boxes. Hence, it can be seen that prototypes can be quite crude in terms of their construction and in comparison to the finished product.

There are many advantages attached to the use of prototypes when assessing products and systems. The primary one stems from the fact that they provide a tangible means by which users can evaluate the product. This results in improving the quality of the feedback that can be collected from the prospective users. A decade or so ago, rapid prototyping was popular. This conveys the idea of the technique being a quick means of helping assess the design of a product during its development. A further advantage of prototyping is that it is more efficient than evaluating the finished product since there is still an opportunity to make changes. This may not be possible for practical and cost reasons, if the final product is being assessed. Hence, the use of prototypes provides a very flexible approach [49]. In terms of disadvantages, it is suggested that it is only possible to prototype products when the task has been decided and can be fully described. Carroll and Rosson [50] illustrated this problem by referring to the task–artefact cycle. The design of a product will influence how it is used, and the user's goals when they use the product. Applying this during prototyping means that the developing design will determine the final design, and a vicious circle may be entered. Scenario analysis or storyboarding has been suggested as a way of providing a flexible means of trying to prevent the development of this circle. A storyboard 'consists of part of the interface mounted onto card or paper' ([51], p. 103). It is particularly appropriate when the design is past the embryonic stages. Card/paper is less intimidating than a screen prototype, and comprises a cheap and efficient way of allowing designers and prospective users to discuss design issues.

Another means of representing designs are state space and transition diagrams. These allow the operational states of a device, and legal/illegal/impossible operations to be represented in diagrammatic form. Often, matrices are used. These types of diagrams allow predictions concerning operation to be made. These can be used as part of an error analysis, and can provide the basis for addressing the redesign of a product or system (see [52]).

Prototyping is often used in conjunction with walkthrough and talkthrough techniques. Walkthroughs are when the intended users 'walk through' using the product or system (see [53]). The benefit of a walkthrough is that it allows using the interface to be simulated, and specific tasks to be considered. A variation on this is a cognitive walkthrough where the focus of the tasks is on the cognitive aspects of the user–system interaction [54]. Other modifications of the walkthrough technique include the programming walkthrough [55] and the cognitive jogthrough [56]. In the jogthrough technique, the evaluation session is recorded on video-tape allowing the evaluation to be conducted in a much shorter time. Lewis and Wharton [54] reported that about three times more problems could be found in the same amount of evaluation time by adopting the jogthrough technique. Finally, an extension of the basic walkthrough is the pluralistic walkthrough [57]. In this type of walkthrough, the practical evaluation is carried out by a number of different groups of individuals who have been involved in the system development, e.g. usability experts, users and the system developers.

Walkthroughs are often accompanied by talkthroughs, where users provide a verbal protocol as they are carrying out the task. Since it is impossible to know what a person is thinking, a talkthrough where individuals 'think aloud' is probably the closest we can get. The intention of a talkthrough is that users should be unconstrained in what they say, and they are not prompted or asked specific questions. Verbal protocols have frequently been used in problem-solving experiments in order to increase our understanding of how people solve problems. From this body of work, a number of implications arise when considering this technique in human factors research. The first is that individuals can exhibit very different behaviours but similar verbal reports [58]. The second concerns disruption to the task; having to talk whilst carrying out a walkthrough, for example, will probably result in a longer completion time. However, evidence from the problem-solving literature suggests that thinking aloud results in more accurate problem-solving with fewer errors being made. A study by Wright and Converse [59] supported this. They found that individuals taking part in a usability study made 20 per cent fewer errors when providing a verbal protocol than those who were working silently. A third issue relates to the embarrassment some individuals might feel whilst talking aloud when carrying out a task in an experimental setting. A fourth point concerns bias. Often individuals will provide comments that they anticipate the experimenter wants to hear. As a general rule, people generate factual statements rather than opinions when carrying out a talkthrough. Like video data, verbal protocols will result in a lot of data being produced. The time needed for analysis of this output needs to be taken into account when using this tool.

4.4.3.3 Fitting trials and mannequins

Although the emphasis of the methods covered in this chapter has been on measuring the cognitive attributes of the user, the increasing use of CAD technology for physical measurements is an important tool in the human factors toolkit. Fitting trials are experimental investigations of the relationships of the dimensions of a product with those of the user. These are increasingly being carried out using CAD technology and mannequins. The mannequins allow different dimensions, e.g. height, arm reach, to be manipulated in order to see how these fit within a workspace for a given sector of the population and their particular anthropometric characteristics. One of the best-known of the mannequins is SAMMIE (System for Aiding Man–Machine Interaction Evaluation) that was developed at Loughborough University. The web site address is: http://www.lboro.ac.uk/departments/cd/docs_dandt/research/ergonomics/sammie_old/sammie.html). Others include Jack and Combiman [60].

The main advantage of carrying out fitting trials using CAD is having the capability to vary the dimensions of the mannequin or the workspace at the touch of a button. Thus, the technique is fast, extremely accurate and flexible. The disadvantages stem from the fact that experienced personnel are needed to operate the

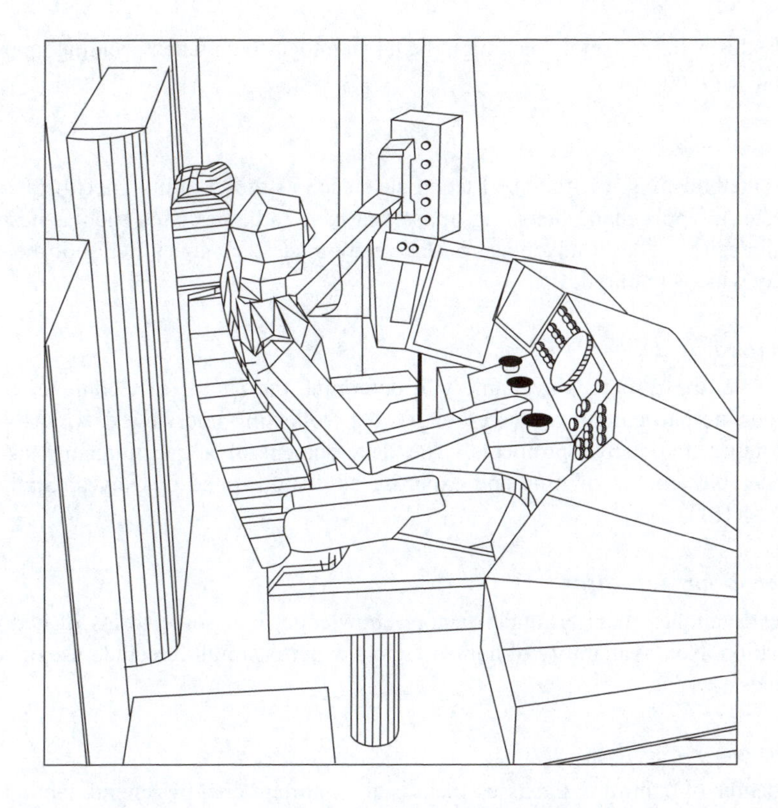

Figure 4.1 *Example of a SAMMIE display. (Reprinted with the permission of SAMMIE CAO Ltd.)*

CAD system; hence, costs can be high. Figure 4.1 shows an example of a SAMMIE display.

4.5 Which method to use?

A wide range of methods exists for assessing and evaluating the design of products and systems from a user perspective. The most commonly used ones have been covered in this chapter, but there are others, e.g. functional analysis, card sorts, repertory grid, etc. The issue concerns which method to choose, i.e. how do we select the one that is most appropriate from the human factors toolkit.

The answer to this lies in considering both the particular methods available and the characteristics of the product or system in conjunction with any requirements and constraints imposed by the assessment. It is likely that this final point will be a major factor in determining which method is appropriate. For example, if there is little time or resource available, a checklist approach would be more appropriate than developing a bespoke questionnaire. If there are no human factors specialists available to undertake the evaluation, this may rule out some techniques. In order to clarify this, a list of factors that might be taken into account when deciding upon the method is given below.

Time available

Some methods, e.g. off-the-shelf techniques such as the usability checklists, are a lot faster to apply than others, e.g. those that have to be designed such as bespoke questionnaires. The disadvantage is that ready-made techniques may not measure exactly what is required.

Resource availability

Like time, the resources available will determine the choice of technique. Some methods are particularly resource-hungry, e.g. collecting interview data, analysing video-tapes and verbal protocols. The development of a questionnaire takes a considerable amount of time and expertise as demonstrated by Noyes, Starr and Frankish [61].

Human factors expertise

Some techniques require human factors knowledge, e.g. some forms of heuristic evaluation. Non-availability of human factors expertise might preclude use of some methods.

Laboratory versus field

The degree of control in the assessment is an important one. In general, laboratory-based tasks can be more tightly controlled than those that take place in the field. The disadvantage is the loss of ecological validity, which may be important to the task being considered. An interesting compromise in this respect was the CAFÉ OF EVE (Controlled Adaptive Flexible Experimental Office of the Future in an Ecologically Valid Environment: [62]). This was a 'living research and development' laboratory for the design and evaluation of advanced office systems. It was set up in order to combine the best of laboratory and field approaches. However, it entailed high set up and running costs, and, as the workers knew they were in a research environment, some loss of ecological validity (i.e. the Hawthorne effect).

Participatory influence

The extent to which the researcher influences the sequence of events and the outcome of the assessment needs to be considered. In some techniques, e.g. observation studies, the presence of the experimenter can influence people's behaviour and subsequent results.

Bias

Some methods are more subject to bias, e.g. focus groups, where the use of stooges can influence the outcome of the discussions. The extent to which bias is influencing the results might therefore be a consideration.

Reliability and validity

Reliability is a measure of the extent to which the data collected from the same or similar group of people in similar circumstances would yield the same finding. It is usually expressed on a numerical scale from 0 (unreliable) to 1 (reliable).

Validity is the extent to which the data collected is actually measuring what you as the experimenter intended it to measure. For example, people completing a questionnaire may interpret the questions in different ways when they answer them.

The extent to which reliability and validity will need to be taken into account will depend on the application.

Ethical considerations

There are ethical issues to be considered when collecting data about users: under the UK Data Protection Act of 1998, all information to be used for research purposes needs to have the informed consent of the participants. This has important implications for the use of observation and similar studies.

Although the above list may look daunting, the important message from the human factors viewpoint is to take the user or the prospective user into account in the design and evaluation process. Many designs in the built world have not done this. It is suggested that even minimal input from a human user will enhance the design in the longer term, and accrue benefits for future users.

4.6 Exercises

4.6.1 Exercise 1: heuristic evaluation

Heuristic evaluation requires that a small group of testers (evaluators) examine an interface and judge its compliance with recognised usability principles (heuristics). Select an interface, e.g. a web site, and act as an evaluator to assess the extent to which the design meets the following ten heuristics by answering the following questions.

1. Visibility of system status
   Are users kept informed about system progress with appropriate feedback within reasonable time?
2. Match between system and the real world
   Does the system use concepts and language familiar to the user?
3. User control and freedom
   Can users do what they want when they want?
4. Consistency and standards
   Do design elements such as objects and actions have the same meaning in different situations?
5. Error prevention
   Can users make errors which good design would prevent?
6. Recognition rather than recall
   Are design elements such as objects, actions and options visible?

7. Flexibility and efficiency of use
   Can users customise frequent actions or use short cuts?
8. Aesthetic and minimalist design
   Do dialogues contain irrelevant or rarely needed information?
9. Recognition, diagnosis and recovery from errors
   Are error messages expressed in plain language; do they accurately describe the problem and suggest a solution?
10. Help and documentation
   Is appropriate help information supplied, and is this information easy to search and focused on the user's tasks?

When you have answered all the questions, assess the extent to which your interface meets basic usability requirements.

4.6.2 *Exercise 2: checklists*

Checklists have been used extensively in the evaluation of products and systems. One of the simplest (and shortest) is SUS (System Usability Scale: [12]). Select two systems, e.g. search engines, and compare their usability by applying the SUS checklist in Table 4.3.

The SUS is intended to be completed fairly quickly, i.e. record your immediate response rather than thinking about items for a long time. If you feel you cannot respond to an item, mark the centre point of the scale.

Table 4.3 SUS (System Usability Scale)

	Strongly disagree				Strongly agree
1. I think I would like to use this system frequently.	1	2	3	4	5
2. I found the system unnecessarily complex.	1	2	3	4	5
3. I thought the system was easy to use.	1	2	3	4	5
4. I think that I would need the support of a technical person to be able to use this system.	1	2	3	4	5
5. I found the various functions in this system were well integrated.	1	2	3	4	5
6. I thought there was too much inconsistency in this system.	1	2	3	4	5
7. I would imagine that most people would learn to use this system very quickly.	1	2	3	4	5
8. I found the system very cumbersome to use.	1	2	3	4	5
9. I felt very confident using the system.	1	2	3	4	5
10. I need to learn a lot of things before I could get going with this system.	1	2	3	4	5

To calculate the SUS score, first, sum the contributions from each item. Each item's score contribution will range from 0 to 4. For items 1, 3, 5, 7 and 9, the score contribution is the scale position minus 1. For items 2, 4, 6, 8 and 10, the contribution is 5 minus the scale position. Multiply the sum of the scores by 2.5 to obtain the overall value of usability for a particular system. This final value will lie in the range 0 to 100.

4.7 Acknowledgements

Grateful thanks are extended to Chris Baber and Jurek Kirakowski for their input to this chapter.

4.8 References

1 NOYES, J. M.: 'QWERTY – the immortal keyboard', *Computing and Control Engineering Journal*, 1998, **9**, pp. 117–122
2 PHEASANT, S. T.: 'Bodyspace: anthropometry, ergonomics and the design of work' (Taylor & Francis, London, 1996, 2nd edn.)
3 BIAS, R. G., and MEYHEW, D. J.: 'Cost-justifying usability' (Academic Press, Boston, MA, 1994)
4 STANTON, N. A., and YOUNG, M. S.: 'What price ergonomics?', *Nature*, 1999, **399**, pp. 197–198
5 RAUTERBERG, M., and STROHM, O.: 'Work organisation and software development', *Annual Review of Automatic Programming*, 1992, **16**, pp. 121–128
6 HART, S. G., and STAVELAND, L. E.: 'Development of NASA-TLX (Task Load Index): results of empirical and theoretical research', in HANCOCK, P. A., and MESHKATI, N. (Eds): 'Human mental workload' (North-Holland, Amsterdam, 1988)
7 REID, G., and NYGREN, T.: 'The subjective workload assessment technique: a scaling procedure for measuring mental workload', in HANCOCK, P., and MESHKATI, N. (Eds): 'Human mental workload' (North Holland, New York, 1988)
8 TATTERSALL, A. J., and FOORD, S. F.: 'An experimental evaluation of instantaneous self-assessment as a measure of workload', *Ergonomics*, 1996, **39**, pp. 740–748
9 OPPENHEIM, A. N.: 'Questionnaire design, interviewing and attitude measurement' (Pinter, London, 1992, 2nd edn.)
10 NIELSEN, J.: 'Heuristic evaluation', in NIELSEN, J., and MACK, R. L. (Eds): 'Usability inspection methods' (Wiley, New York, 1994)
11 NOYES, J. M., and MILLS, S.: 'Heuristic evaluation: a useful method for evaluating interfaces?' (submitted)
12 BROOKE, J.: 'The System Usability Scale' (Digital Equipment Co. Ltd., Reading, 1986)

13 CHIN, J. P., DIEHL, V. A., and NORMAN, K. L.: 'Development of an instrument measuring user satisfaction of the human-computer interface'. Proceedings of CHI '88 (ACM, New York, 1988) pp. 213–218
14 KIRAKOWSKI, J., and CORBETT, M.: 'Measuring user satisfaction', in JONES, D.M., and WINDER, R. (Eds): 'People and Computers IV' (Cambridge University Press, Cambridge, 1988) pp. 329–430
15 KIRAKOWSKI, J., and CORBETT, M.: 'SUMI: the software usability measurement inventory', *British Journal of Educational Technology*, 1993, **24**, pp. 210–214
16 HULZEBOSCH, R., and JAMESON, A.: 'FACE: a rapid method for evaluation of user interfaces', in JORDAN, P. W., THOMAS, B., WEERDMEESTER, B. A., and McCLELLAND, I. L. (Eds): 'Usability evaluation in industry' (Taylor & Francis, London, 1996) pp. 195–204
17 MACLEOD, M., BOWDEN, R., BEVAN, N., and CURSON, I.: 'The MUSiC performance measurement method', *Behaviour and Information Technology*, 1997, **16**, pp. 279–293
18 KIRAKOWSKI, J., and CIERLIK, B.: 'Measuring the usability of web sites'. Proceedings of Human Factors and Ergonomics Society (HFES) Conference, Chicago, IL (HFES, Santa Monica, CA, 1998)
19 WILSON, V.: 'Focus groups: a useful qualitative method for educational research?', *British Educational Research Journal*, 1997, **23**, pp. 209–224
20 McCLELLAND, I. L., and BRIGHAM, F. R.: 'Marketing ergonomics – how should ergonomics be packaged?', *Ergonomics*, 1990, **33**, pp. 519–526
21 AINSWORTH, L. K.: 'Task analysis', in NOYES, J. M., and BRANSBY, M. L. (Eds): 'People in control: human factors in control room design' (IEE, Stevenage, 2001) pp. 117–132
22 DRURY, C. G.: 'Methods for direct observation of performance', in WILSON, J. R., and CORLETT, E. N. (Eds): 'Evaluation of human work: a practical ergonomics methodology' (Taylor & Francis, London, 1995) pp. 45–68
23 ANNETT, J., and STANTON, N. A. (Eds): 'Task analysis' (Taylor & Francis, London, 2000)
24 SHEPHERD, A.: 'HTA as a framework for task analysis', *Ergonomics*, 1998, **41** (11), pp. 1537–1552
25 SHEPHERD, A.: 'HTA as a framework for task analysis', in ANNETT, J., and STANTON, N.A. (Eds): 'Task analysis' (Taylor & Francis, London, 2000) pp. 9–24
26 CARD, S. K., MORAN, T. P., and NEWELL, A.: 'The psychology of human-computer interaction' (LEA, Hillsdale, NJ, 1983)
27 PAYNE, S. J., and GREEN, T. R. G.: 'Task-action grammars: a model of the mental representation of task languages', *Human-Computer Interaction*, 1986, **2**, pp. 93–133
28 DIAPER, D.: 'Task analysis for knowledge descriptions (TAKD); the method and an example', in DIAPER, D. (Ed.): 'Task analysis for human-computer interaction' (Ellis Horwood, Chichester, 1989)

29 MILITELLO, L. G., HUTTON, R. J. B., and MILLER, T.: 'Applied cognitive task analysis' (Klein Associates Inc., Fairborn, OH, 1997)

30 DIAPER, D., McKEARNEY, S., and HURNE, J.: 'Integrating task and data flow analyses using the pentanalysis technique', in ANNETT, J., and STANTON, N. A. (Eds): 'Task analysis' (Taylor & Francis, London, 2000) pp. 25–54

31 SALVENDY, G. (Ed.): 'Handbook of human factors and ergonomics' (Wiley, New York, 1997, 2nd edn.)

32 KIRWAN, B., and AINSWORTH, L. K.: 'A guide to task analysis' (Taylor & Francis, London, 1992)

33 NOYES, J. M.: 'Human error', in NOYES, J. M., and BRANSBY, M. L. (Eds): 'People in control: human factors in control room design' (IEE, Stevenage, UK, 2001) pp. 3–15

34 SWAIN, A. D., and GUTTMANN, H. E.: 'A handbook of human reliability analysis with emphasis on nuclear power plant applications', Nureg/CR-1278 (USNRC, Washington DC, 1983)

35 WILLIAMS, J. C.: 'HEART – a proposed method of assessing and reducing human error'. Proceedings of the 9th Symposium on *Advances in reliability technology* (University of Bradford, UK, 1986)

36 EMBREY, D. E.: 'SHERPA: a systematic human error reduction and prediction approach'. Proceedings of the International Topical Meeting on *Advances in human factors in nuclear power systems* (Knoxville, TN, 1986)

37 WHALLEY, S. P.: 'Minimising the cause of human error'. Proceedings of the 10th Symposium on *Advances in reliability technology* (Elsevier, London, 1988)

38 REASON, J. T.: 'Human error' (Cambridge University Press, Cambridge, 1990)

39 EMBREY, D. E.: 'Quantitative and qualitative prediction of human error in safety assessments'. Proceedings of IChemE Symposium Series No. 130, London, 1992

40 BABER, C., and STANTON, N. A.: 'Task analysis for error identification: a methodology for designing error-tolerant consumer products', *Ergonomics*, 1994, **37**, pp. 1923–1941

41 INTERNATIONAL STANDARDS ORGANISATION.: 'Guidance on usability, ISO 9241' (International Organisation for Standardisation, Geneva, 1997)

42 PICARD, R. W.: 'Affective computing' (MIT Press, Cambridge, MA, 1997)

43 SHACKEL, B.: 'The concept of usability'. Proceedings of IBM Software and Information Usability Symposium (IBM, New York, 1981) pp. 1–30

44 SILVERMAN, D.: 'Interpreting qualitative data: methods for analysing talk, text and interaction' (Sage, London, 1993)

45 MARSHALL, C., and ROSSMAN, G.: 'Designing qualitative research' (Sage, London, 1989)

46 BLAIKIE, N.: 'Analyzing quantitative data: from description to explanation' (Sage, London, 2003)

47 FIELD, A., and HOLE, G.: 'How to design and report experiments' (Sage, London, 2003)

48 SMITH, J. A.: 'Qualitative psychology: a practical guide to research methods' (Sage, London, 2003)
49 BEAGLEY, N .I.: 'Field-based prototyping', in JORDAN, P. W., THOMAS, B., WEERDMEESTER, B. A., and McCLELLAND, I. L. (Eds): 'Usability evaluation in industry' (Taylor & Francis, London, 1996) pp. 95–104
50 CARROLL, J. M., and ROSSON, M. B.: 'Deliberated evolution: stalking the view matcher in design space', *Human-Computer Interaction*, 1991, **VI**, pp. 281–318
51 FAULKNER, C.: 'The essence of human-computer interaction' (Prentice-Hall, Hemel Hempstead, UK, 1998)
52 STANTON, N. A., and BABER, C.: 'Task analysis for error identification: applying HEI to product design and evaluation', in JORDAN, P. W., THOMAS, B., WEERDMEESTER, B. A., and McCLELLAND, I. L. (Eds): 'Usability evaluation in industry' (Taylor & Francis, London, 1996) pp. 215–224
53 LEWIS, C., POLSON, P., WHARTON, C., and RIEMAN, J.: 'Testing a walkthrough methodology for theory-based design of walk-up-and-use interfaces'. Proceedings of CHI '90, Seattle, WA (ACM, New York, 1990) pp. 235–242
54 LEWIS, C., and WHARTON, C.: 'Cognitive walkthroughs', in HELANDER, M., LANDAUER, T. K., and PRABHU, P. (Eds): 'Handbook of human-computer interaction' (Elsevier, North-Holland, 1997, 2nd edn.) pp. 717–732
55 BELL, B., RIEMAN, J., and LEWIS, C.: 'Usability testing of a graphical programming system: things we missed in a programming walkthrough'. Proceedings of CHI '91, New Orleans, LA (ACM, New York, 1991) pp. 7–12
56 ROWLEY, D. E., and RHOADES, D. G.: 'The cognitive jogthrough: a fast-paced user interface evaluation procedure'. Proceedings of CHI '92 (ACM, New York, 1992) pp. 389–395
57 BIAS, R. G.: 'The pluralistic usability walkthrough: coordinated empathies', in NIELSEN, J., and MACK, R. L. (Eds): 'Usability inspection methods' (Wiley, New York, 1994) pp. 63–76
58 NISBETT, R. E., and WILSON, T. D.: 'Telling more than we can know: verbal reports on mental processes', *Psychological Review*, 1977, **84**, pp. 231–259
59 WRIGHT, P. C., and CONVERSE, S. A.: 'Method bias and concurrent verbal protocol in software usability testing'. Proceedings of the Human Factors Society Annual Meeting (HFS, Santa Monica, CA, 1992) pp. 1220–1224
60 McDANIEL, J. W.: 'CAE tools for ergonomic analysis', in ANDRE, T. S., and SCHOPPER, A. W. (Eds): 'Human factors engineering in system design' (Crew System Ergonomics Information Analysis Center, Wright-Patterson AFB, OH, 1997) pp. 100–140
61 NOYES, J. M., STARR, A. F., and FRANKISH, C. R.: 'User involvement in the early stages of the development of an aircraft warning system', *Behaviour and Information Technology*, 1996, **15**, pp. 67–75
62 GALE, A., and CHRISTIE, B. (Eds): 'Psychophysiology and the electronic workplace' (Wiley, Chichester, UK, 1987)

Further Reading

JORDAN, P. W., THOMAS, B., WEERDMEESTER, B. A., and McCLELLAND, I.: 'Usability evaluation in industry' (Taylor & Francis, London, 1996)

This book comprises 26 relatively short chapters compiled by authors primarily from industry concerning human factors work. It thus provides a unique snapshot of human factors activities and practice with an emphasis on tools and methods used in the industrial setting. A little dated now, but an excellent read.

NOYES, J. M., and BABER, C.: 'User-centred design of systems' (Springer, London, 1999)

As the title suggests, this book focuses primarily on the design and evaluation of computer systems from the perspective of the user. Users are considered in terms of their cognitive and physical attributes and their social needs, and at various stages throughout the design lifecycle. This is essentially a book on the methods available for those designing for humans.

PHEASANT, S.: 'Bodyspace: anthropometry, ergonomics and the design of work (2nd edn.)' (Taylor & Francis, London, 1996)

Bodyspace focuses more on the physical design of workspaces and the anthropometric characteristics of the users than the cognitive or social aspects. The book is in two parts: the first part comprises the bulk of the text while the second part is mainly tables of anthropometric data. This is a classic text, and much needed in terms of filling a gap in the market.

STANTON, N. A., and YOUNG, M. S.: 'A guide to methodology in ergonomics: designing for human use' (Taylor & Francis, London, 1999)

This book describes a range of methodologies for analysing a product or process. Validity and reliability, as well as the costs and benefits of each method are assessed with usability and applicability. If you are looking for a relatively short guide to methods used in ergonomics, this book will meet your needs.

WILSON, J. R., and CORLETT, E. N.: 'Evaluation of human work: a practical ergonomics methodology' (Taylor & Francis, London, 1995, 2nd edn)

The first edition of this book appeared in 1990, and in revised, updated and extended form in the second edition in 1995. It is an extremely detailed and comprehensive compendium of ergonomics methods and techniques covering every aspect of human work. If you are looking for in-depth coverage of a particular Human Factors method, this excellent book will provide it for you.

Chapter 5
Task analysis

Leslie K. Ainsworth

5.1 Introduction

Task analysis is the name given to a range of techniques that can be used to examine the ways that humans undertake particular tasks. Some of these techniques focus directly upon task performance, whilst others consider how specific features of the task, such as the interfaces, operating procedures, team organisation or training, can influence task performance. Thus, task analysis techniques can be used either to ensure that new systems are designed in a way that assists the users to undertake their tasks safely and effectively, or to identify specific task-related features that could be improved in existing systems. The general family of tools known as Task Analysis can contribute substantially valuable methodologies. This chapter should be read in conjunction with the additional information concerning error and its quantification provided in Chapter 8 of this book.

The task analysis process is essentially about understanding the requirements of a person who might undertake a task and then assessing whether the task and the environment, or situation, in which the task is undertaken is designed in a manner that meets these requirements. At the highest level these requirements can often be met by developing a set of task descriptions that can then be used to identify the basic interface requirements. However, task performance seldom depends solely upon the interfaces, because there are many other influences upon performance, such as training, team organisation, communications, workload, time pressure or environmental factors, that can all influence the way in which human operators will behave. Therefore, in order to ensure an acceptable level of performance, regardless of whether this is defined in terms of safety or productivity, it will be necessary to utilise those task analysis methods that will enable the analyst to obtain data on the issues of concern. Where the analyst identifies mismatches between what is provided and what the human operator requires, it will be possible to develop ways to overcome these mismatches.

In some situations the issues of interest can be defined before any task analysis is undertaken. For instance, the purpose of a task analysis may be defined at the outset as being to assess whether the workload imposed by a proposed system will be acceptable. However, in most task analyses the issues of interest, and, therefore, the direction of the study, will only become clear as more data are obtained. Hence, task analysis will usually involve an iterative process of identifying potential weaknesses and then undertaking further analysis in order to gain a better understanding of the underlying issues.

This process is illustrated in the overview model of the task analysis process that is presented in Figure 5.1 and the remainder of this chapter is based upon this structure.

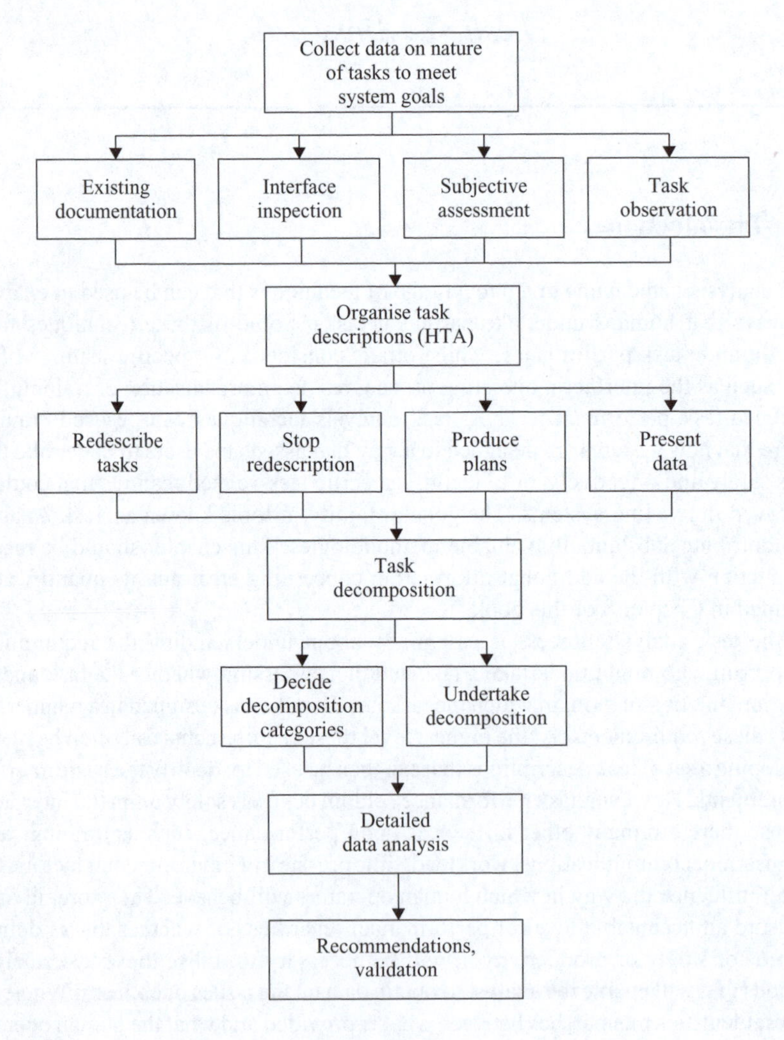

Figure 5.1 An overview of the task analysis process

5.2 Data collection for task analysis

In any task analysis it is first necessary to collect data that will determine how tasks are done and then to supplement these data with more focused information about aspects of the tasks, or the task environment, that are of particular interest. Inevitably this will require that the analyst will have to use a variety of sources. This means that on occasion it will be necessary to resolve potentially conflicting data, or to collate several data items to clarify ambiguities. Thus, the data collection problems are similar to those that may be met in the analysis of any complex system in which many of the details are either difficult to obtain or are open to different interpretations.

However, there are also some aspects of data collection that are particularly important for task analysis and so the following brief summary of some of the main issues is presented, based upon the nature of the information source.

5.2.1 Existing documentation

Even in a new system, a vast amount of documentation is often available about tasks and task interfaces, including such things as functional descriptions, system descriptions, logic diagrams, piping and instrumentation diagrams, interface layout diagrams, training programmes and written procedures. In existing systems this could be supplemented by historical information about past problems, such as operator logs and critical incident reports.

However, whilst this information is useful for task analysis purposes, it is important to recognise where there may be potential inaccuracies. In a review of several task analyses undertaken for a wide range of applications (summarised in [1]) a major source of such inaccuracies was identified as being due to either ignoring many features that could influence human performance, or making implicit assumptions that are not documented. This is particularly true for system descriptions, because these tend to focus on the functional requirements rather than the human requirements. Thus, features that can have significant implications to human performance may either not be documented, or decisions may be made without properly considering their potential impact. For example, whilst a designer is likely to define the electrical specification for a particular switch, other features that may directly affect the way that the switch is used, such as its accessibility or the feedback that it provides, may not be specified. If an inappropriate switch is then installed, an operator may subsequently make errors because the status of that switch is misidentified, or because it is difficult to reach during an emergency.

In order to guard against such misinterpretation, an analyst must always be careful not to extrapolate beyond what is actually specified. Also, where possible, all information should be corroborated from a different source.

Written procedures and training documentation can both be particularly useful for developing task descriptions. However, the analyst must be aware that some steps may be covered very briefly, or omitted altogether. This is particularly likely in training notes, because these will tend to focus upon aspects that tend to cause difficulties for trainees, or that have safety significance.

Operational feedback records and incident reports can provide valuable information about features that have caused past difficulties, and which, therefore, may well merit special attention.

5.2.2 *Interface inspection*

If the control interfaces, or a simulation of them, have already been built, much information about those interfaces can be obtained from making a systematic survey of them. This could be done as part of a walk-through exercise in which particular controls and displays are examined as they become used during a task sequence, but it is often preferable to undertake a systematic inspection of an entire interface, either by working from the top left to the bottom right of each control panel, or by systematically working through a set of VDU display formats.

Such inspections should consider compliance with basic ergonomics guidelines such as the consistency of labelling (terminology, size and relative position), the grouping of items, legibility (of labels and parameter values), coding consistency, the adequacy of feedback, direction of motion stereotypes and control/display compatibility. For important indications, that may be more remote from the main operating position, it may also be necessary to check the sightlines to ensure that both the shortest and the tallest operator have unhindered visibility of such items without having to move from their normal working positions.

For interfaces that have been operational for some time, it can also be useful to undertake an operator modifications survey to look for areas where the operators have felt it necessary to supplement the interfaces by adding Dymo tape labels or marking critical limits on displays.

5.2.3 *Subjective assessment*

A considerable amount of useful task information can be obtained by individual or group discussions with operational personnel or technical experts. However, there is always a risk that such information could represent a distorted, or even inaccurate, view. Therefore, an analyst must take considerable care to ensure that their interpretation of such discussions is accurate and to recognise situations where there are different ways to undertake a task. Problems of misinterpretation can be reduced by providing an opportunity for respondents to comment on an analyst's interpretation, either during a discussion session, or by undertaking a validation of the analyst's interpretation at a later date. The importance of such validation cannot be overstressed because it is surprisingly easy for an analyst either to miss out an important aspect of a task, or to derive an ambiguous or erroneous view of the task by misinterpreting an interviewee's accounts or opinions. It must always be remembered that what may appear straightforward to a highly experienced operator may in reality involve some highly complex cognitive behaviour and decision-making. Thus, whilst there is a tendency to obtain task data from the most experienced operators, in some cases, a less experienced operator may be a better source of information about aspects of the task that eventually become highly skilled and undertaken with little apparent conscious effort.

The best way to identify alternative approaches to undertaking a task is to have discussions with more than one operator and then to compare their approaches. However, care must be taken to select the most appropriate respondents. The author recalls a situation where the most experienced operator described how an additional pump would be brought into operation and this approach was later corroborated by another person on the same shift. However, further investigation with a relative newcomer revealed that there was a better way to do this task, which involved less thermal shock to the system. The experienced operator was not aware of this problem, and so for several years he had continued to add pumps in this way and to train others to adopt the same, suboptimal, approach.

In order to get the most out of group discussions it is advisable to structure such discussions upon a particular procedure, or to undertake them as a debrief session after a particular task has been completed. It is also helpful if those attending group discussions can be pre-warned of the topics that will be covered. In any discussion group there must be an appropriate mix of individuals with the necessary expertise and, if there is to be a series of discussion sessions, it will be helpful if the same individuals can be used. To encourage discussion whilst avoiding certain views predominating, only one expert should generally be permitted from each area of expertise. There should also not be too great a difference between the managerial levels of the members of a discussion group, otherwise, there is a strong risk that the views of senior management will dominate.

5.2.4 Task observations

Task performance can be observed directly during walk-throughs, simulator trials, or whilst tasks are being carried out on the real plant. Walk-throughs can be undertaken in a similar manner to the actual tasks, apart from actually intervening with the plant, or else they can be interrupted by the analyst so that detailed questions can be asked. All of these data collection methods should provide the analyst with sufficient information to develop a comprehensive set of task descriptions and to identify the controls and displays that are used. They can also provide a way to identify potential problems, such as difficulties locating particular information. During any task observations the analyst must be aware of ergonomics guidelines and note any situations where these are not being met. Sometimes these will be obvious, because an operator will experience some difficulties, but in other cases this may be more difficult because an operator has learned to overcome such deficiencies. Nevertheless, even when operators experience no apparent difficulty with a feature that does not comply with ergonomics guidance, this should still be noted, because these features may cause difficulties to less experienced personnel, or to anyone under more stressful conditions.

The debriefs or the walk-through questions also provide a opportunity to explore the underlying decision-making and reasoning, and also to identify alternative sources of information, or corroborating sources of information, such as upstream flow meters to confirm a valve position.

5.3 Hierarchical Task Analysis (HTA)

Hierarchical Task Analysis (HTA) provides a convenient way to identify, organise and represent the constituent tasks and sub-tasks that are involved in a complex activity. The starting point of an HTA is a high-level description of the main goal of an activity. The analyst then redescribes this main goal in greater detail as a small, but comprehensive set of sub-goals. This redescription process is then continued to develop a hierarchically organised set of task descriptions, such that at the lowest levels of this hierarchy the tasks are described in sufficient detail for the analyst.

It must be stressed that the main purposes of HTA are to provide an analyst with a framework to understand the relationships between different task elements and to ensure that no constituent tasks are missed. In particular, it must be stressed that the hierarchical structure does not necessarily imply that the tasks will be undertaken in a hierarchical way. This means that HTA can be applied to a series of steps that are always undertaken in an identical linear sequence, just as effectively as it can be used for a complex diagnostic task. Thus, it is possible that a complex task could be redescribed into several different hierarchies. Therefore, providing that they correctly reflect the relationships between the tasks that are described, it is permissible for different analysts to produce different HTAs for the same task.

5.3.1 The redescription process

In order to avoid unnecessary complexity within an HTA, it will often be preferable to produce a set of HTAs rather than to attempt to cover all situations within a single HTA. Therefore, this has to be considered before any redescription is started. Generally speaking it is recommended that separate HTAs should be produced in the following situations:

- Different scenarios.
- Different operational modes, except when there is specific interest in tasks undertaken during the transition between operational modes.
- When operational and maintenance tasks are undertaken independently of each other on similar equipment.
- When tasks are undertaken by different persons, providing that these tasks require little interaction or communication with others.

For each individual HTA the main goal is first defined; then this is broken down into a set of all the sub-goals that must be fulfilled in order to ensure that the main goal is met. In some situations all of these sub-goals will have to be met, but in other situations some of the sub-goals may only apply in certain conditions, or they may even be mutually exclusive. For instance the goal of controlling temperature could be redescribed into three sub-goals for maintaining the current temperature, increasing it or decreasing it. The sub-goals are subsequently redescribed in a similar way and this is continued to produce a hierarchy consisting of sub-goals at the higher levels and task descriptions at the lower levels, with each of these items given a unique hierarchical number.

In order to ensure that no important tasks are missed, it is recommended that each superordinate goal should not be broken down into more than seven or eight sub-tasks. This should ensure that an analyst can mentally visualise all the main goals/activities that might be required to successfully achieve the superordinate goal. Thus, whilst in some conditions it may not be necessary meet all of the subordinate goals, an analyst should check that for each level of redescription the superordinate goal can always be fully met by some combination of the subordinate goals at that level.

Each task description should consist of a short, but clear and unambiguous summary of the goals or tasks that would be sufficient to identify the essential elements of that task. Hence, general statements such as *Open valve* should be avoided and better identification of the valve or flowline should be provided. It is also necessary to ensure that the wording for each task description is unique. In particular, when a task is redescribed, none of the redescribed items should use the same wording as the original. However, it is acceptable to repeat the same wording between some of the lowest-level elements when they imply very similar tasks and where these are relatively straightforward.

At this juncture it is necessary to warn analysts to resist the temptation to blindly base an HTA directly upon an existing hierarchical structure, such as a written procedure. This is because HTAs developed in this way would use an identical framework to the written procedures, so that any errors or limitations in the procedures would be repeated. Thus, any such HTA would be unlikely to identify any important steps or actions that were not specifically mentioned in the source procedures. The author has seen many examples of HTAs produced in this way that have then been used to prove that the procedures are comprehensive, with no appreciation of the circularity of this argument.

Whilst the analysts must be permitted some flexibility in the HTA process, it is appropriate to provide some guidance about the way that task redescriptions can be organised. The following approaches to task redescription can be used:

- The most useful way to redescribe a superordinate task is by identifying the functions and sub-functions that have to be completed.
- At the lower levels in an analysis the different systems and equipments that are used can often serve as a helpful way to redescribe tasks (e.g. it may be useful to split the communication function dependent on the equipment used, such as phone, pager or public address).
- Once a set of redescriptions has been developed at a particular level, it is often helpful to arrange them into the sequential order in which they would normally be undertaken.
- In complex, multi-person tasks it is often helpful to keep tasks undertaken by the same person, or at similar locations, close to each other within an HTA.

5.3.2 *Stopping rules*

As the redescription process continues there comes a point at which it is no longer fruitful to redescribe the tasks in any greater detail. This level depends upon the requirements of the analyst and it is often very clear when this point has been reached. Indeed, as task analysis is an iterative process, an analyst might well take a relatively arbitrary decision about where to stop redescription and then amend the HTA by undertaking further redescription at any points where subsequent analysis suggested that this might be useful.

Alternatively, an analyst could set criteria in advance to decide where redescription should be stopped. When HTA was first developed by Annett and Duncan [2] it was intended as a method of examining training requirements and so they suggested that redescription should only continue whilst the consequences of making an error were considered critical. Therefore, they argued that when there was a very low probability that an action was required or when the system cost of an error was low, no further redescription was needed. This was formalised into what is now known as the $P \times C$ Rule, which stated that redescription should cease at the point at which the probability of an error and the cost of its potential consequences fell below a particular criterion level.

Other stopping rules are directly related to the subsequent purpose of the analysis. These criteria include continuing the redescription to the point at which:

- The main interfaces or functional elements within an interface can be identified.
- The tasks have sufficient detail to enable a workload assessment to be undertaken.
- The underlying knowledge and skills can be defined in sufficient detail. Thus for selection and manpower planning less detail is required than for developing a training programme.

5.3.3 *Plans*

At the end of the redescription process the task descriptions that have been obtained should provide a comprehensive picture of the tasks that are involved. However, in most situations, it is also necessary to understand how the constituent tasks are related to each other. For this purpose HTA plans are produced. Each time a superordinate item is redescribed, the relationships between all its immediately subordinate items is investigated and a separate plan is developed that explains the conditions necessary for undertaking that section of the plan. Thus, for example, if 1.2 was redescribed into three tasks, Plan 1.2 would describe the relationships between 1.2.1, 1.2.2 and 1.2.3. The simplest plan is for the tasks to be undertaken as a simple linear sequence, with no constraints, other than that the previous steps have been completed. This is so common that, in order to avoid cluttering an HTA, most analysts do not specifically present these plans where all the tasks are undertaken as a simple linear sequence.

In other cases certain conditions may have to be met before a task can be left (e.g. 'continue until the temperature reaches 100°C', or 'continue for 10 mins'), whilst in other situations it may be necessary to undertake different actions dependent upon the situation (e.g. 'if the pressure rises above 5 bar, do this, otherwise do something else').

All plans must involve at least one of the following structures and many may be a hybrid of them.

- Simple linear sequence.
- Constrained linear sequence, where progress to the next element depends upon particular conditions being met.
- Unordered list, where a person is free to undertake the tasks in any order.
- Conditional or free choice branching, where at each branch point a person must decide which branch to follow next.
- Condition attainment looping, where a task is continued until certain conditions are met.
- Continual looping, such as monitoring and control tasks that are undertaken intermittently in parallel with other tasks.
- Concurrent tasks, where a person has to attend to two or more tasks over a period of time.

If an analyst identifies tasks that are being undertaken as an unordered list, it is suggested that a check should be made to ensure that these should really be undertaken in this way. In the author's experience, even when operators consider that tasks can be undertaken in any order, there are sometimes good reasons for adopting a specific approach.

5.3.4 Presentation of HTA information

It is extremely helpful, both for the analyst and for anyone reading an HTA report, to produce HTA diagrams like the one presented in Figure 5.2. This shows the hierarchical relationships between the tasks and makes it easier to see how sets of tasks interrelate. However, the value of HTA diagrams will be limited if they are too complex to enable the information on them to be readily assimilated. Therefore, in a complex HTA it may be helpful either to reduce the amount of supplementary information that is shown, to split the diagram on to several pages, or to remove the lowest level elements. However, the latter approach should only be taken when the lowest level elements are trivial or repetitive (e.g. if most communications tasks split into similar elements for using either the phone or a radio).

Where HTA is used, an HTA diagram should be considered as the reference point for the task analysis. Therefore, all other references to tasks shown in the HTA diagram should use identical wording. If possible, it will also be extremely helpful to show the hierarchical numbering at the top of each box on the diagram. However, this may not be practical if there are five or more levels, and in this case it may be sufficient just to show the numbering for the top-level boxes on each page.

Considerable flexibility is available in the way that HTA data are presented, especially if colour is used. However, whilst there are no firmly established conventions, the following are recommended.

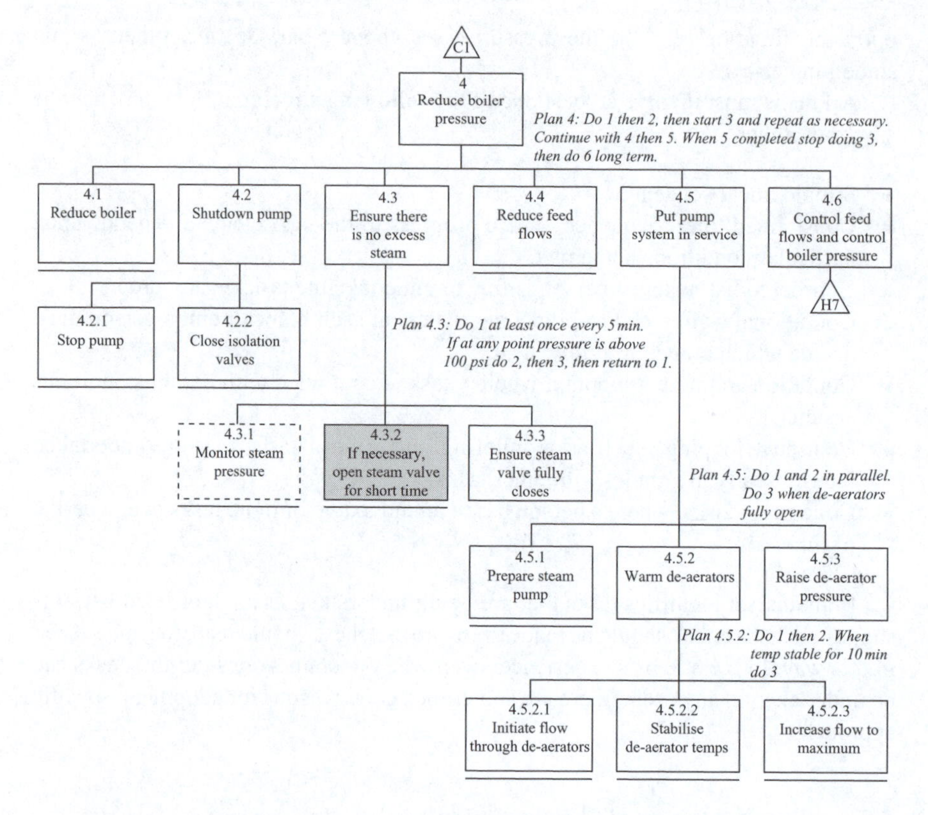

Figure 5.2 An HTA diagram

- Where plans are shown, these can be presented as plain text, as conditional logic statements, or as flow diagrams. However, for most audiences, plain text is preferred. In order to differentiate plans from other information, they should be presented in italic text. It can also be confusing to present plans within boxes and so this should not be done.
- Shading and colouring the text boxes can be used to differentiate task features that an analyst deems particularly important. Such features could include; safety critical tasks, tasks done by particular persons, tasks done at particular locations, tasks for which recommendations are made, tasks where potential errors are likely, or tasks that share particular characteristics (e.g. decision-making). However, no more than three classes of feature should be shown in this way.
- Repetitive tasks, such as monitoring, can often be shown very effectively by drawing their boxes with pecked lines.
- It is useful to indicate those tasks where no further redescription has been made; this is commonly indicated by drawing a horizontal line below that box. This serves to differentiate these from task boxes where further redescription was carried out.

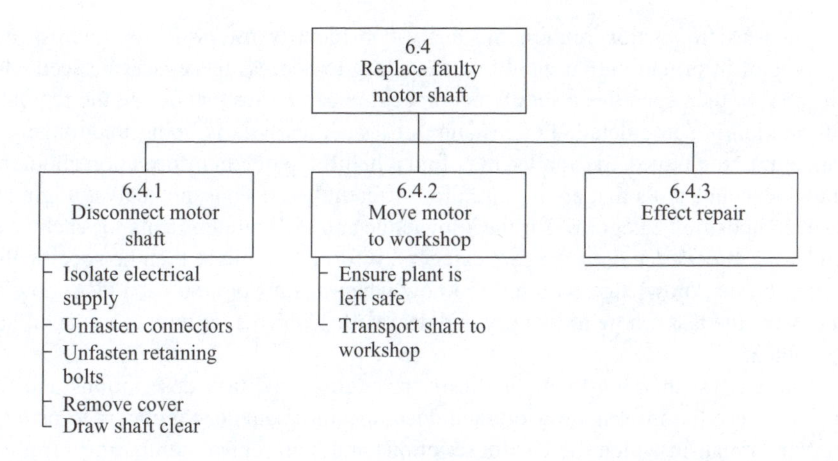

Figure 5.3 Ladder style of HTA diagram

- Where further redescription is shown on another page, there should be a triangular box below that box with a clear reference to the page and the entry position on the next page. Then, on that next page, another triangular box should indicate the location of the superordinate tasks and the lowest level box from the previous page should be repeated.

Although HTA diagrams normally follow the box structure as presented in Figure 5.2, an alternative structure, known as a ladder, can be used in some cases to show the lower levels, so that more information can be shown. An example of this is given in Figure 5.3.

HTAs can also be presented in a tabular format (see [3] for further details).

5.4 Task decomposition

At the end of the HTA stage an analyst will have a set of task descriptions that are linked together by plans which indicate the conditions necessary for undertaking particular tasks. Thus, the analyst knows the basic structure in terms of what has to be done and how the constituent tasks are organised. However, the HTA itself provides no information about such concerns as task feasibility, task safety or other potential problems that might be encountered. To examine these it is first necessary to obtain more information about the context, or environment, in which the tasks are undertaken. For these purposes, the task context can be taken as any features that may influence performance. This includes items that are directly used to undertake the tasks (such as the controls, displays, job aids or other tools), through to less direct influences (such as the procedures provided, training, environmental conditions, or the way that work is organised and managed). Task decomposition provides a structured way to collect and collate such information based upon a set of task descriptions.

Task decomposition requires an analyst to identify the issues of interest and the type of information that would be necessary to address these issues effectively. The analyst then specifies a set of decomposition categories that define the required information is some detail. For instance, if a task analysis is being undertaken to look at training issues, the analyst may find it helpful to obtain information about the knowledge and skills that are required for different tasks. Thus, an analyst might set up decomposition categories for the knowledge and skill requirements for each task. Under the *knowledge requirements* category, information might then be specifically sought on the knowledge required to know when to start or stop each task, how to undertake the tasks, how to identify and deal with problems, or how to use ancillary equipment.

The analyst then works methodically through all the task descriptions and for each of these the information for each decomposition category is recorded using a tabular format, in which the task descriptions and hierarchical numbering form the first column, with the other decomposition categories in the subsequent columns. Thus, the task decomposition process merely provides a template to ensure that the relevant data are collected for the constituent tasks. The completed template can then be used as a convenient way to represent this information and as the basis for further analysis. It should be noted that, because a tabular structure is used, task decomposition is also known as tabular task analysis. However, this should not be confused with other tabular representations that are used for task analysis, such as tabular HTA.

Although this is a very simple approach, it can prove to be a very effective way to focus attention upon the issues of concern. For instance, in a survey of several different task analyses, Ainsworth and Marshall [1] found that those that used six or more formal decomposition categories were considerably more insightful than those that involved less task decomposition. Thus, the analyses that had made more use of task decomposition identified considerably more specific issues of concern and provided more detail, whilst the other analyses were characterised by recommendations that tended to be limited to a few generic statements.

5.4.1 Types of decomposition information

Data collection can be very resource intensive and so the decomposition categories must be carefully selected so that they focus on information that is likely to contribute to the specific analysis, without wasting resources collecting data that will not be used. This means that it is not appropriate to propose the use of a standard set of decomposition categories to be applied to most task analyses.

Nevertheless, it is possible to obtain some guidance on potential decomposition categories by considering those that have been used in other studies. By far the most widely used decomposition information has been that which describes the interfaces that are, or that should be, provided. This includes data about the controls, displays, job aids or communications that are provided. The amount and type of data that are collected under these headings will depend upon the analyst's requirements. This could range from being a simple check that appropriate interface facilities have been provided, to a more comprehensive assessment of these facilities against appropriate ergonomics guidelines. From a practical standpoint, it is recommended that if a wide range of issues are to be considered for different displays and/or controls, the entry in the decomposition table should generally be limited to an identifying label. Then any discrepancies that are of concern to the analyst can be noted in a *Comments* column, rather than listing all the acceptable features that were assessed.

Other potentially useful decomposition information includes descriptions of the wider context in which the task is carried out, such as the task location, the person responsible, or adverse environmental conditions and other hazards. It can also be useful to look directly at some of the psychological issues, such as task initiation cues, feedback or decision-making. However, analysts are warned against expecting too much by recording the task initiation cues, because in many cases this will merely be the completion of the preceding task.

It is also possible to use the task decomposition to record various measures of task performance. These can be derived from direct measurements or observations or they can be based upon some subjective estimate either by the analyst or some other person. Particularly useful performance measures include the time required to complete a task, errors made (or potential errors identified), and knowledge required for the task or skills required. For such data, the analyst should also record the source of that data. If the data are based on the performance or judgements of more than one person, the analyst should carefully consider how variability is to be dealt with. For task analysis purposes, it must be stressed that it is often more useful to know the range of performance than the mean value.

5.4.2 Collection of task decomposition data

There are two basic approaches that an analyst can adopt in order to collect decomposition information. The first of these is to produce most of the task descriptions and then to return to collect specific decomposition data for each task description in turn. Alternatively, all the decomposition data can be collected whilst the task descriptions are being developed. The latter approach is usually considered to be more efficient when a lot of different information is being obtained from a single source. However, the former approach is useful when several different sources are being used, because an analyst can then work methodically through each data source.

In practice either approach is likely to involve some iteration. Therefore, the main aims should be to reduce this by careful planning and to make provision for collecting further data, or clarification, at a later time.

5.5 Data analysis

5.5.1 Introduction

There are a wide range of methods that can be used to assist an analyst to examine the data that have been collected and either to identify any potential problems or to assess whether particular criteria can be met. Each of these analysis methods is appropriate for different situations and it is beyond the scope of this chapter to provide detailed descriptions of each method. Readers who do wish to read more detailed descriptions of the methods are referred to Kirwan and Ainsworth [4] which provides a more comprehensive discussion of several task analysis techniques together with some case studies that show how they were used.

In order to provide the analyst with an indication of the way that potential issues can be examined and the choices that are possible, seven general application areas have been chosen and for each of these, some potential analysis methods are described.

It must be stressed that whilst the methods have been described in the context of specific applications, several of the methods could also be applied very effectively to help resolve other issues. It must also be emphasised that often the underlying issues are so evident from the basic data, that there is no need to undertake further analysis of the decomposition data in order to clearly identify potential human factors problems.

5.5.2 Analysis of the working environment

Task performance on both physical and mental tasks can be influenced by the environmental conditions under which the tasks have to be undertaken, and in more hostile environments there is also a health and safety implication. Therefore, it is often necessary to obtain some information about lighting, noise levels, or thermal conditions in which tasks may have to be undertaken. For some specific tasks it may also be necessary to consider other potential environmental hazards, such as vibration, motion or radiation.

Providing that there is a system already in existence, the following environmental assessments can be made:

- Environmental surveys to establish the thermal conditions, such as air movement, humidity and temperature by taking the appropriate measurements, such as wet and dry bulb temperatures, radiant temperatures, etc.
- Lighting surveys to measure the lighting levels for particular tasks and to identify potential sources of glare. For many applications it will also be necessary to assess the adequacy of emergency illumination. It should be noted that these measurements can be taken in a pre-operational environment.
- Noise surveys to measure the background sound level in dB(A) under typical working conditions to ensure that hearing protection will be provided if there is a risk to health. Such surveys should also measure the sound levels of communications and of auditory alarm signals to ensure that these are sufficiently above the background.

Alternatively, the analyst will have to rely upon statements about the proposed environmental conditions, or else to define the conditions that should be provided. If necessary, this information can then be assessed against guidance provided in ergonomics checklists. The checklists provided in Section 5.8 of MIL STD 1472 [5] and in Sections 12.1.2 and 12.2.5 of NUREG-0700 [6] provide very comprehensive guidance information about the lighting conditions, noise levels or thermal conditions. More comprehensive design guidance is provided in Chapter 9 of this book.

5.5.3 Analysis of the workstation design

To check that workers can be accommodated at their workstations and that they will be able to reach all the necessary controls, it will first be necessary to identify the relevant dimensions. There may be two sets of dimensions that will be identified. The first of these will be the general workstation dimensions that must be considered in order to ensure that the largest (95th percentile) users can fit into the workplace and the smallest (5th percentile) will be able to reach common controls, such as a mouse or keyboard. In many cases the task descriptions will also identify that other controls or displays may be required, and it will then be necessary to ascertain whether these can be used from the normal operating positions. The analyst must then obtain the relevant workplace measurements, either from drawings that are provided, or by direct measurement. The workplace dimensions can then be compared with standard anthropometric data, such as Pheasant [7], after adding allowances for clothing, such as increasing standing height by 25 mm for males wearing shoes. At the same time, the operating forces on the controls can be checked against ergonomics guidance that is provided in Pheasant and in other standard ergonomics texts.

If any of the tasks require manual handling, the NIOSH Lifting Equation (see Waters *et al.* [8]) can be used to assess whether these tasks can be undertaken safely. More comprehensive design guidance is given in Chapter 9 of this book.

5.5.4 Analysis of information requirements

An understanding of the information requirements for each task may be useful either to define the interface requirements, or to identify the knowledge and skill requirements.

Perhaps the most effective way to examine these issues is to produce some form of diagrammatic representation of selected tasks that indicates the relationships between particular parameters or activities. This can be done by drawing process charts, but these tend to be too coarse-grained for most task analysis purposes. Other possibilities are to produce input–output diagrams [9] or Petri nets [10]. The former use block diagrams to show the information transitions that occur during tasks, whilst the latter examine the possible state changes that could occur in key parameters and trace their impact. However, this section will be limited to short descriptions of information flow diagrams and functional block diagrams.

Information flow diagrams, which are also known as decision–action diagrams, or decision–action–information diagrams, depict the information flows and transformations in a system. On these diagrams, actions are presented in rectangles and decisions are shown within diamonds. After each decision box the possible outcomes are labelled on the exit lines, usually as binary choices, though they can also be multiple choices. An example of a decision–action diagram is presented in Figure 5.4 to illustrate this. This particular example is based upon a continuous process control task and so it has a closed loop structure. However, for other applications the diagram may well be much more linear.

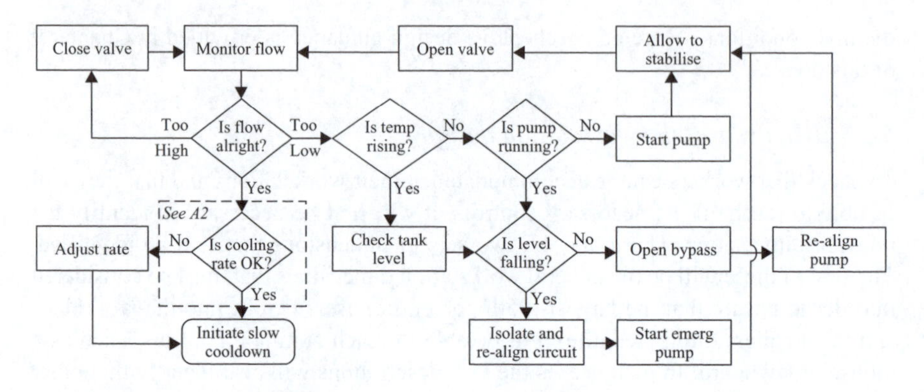

Figure 5.4 Decision–action diagram

These diagrams provide an indication of the information requirements and they are particularly useful because they focus attention upon decision-making activities. Wherever an analyst identifies any decision-making activity, it is recommended that this should be examined carefully to ascertain whether all the parameters that could influence a particular decision have been identified. If there are other important influences, these can either be added to the diagram, or else the component decisions can be shown elsewhere. Where it is appropriate to provide a separate diagram for a particularly complex decision, it is recommended that the relevant decision box on the parent diagram should be surrounded by pecked lines and a reference to the more detailed diagram should be given, as is illustrated on the 'Is cooling rate OK?' box in Figure 5.4.

Functional block diagrams provide a representation of the main functional requirements, in which the different functional requirements are presented in separate boxes, with arrows linking the boxes to show the sequential relationships between the functions and subfunctions that have to be undertaken to complete a task. AND and OR gates are used on these diagrams to describe more complex sequential relationships. An example of a simple functional block diagram is shown in Figure 5.5.

Each function represented on the diagram should be uniquely numbered, with lower level subfunctions being hierarchically related to their parent function. It will seldom be fruitful to proceed past four levels in this hierarchy. It is important to stress that this numbering is based upon functional relationships and as such it may differ from the normal HTA numbering. For this reason, the numbers are separated from the rest of the box by a horizontal line.

For some purposes it will not be possible to present all the functional blocks on a single page and so a set of diagrams will be necessary. Such sets of diagrams can be organised either sequentially or hierarchically. In the former case, the functions that are undertaken will be shown on the first diagram and subsequent diagrams will show functions that are undertaken later. For a hierarchical arrangement, the main functions will be shown on the first diagram and each of the main functions will then be shown in greater detail elsewhere.

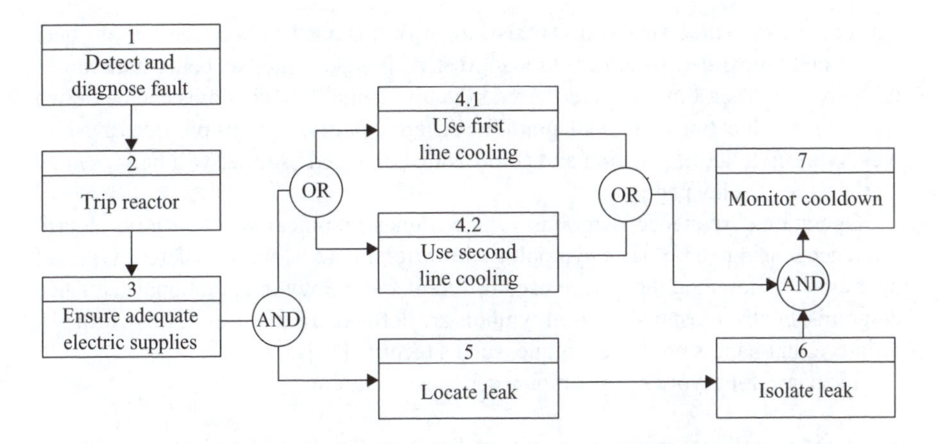

Figure 5.5 Functional block diagram

5.5.5 *Analysis of task sequence*

Link analysis and operational sequence diagrams are specifically used to examine task sequences in terms of the use of interfaces or communications links. Both techniques can be used either to present information schematically, or they can be based upon more accurate plans or drawings that maintain the topographic relationships between items.

Link analysis was originally used to investigate the social relationships between people in group situations. This was done by representing each person as a circle on a schematic diagram and then drawing a separate line between the appropriate circles each time that two people interacted with each other in some way. Thus, the number of lines provided an indication of the frequency of interpersonal interaction or communication between individuals. Clearly, this technique can be applied directly to working situations to look at communications between team members during various tasks. It can also be adapted to assess the relationships between different instruments or parts of an interface, by using a schematic or topographical diagram of the interfaces and then drawing a link between items on this drawing when the focus of attention changes. This is known as a frequency link analysis diagram, but it is of limited use because it is not possible to determine the task sequence from such diagrams. Therefore, such diagrams are open to misinterpretation. For example, the task context may mean that it is more important to keep together two components with relatively weak frequency links that are used during an emergency, rather than other components with much stronger links that are always used in less critical situations with no time stress.

In order to examine the sequential relationships a continuous line can be drawn to indicate the changes of attention that occur in a task. This is known as a spatial link diagram and it enables an analyst to directly trace the changes between different interface elements. This type of analysis is particularly useful for identifying where there is a wide physical separation between interface items that are accessed sequentially that could cause time stress. It can also identify mismatches between the physical layout and task order, which could lead to steps being omitted or being undertaken in the wrong order. For computer-based systems, spatial link diagrams can be drawn with separate locations on the diagram for different pages, and this provides a useful way to ensure that information and controls that are used together will be presented on the same display page.

Operational sequence diagrams provide other graphical ways to look at task sequences, using some basic symbology to differentiate between different types of operation. Different authors have used different symbology in operational sequence diagrams, but the most widely used symbols are defined in Kirwan and Ainsworth [4], or in original papers on the technique, such as Kurke [11].

There are four types of operational sequence diagram:

- Basic operational sequence diagrams. These categorise each task into one of five types of activity corresponding to the activities listed in the basic symbol set. Then these are represented as a vertical flow chart using these symbols. This type of representation is sometimes useful for tasks that are undertaken at a number of different locations, but it generally adds little to many applications, such as control room-based tasks.
- Temporal operational sequence diagrams. Effectively these are basic operational sequence diagrams on which the vertical axis is also a timescale. It is considered that timelines are a better way to show this information and so these are not recommended.
- Partitioned operational sequence diagrams. These consist of separate operational sequence diagrams for different aspects of a task that are presented together. For example, a two-person task could be presented as a vertical operational sequence diagram for each person, with a centre column being used to indicate interactions between them. Tainsh [12] has developed a method that he called job process charts, which essentially uses this approach to look at the relation between computer tasks and human tasks. In job process charts, the outside columns are used for computer tasks and human tasks, whilst the central column shows the interaction between the two.
- Spatial operational sequence diagrams. These are simply spatial link diagrams to which operational sequence symbols have been added.

In order to produce any form of link diagram or operational sequence diagram it is first necessary to collate the information in some way. Normally this would mean that task decomposition tables could be used directly as the data source. However, it can be helpful to organise this information in some way prior to producing the diagrams. Information sequence tables provide a tabular way to do this, in which the different columns of a table are used to group the information that comes from different sources. Such diagrams can provide a particularly effective way to look at computer-based workstations that comprise several screens, so that screen usage can be optimised. An example of how this approach could be used to examine the usability of a three-screen workstation is shown in Table 5.1.

Table 5.1 Information sequence table

Task	Key presses	Screen 1	Screen 2	Screen 3
1.1.1 Check inventory	1	R100	(unassigned)	Alarms
1.1.2 Ensure conditions stabilising	3	R101	T100 T1011	Alarms
1.1.3 Isolate RHR	1	R102		Alarms
1.1.4 Start four emergency pumps	1	R102	T1011 T1012	Alarms

5.5.6 *Analysis of temporal and workload issues*

Two persistent issues in task analyses are whether personnel will have sufficient time to successfully complete all of the necessary actions and whether they will have the capacity required to perform their tasks reliably. These two issues are assessed by timelines and workload analyses respectively.

The starting point for the development of any timeline is the collection of data about the time required for each task. This can be done during the task decomposition stage by directly observing task performance, or by obtaining subjective assessments. In the former case, it will be necessary to ensure that the timings are based upon realistic situations. In particular, there is a risk that if operators are timed whilst they undertake actions that they are prepared for, the resultant times may well be much shorter than would be required when there is some uncertainty, or when there are interruptions from other tasks that are competing for attention. Thus the analyst must endeavour to ensure that the tasks are undertaken in a realistic manner and that where necessary allowances are made for any interruptions or problems that may occur. If subjective estimates of time are being used, it will be necessary to select people with sufficient understanding of the tasks to be able to provide realistic assessments and, if it is feasible, an attempt should be made to validate some of the time estimates against direct measurements. Yet another way to estimate task times is to use standard data, either from psychological models or by timing representative tasks. For instance, rather than timing operators opening every valve on a plant, an analyst may use the time taken to open one specific valve and use the same value throughout a task analysis.

A final consideration that should be made when determining realistic task times, is how to make allowances for the effect of adverse environmental conditions, such as darkness, cold or wind upon the time required. The impact of such conditions is illustrated by Umbers and Reiersen [13] who noted that a combination of adverse conditions could increase task times by as much as 70 per cent. Therefore, it has been proposed that task times should be multiplied by an appropriate factor for particular environmental conditions. In Table 5.2, adapted from [14], some weighting factors are given.

Table 5.2 Adverse conditions factors (adapted from [14])

Environmental factor	Type of task	
	Simple manual	Walking
Sub zero temperatures, ice and snow	2	1.5
High winds	1	1.5
Darkness	1	2
Wearing breathing apparatus	1.25	2

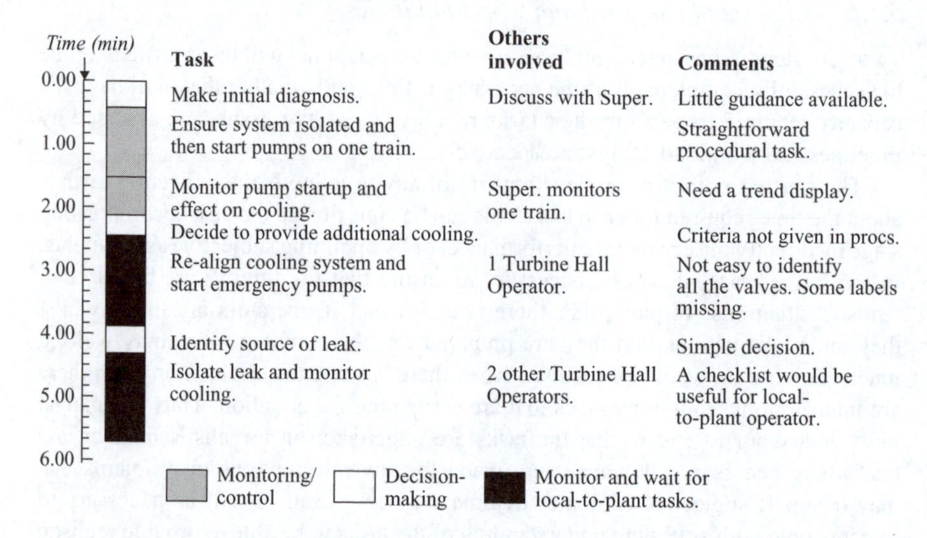

Figure 5.6 Vertical timeline (from unpublished data by Ainsworth)

A timeline is basically a bar chart for either single-person or team-based tasks, in which the length of the bars is directly proportional to the time required for those tasks. Figure 5.6 shows a timeline in which the time axis is presented vertically. This has the advantage that ancillary information can be presented as text alongside the timeline, but adding this text also means that the timescale must be relatively coarse.

It is, however, more common to present timelines horizontally. For a single person the tasks for a horizontal timeline are generally listed along the vertical axis, so they appear as a rising, or falling, series of bars, as depicted in Figure 5.7.

For team-based tasks, the horizontal timeline can be modified by presenting all the tasks undertaken by the same person at the same level (see Figure 5.8 for an example). This then enables an analyst to examine both individual tasks and the interactions between team members. In such situations there are often critical points

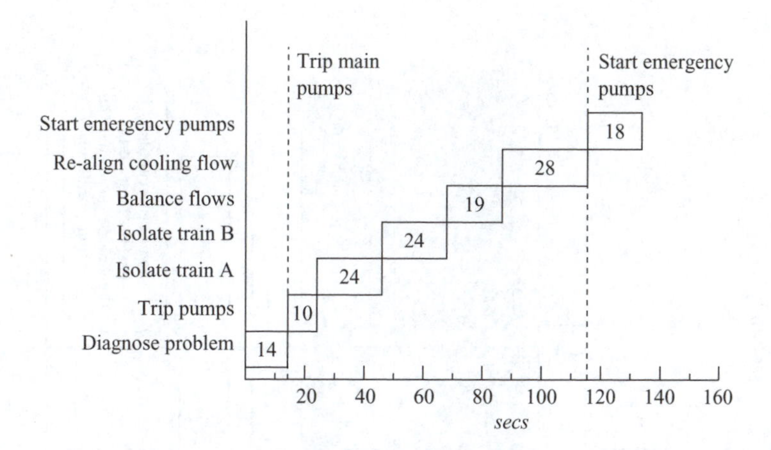

Figure 5.7 Horizontal single-person timeline (from unpublished data by Ainsworth)

where one team member must complete a particular task before others can continue with their tasks. Therefore, to highlight such dependencies, such situations are shown by using arrows to link the respective bars on the timeline. Where team members work at different locations, it can also be useful to present tasks undertaken at the same location adjacent to each other in the timeline.

Clearly, timelines are particularly useful for assessing time-critical tasks. Therefore, it is recommended that the occurrence of time-critical events should be shown on timelines by vertical lines. This will assist an analyst to identify any time-critical events that cannot be completed within the required time. In such situations there are four strategies that can be adopted to reduce the time required to achieve the required system states.

- Try to avoid less important tasks during time-critical periods.
- Provide job aids or additional automation.
- For team-based tasks, re-organise or re-allocate the tasks to reduce waiting times by other team members.
- Increase the staffing levels for these tasks.

Whilst timelines represent the time required to undertake tasks, they do not provide any indication of the amount of effort, or the workload, involved. There are several subjective methods that can be used to assess workload, the most widely used of which is probably the NASA Task Load Index (TLX) [15].

NASA TLX is based upon subjective ratings made on six different dimensions of workload, namely mental demand, physical demand, temporal demand, effort, performance and frustration level. In the first stage of any assessment the subjects are presented with descriptions of each of these dimensions and they are asked to make pairwise comparisons of the importance to workload of each possible pair of dimensions. This enables individual weightings to be developed for each dimension. After undertaking a task, the respondents are then asked to rate the workload that

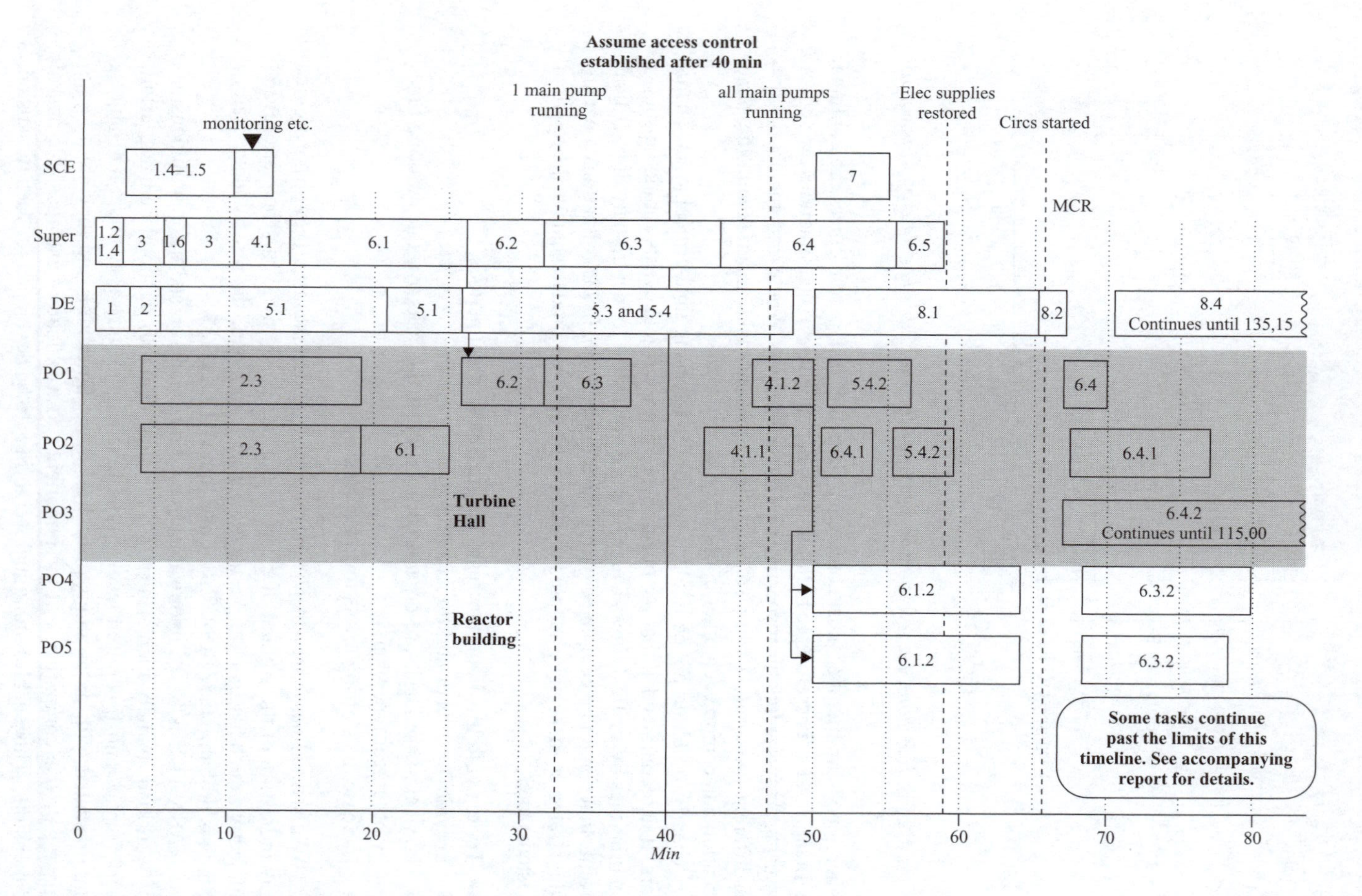

Figure 5.8 Multi-person timeline (from unpublished data by Ainsworth)

they experienced on a scale of one to a hundred on each of the six dimensions. The workload ratings for each dimension are adjusted by the weightings and an overall workload index is then calculated as a percentage.

Unfortunately, for task analysis purposes, NASA-TLX was developed to assess relatively large tasks, or sets of tasks, shortly after they have been completed. Some analysts have used them as a predictive tool by conducting the ratings on the basis of detailed task descriptions rather than actual task performance, but such use is questionable, especially for the performance and the frustration level dimensions. Nevertheless, where NASA-TLX is appropriate, it provides a useful and widely validated indication of workload.

An alternative to NASA-TLX is the Subjective Workload Assessment Technique (SWAT) [16]. This is a rating scale method that assesses workload on three dimensions, known as time load, mental effort load and stress load. Each of these dimensions is a three-point rating scale with detailed verbal criteria provided. The scales are standardised by asking the subjects to compare all the different combinations of these criteria. After task completion, each subject is then asked to rate the effort involved using the same three-point rating scales for each dimension. SWAT has similar limitations to NASA-TLX. It is claimed that it is more sensitive to small changes in workload than NASA-TLX, but the ratings take longer to obtain.

5.5.7 Analysis of errors

Before discussing methods of identifying, or assessing, potential sources of human error by task analysis, it is necessary to sound a cautionary note. Task analysis can certainly be used to identify task features, such as the procedures or interface design which directly influence the ease and accuracy of performance. Thus, task analysis can provide some valuable insights into potential human errors whilst undertaking critical tasks. However, there are also other factors, such as staffing levels, personnel management or the safety culture, that can exert an impact at the point where operators select particular procedures or strategies. These may well predispose a person to a different set of human errors, such as taking inappropriate actions, misdiagnosing the available information, acting upon incomplete information, or failing to check alternative information sources. These issues are generally better addressed by taking a more holistic view of error causation, rather than from task analysis. Therefore, it is necessary to be aware of this so that any task analysis to identify specific human errors can be supplemented by a consideration of the psychological mechanisms that could induce other forms of inappropriate or incorrect behaviour.

Perhaps the most straightforward way to identify where there is a potential for human error is to undertake a decompositional human error analysis. This starts by attempting to identify potential human errors during the decomposition process by observing task performance and by discussions with experienced personnel. Because this part of the analysis is directly based upon operational experience, at the end of this process the most likely human errors should have been identified. This chapter provides some summary information on the quantification of errors. For a detailed exposition of this topic the reader is directed to Chapter 8.

However, it is also necessary to consider less likely errors. Therefore, the analyst must expand this experiential description of the errors by postulating other potential error mechanisms, which is also termed Predictive Human Error Analysis (PHEA). This is very vulnerable to biases from the analyst. In particular, many analysts find it difficult to consider the possibility of some less likely errors. This is not surprising, because the most obvious errors will have already been identified, so that the errors postulated during PHEA will largely comprise other very rare errors, whose probability of occurrence is so low that even highly experienced operators may never have encountered them. The tendency to dismiss potential errors will be particularly marked when they depend upon two or more relatively rare events occurring concurrently.

The danger of dismissing potential errors is exemplified by the assessment of an interlock system that is meant to prevent a dangerous action. In such a case, the analyst may be tempted to dismiss the possibility that the operator would override the interlock, because this would require a conscious effort by well trained operators who were aware of the situation. Furthermore, it may require permission from a supervisor to manually override this interlock. However, whilst this reasoning may be true, it ignores the fact that the most likely way for the interlock to be 'overridden' might be for the interlock to fail and then for this failure to go undetected by an operator.

In order to minimise the risk of ignoring potential human errors, analysts must be rigorous in postulating errors solely upon the logic and the rationality of the situation, with no consideration, at this stage, as to the actual likelihood of the error occurring.

This identification of the potential error mechanisms is best done by developing detailed human error analysis tables from the initial task decompositions. For each task description, these tables should list the potential problems in a way that permits the analyst to identify potential error mechanisms. For this purpose, the following headings will be useful:

- Cognitive issues. Understanding, expectation, diagnostic cues, focus of attention, recognition, response planning, response success monitoring, perceived task importance.
- Task execution. Task cues, successful action feedback (i.e. control actions have worked), functional success feedback (i.e. control actions have achieved the desired result on the wider process).
- Interface design. Ease of identifying controls and displays, labels, appropriateness of displays, controls following standard or stereotypical formats, circumstances where the information becomes unavailable.
- Procedural issues. Procedural coverage of scenarios, clarity of instructional presentation, matching of content to requirements for recollection and management of task difficulty.
- Workload. Time available to perform the task, time required to perform the task, difficulties in performing the task (excluding issues addressed above), possibility of performing concurrent tasks.
- Team design. Clarity of responsibilities for task execution, team assembly delays (e.g. due to mustering, or call-out times).

- Communications. Availability of communication systems, clarity of communications, use of communication protocols.

Some analysts, particularly those with no psychological background, may find it preferable to substitute key words, such as those used for Human HAZOP (see below), for the human error decomposition categories. Similarly, some analysts may prefer to use error decomposition categories derived directly from a human error model, such as the Generic Error Modelling System (GEMS) proposed by Reason [17].

It will not be possible, within an error analysis table, to provide comprehensive descriptions of both the potential causes of human error and the consequences for risk. Therefore, this table should be limited to describing the potential causes, and each human error should then be referenced using the HTA number, so that it will be possible to refer back to other information. Where necessary, letters should be added to this number to differentiate between different error mechanisms related to the same task description. It must also be noted that whilst many human errors will be specific to the deepest hierarchical levels, some error mechanisms, such as those associated with cognitive or attentional cues, will exert a more widespread effect. Therefore, such errors are best reported at the higher levels.

Finally, the analyst must make an assessment of the impact of each postulated human error. This should be a qualitative assessment that involves tracing through the task analysis from the error to examine its consequences. The analyst must then decide whether to delete any of the postulated human errors. For example, if an operator fails to send someone to unlock a valve, this may be detected later when there is no flow through a circuit, at which point the error could be corrected. However, if establishing this flow is time-critical, such an error might still merit further consideration.

Another approach to identifying human errors is to use a table-top discussion method known as a hazard and operability (HAZOP) study. This will involve setting up a meeting of specialists with some understanding of the technical and operational requirements, who will then systematically work through a particular scenario or procedure and attempt to identify potential errors and failure modes. At each step in the scenario or procedure, the respondents are prompted to identify different types of problem by using appropriate guide words. For human errors, guide words such as *too soon*, *too late*, *too little*, *too much* or *wrong order* are used.

When the key events or potential human errors have been identified, either through task analysis or from a general reliability analysis, it can be useful to produce a tree diagram such as an event tree or a fault tree. These can aid the analyst to understand the consequences of errors and will also be useful if the analyst later wishes to undertake any quantified risk assessments.

Event trees start with a specific initiating event or human action, then branches are drawn to the subsequent events that either lead to recovery or to a fault situation. Normally, but not always, there is a binary split at each node of the event tree, with one path leading to success and the other representing a failure mode. By convention, they are usually drawn from left to right from the initiating event, with the success paths being shown above the failure paths. Thus, event trees provide a chronological view

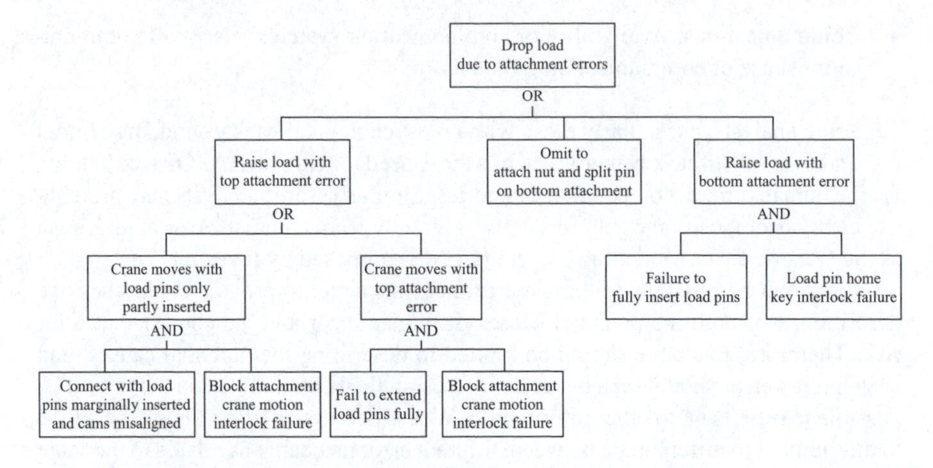

Figure 5.9 Fault tree (from unpublished data by Ainsworth)

of the range of situations that could occur after any initiating event. Normal event trees can include both technical consequences and human interactions. However, it is also possible to produce human event trees, which are sometimes known as Operator Action Event Trees. These show only the temporal and logical dependencies that could arise between human interactions.

Fault trees (Figure 5.9) provide a view of the way that hardware faults and human errors can lead to an undesirable outcome. They are usually in a similar format to that used for HTAs. Thus, they trace backwards from the undesirable situation to identify all the potential causes that might contribute to it. AND and OR logic gates are shown on the diagrams to indicate how these different technical faults or human errors are combined in order to cause the final outcome. They are directly useful for quantified error assessments, but they are also useful for qualitative analyses.

5.5.8 *Analysis of cognitive processes*

The analysis and documentation of cognitive task performance can be very time consuming. Therefore, it is recommended that an analyst should only undertake detailed cognitive analysis after undertaking a general task analysis to understand the context in which cognitive processing is required, and then deciding where such effort is justified.

Where potential cognitive issues are identified, in many cases the psychological issues of concern will be evident and there will be sufficient information to enable an analyst with some psychological background to assess whether personnel are likely to experience any difficulties though in some cases it may be helpful to produce a decision–action diagram that splits some of the complex tasks sufficiently to identify the main information flows and transformations. When assessing potential cognitive issues, the analyst should be looking for tasks, or aspects of tasks, that:

- Impose high, or unnecessary, long-term or short-term memory loads.
- Require the collation of data from different sources, especially if these are widely separated spatially or temporally.
- Involve complex processing of data before it can be used.
- Involve simple processing of data that has to be done under time stress.
- Involve complex diagnostics.
- Require decision-making where there are multiple, and possibly conflicting, criteria.
- Involve incomplete or ambiguous data.
- Involve predicting future conditions or trends, especially where the changes are not linear.

Where such issues as these cannot either be identified or properly understood because the underlying cognitive tasks are more complex, it may be appropriate to obtain further information about the cognitive processes that are involved. However, at this juncture, it is also necessary to give warnings about two potential problems with cognitive data that must be considered before further cognitive data are obtained. First, it must be recognised that just because a task imposes heavy cognitive demands, it does not necessarily follow that a cognitive analysis is justified. For instance, a person may acquire effective cognitive skills, such as decision-making, through practice sessions with effective feedback. Second, in any cognitive analysis it may be difficult to tease out all the underlying cognitive issues, particularly where specific sequences of behaviour have become so well learned that they are undertaken with little conscious effort.

However, if further information is required about the cognitive content of complex tasks, this can be obtained by various knowledge elicitation techniques. The most straightforward of these is verbal protocol analysis, where a person is asked to verbalise without interruption whilst undertaking a cognitive task. This can be very effective, but is limited to tasks where it is possible to verbalise directly without any processing of the information. Hence, verbal protocols should only be obtained for tasks where the verbalisation does not affect task performance and where information is being retrieved from short-term memory in verbal form. Thus, it is not appropriate for tasks that have been coded in a visual form, or where tasks have become subconscious or 'automated' due to overlearning.

Another useful way to collect cognitive information is the withheld information technique. This involves asking someone to diagnose a particular situation when some of the available information is systematically withheld. The analyst can then determine the effects of different items of information.

Finally, it is necessary to discuss methods that are known as cognitive task analysis, but which are better considered as cognitive modelling. The first stage in these models is to use task descriptions and task decomposition data to identify the component task elements. Then the cognitive tasks are modelled on some parameter, usually the execution time. The most widely known of these cognitive task analysis methods is the Goals, Operators, Methods and Selection Rules (GOMS) family of techniques that have been developed from the work of Card, Moran and Newell [18]. The constituent elements of these models are defined below:

- Goals. Goals and subgoals define the aims that are to be accomplished.
- Operators. These are the perceptual, cognitive or motor actions that are performed to accomplish the goals.
- Methods. These are the sequences of operators that are invoked.
- Selection rules. If there is more than one method to accomplish a goal or subgoal, selection rules are used to choose between these.

Thus, for example, one method to satisfy the goal of increasing the flow through a system might involve the following operators.

MOVE MOUSE to an appropriate valve symbol
DOUBLE CLICK to highlight control plate for that valve
MOVE MOUSE to the Open command
CLICK MOUSE BUTTON to initiate opening of valve.
PERCEIVE FLOW

The simplest GOMS technique is the keystroke-level model. This is limited to defining the operators at the level of individual keystrokes (or mouse movements) using six different types of operator that were specifically chosen for computer-based tasks. The time estimates for these operators have since been refined [19]. The keystroke-level model also defines rules for deciding when the mental preparation operator should be used.

The other GOMS techniques, and most of the other cognitive task analysis methods, are generally based upon more comprehensive underlying assumptions and are also intended for wider application than the human–computer interface issues for which the keystroke model was proposed. However, most of the cognitive task analysis methods that are currently available assume error free performance by skilled personnel who are not operating under any stress, and they also make no effective allowances for fatigue. These are serious weaknesses, that lead many to question whether the effort of undertaking a cognitive task analysis can ever be justified.

5.6 Reporting task analysis

When all the required analyses have been completed the analyst should produce a report that clearly explains how the task analysis was undertaken and which provides some guidance that assists readers, who may be unfamiliar with task analysis, to interpret the findings. The report must also clearly specify any assumptions that have been made, especially when the task analysis is based upon a specific scenario. These could include assumptions about the availability of particular plant items, the status of other tasks, the availability of personnel or the errors that might be made.

Where an analyst identifies any potential mismatches between what the users require and what is provided for them, the analyst must provide recommendations for overcoming such difficulties. It is important that such recommendations are written in a manner that assists the decision-makers to focus very quickly upon the issues that concern them. Hence, in large systems, it is incumbent on the analyst to group different recommendations together under appropriate subheadings, so that the different individuals can rapidly identify those recommendations for which they have responsibility.

Each recommendation should outline the reason for the recommendation and should then provide a potential solution. In many cases this will provide the decision-maker with sufficient information to agree to implement a particular change. However, if it is considered that a particular recommendation is costly or difficult to implement, there should be sufficient information available to enable a reader to obtain more comprehensive insights into the problem. They should then be able to assess the importance of overcoming that particular problem and may well be able to suggest alternative ways of overcoming it.

Finally, after agreements have been made about the way that each recommendation will be implemented, these should be recorded, and then at a later date these aspects of the system should be validated to ensure that the recommendations have been implemented in a manner that fully complies with ergonomics guidance.

5.7 Exercises

1. Produce an HTA diagram and a task decomposition for preparing N copies of a report from a word processor file in a double-sided format using a printer that only prints single-sided. Each report is to be stapled within light card covers, with the front cover printed from a separate file.
2. Using an actual video recorder, ask someone to record a specific television programme and monitor how this is done. Then produce an HTA and a task decomposition that identifies the interfaces used and any difficulties experienced. Use this to prepare a written procedure to enable someone with no experience to undertake this task. Compare this to the written instructions that are provided by the manufacturer.
3. Undertake a task analysis for the preparation of breakfasts for two people who each have different requirements. Produce a timeline for one person undertaking all the work and another timeline on the assumption that two people are involved.
4. Undertake a task analysis for selecting and booking a holiday for two people, who have some flexibility about their departure date, using the internet. Assume that this will require separate bookings for the flight, car hire and accommodation and that at least two prices will be compared for each of these items. Then, identify potential errors and suggest ways that the internet interfaces could help to prevent these.

5.8 Answers

1. There are three basic ways that this task could be undertaken, but for this example it is assumed that all the even pages are printed first and then these are turned

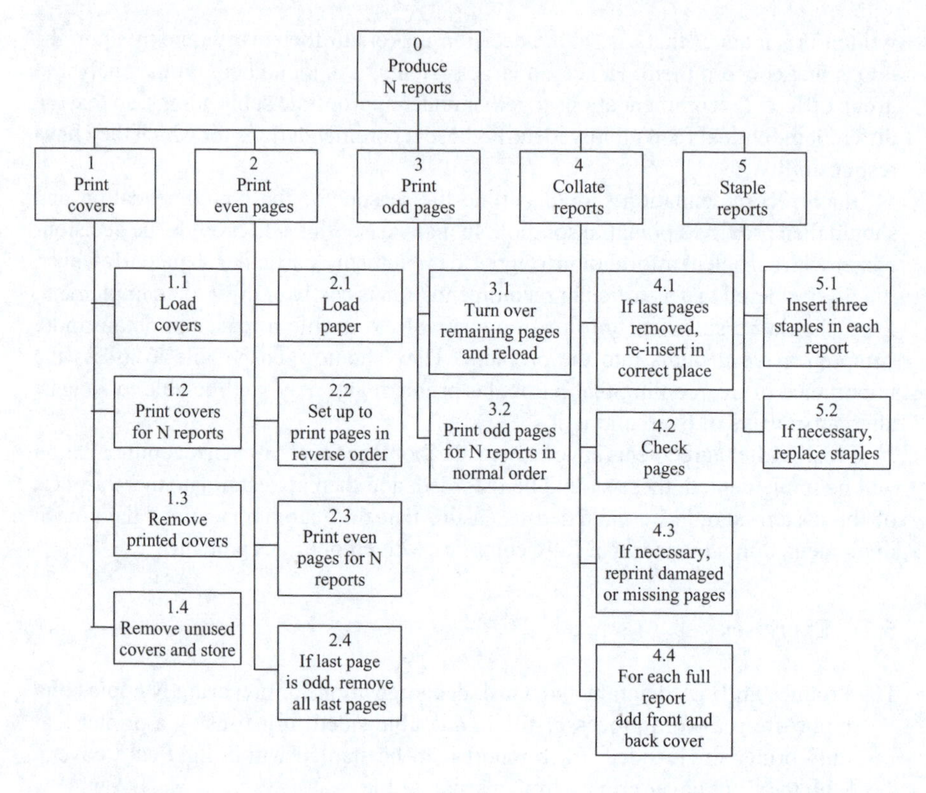

Figure 5.10 Report preparation

over and re-inserted into the printer. No plans are given, because the tasks would all normally be undertaken sequentially. However, some people may wish to undertake task 1 after completing tasks 2 and 3. An appropriate task decomposition might consider the word processor interface and any particular problems that might be encountered. The biggest difficulty is that if there is an odd number of pages, there will be one less even page to print, so unless the last odd page is removed, or a blank even page is inserted, the pagination will be wrong. For example, the first page of the second report will be printed on the back of the last page of the first report. Hence the operators will require good cues to ensure that this problem is avoided (Figure 5.10).

2. The essential elements of this exercise are to define the different ways that this task can be undertaken and to identify aspects of the interface that could cause difficulties to someone who did not use these facilities frequently. The decomposition table could usefully include information on the knowledge required, the nature of the interfaces and dialogues used, and potential errors. The analysis should show that individuals can experience difficulties that are often not anticipated by the designers. For example, what happens if the viewer wishes to use the video for something else before the recordings have been made, or what happens if two programmes follow each other immediately, on different channels?

3. This is a straightforward example. For the two-person timeline the analyst should be careful to ensure that the equipment and space are available before starting each task. It can also be informative to repeat parts of the timelines using different assumptions. For example, if toast is wanted at the end of the meal, the time that toasting can be started will depend upon whether the diners wish to have warm or cooled toast.
4. This exercise is aimed at demonstrating the wide range of errors that can be made on relatively straightforward data entry and decision-making tasks. Errors could be defined as selecting an inappropriate option (e.g. not the cheapest deal), or finding that the options required by the person making the booking were incorrect (e.g. wrong dates, car to be picked up from wrong location, wrong type of accommodation). Consider how the interfaces that are provided could assist the users to avoid such errors. For example, should the person be told the total duration of the holiday to prevent errors such as booking a day trip for the same day on different months. In this particular example, there will be much scope for error by the person making the bookings, because three different bookings will be made, so consider how the booking confirmations could be produced in a way that reduced the potential for errors booking the other items. Also, consider how the booking systems could support the users, by anticipating responses or by providing additional options, such as special meals.

5.9 References

1 AINSWORTH, L. K., and MARSHALL, E. C.: 'Issues of quality and practicability in task analysis: preliminary results from two surveys', in ANNETT, J., and STANTON, N. (Eds): 'Task analysis' (Taylor and Francis, London, 2000)

2 ANNETT, J., and DUNCAN, K. D.: 'Task analysis and training design', *Occup. Psychol.* 1967, **41**, pp. 211–221

3 SHEPHERD, A.: 'Hierarchical task analysis' (Taylor and Francis, London, 2001)

4 KIRWAN, B., and AINSWORTH, L. K.: 'A guide to task analysis' (Taylor and Francis, London, 1992)

5 US DEPARTMENT OF DEFENSE: 'Design criteria standard, human engineering' (Washington: Department of Defense, MIL STD 1472F, 1999)

6 O'HARA, J. M., BROWN, W. S., LEWIS, P. M., and PERSENSKY, J. J.: 'Human–system interface design review guidelines', Washington: US Nuclear Regulatory Commission, Rept. No. NUREG-0700, 2002

7 PHEASANT, S.: 'Bodyspace' (Taylor and Francis, London, 1996)

8 WATERS, T. R., PUTZ-ANDERSON, V., GARG A., and FINE, L. J.: 'The revised NIOSH lifting equation', *Ergonomics*, 1993, **36**, pp. 749–776

9 SINGLETON, W. T.: 'Man–machine systems' (Penguin Books, Harmondsworth, 1974)

10 MURATA, T.: 'Petri nets: properties, analysis and applications', *IEEE Proc.*, 1989, **77** (4), pp. 541–580

11 KURKE, M. I.: 'Operational sequence diagrams in systems design', *Human Factors*, 1961, **3**, pp. 66–73
12 TAINSH, M. A.: 'Job process charts and man–computer interaction within naval command systems design', *Ergonomics*, 1985, **28** (3), pp. 555–565
13 UMBERS, I. G., and REIERSEN, C. S.: 'Task analysis in support of the design of a nuclear power plant safety system', *Ergonomics*, 1995, **38** (3), pp. 443–454
14 UMBERS, I. G.: (2003) Personal communication
15 HART, S. G., and STAVELAND, L.: 'Development of NASA-TLX (Task Load Index): results of empirical and theoretical research', in HANCOCK, P. A., and MESHATI, N. (Eds): 'Human mental workload' (Elsevier, Amsterdam, 1988) pp. 139–184
16 REID, G. B., and NYGREN, E.: 'The subjective workload assessment technique: a scaling procedure for measuring mental workload', in HANCOCK, P. A., and MESHATI, N. (Eds): 'Human mental workload' (Elsevier, Amsterdam, 1988) pp. 184–218
17 REASON, J.: 'Human error' (Cambridge University Press, Cambridge, 1990)
18 CARD, S. K., MORAN, T. P., and NEWELL, A.: 'The psychology of human–computer interaction' (Laurence Erlbaum Associates, New Jersey, 1983)
19 OLSON, J. R., and OLSON, G. M.: 'The growth of subjective modelling in human-computer interaction since GOMS', *Human–Computer Interaction*, 1990, **5**, pp. 221–265

Chapter 6
Automation and human work

Erik Hollnagel

6.1 Introduction

The word automation, which comes from Greek, is a combination of *auto*, 'self', and *matos*, 'willing', and means something that acts by itself or of itself. This is almost identical to the modern usage where automatic means something that is self-regulating or something that acts or operates in a manner determined by external influences or conditions but which is essentially independent of external control, such as an automatic light switch. The following quotes illustrate the various modern meanings of the term:

> Automation is defined as the technology concerned with the application of complex mechanical, electronic and computer based systems in the operation and control of production.
>
> ([1], p. 631)

> 'Automation' as used in the ATA Human Factors Task Force report in 1989 refers to 'a system or method in which many of the processes of production are automatically performed or controlled by self-operating machines, electronic devices, etc.'
>
> ([2], p. 7)

> We define *automation* as the execution by a machine agent (usually a computer) of a function that was previously carried out by a human.
>
> ([3], p. 231)

Automation is often spoken of as if it was just a technology and therefore mainly an engineering concern. Automation is, however, also an approach to work and therefore represents a socio-technical intervention that needs careful consideration. In order to understand how automation relates to human work, it is useful to consider two simple examples, both taken from the history of technology rather than the present day.

6.1.1 Precision and stability

The first example is a self-regulating flow valve for a water clock (clepsydra), which was invented about 270 B.C. by Ktesibios (or Ctsesibius) of Alexandria. The water clock itself was invented about 1500 B.C. by the Egyptian court official Amenemhet ([4], p. 124). The principle of a water clock is that water flows into (or out of) a vessel at a constant rate, whereby the level of water can be used as a measure of the passage of time. In its simplest form, a water clock is a container in which water runs out through a small hole in the bottom. The water clock is filled at the beginning of a time period and the level of water indicates how much time has passed. In more sophisticated versions the water runs into a container instead of out of it. The water clock that Ktesibios improved was even more advanced. In this 'machine' water fell from a level and hit a small scoop-wheel, thereby providing the force to drive a shaft. Here the precision of the clock depended critically on having a constant trickle or flow of water, which in turn depended on the pressure or the level of water in the reservoir. Ktesibios solved this problem by designing a self-regulating flow valve (Figure 6.1), which produced the desired flow of water.

The second example is almost 2000 years younger and far better known, even though a version of the self-regulating valve is used by billions of people every day. The example is the flying-ball governor that we know from James Watt's 'classical' steam engine. Watt did not invent the steam engine as such, Thomas Savery had done that already in 1698, but he significantly improved the existing models in several ways, one by adding a separate condenser (1756) and the other by introducing the conical pendulum governor (1769) (see Figure 6.2).

The purpose of the governor was to maintain the constant speed of the engine by controlling the inflow of steam. This was done by having a pair of masses (balls) rotating about a spindle driven by the steam engine. With an increase in speed, the masses moved outward and their movement was transmitted to the steam valve via

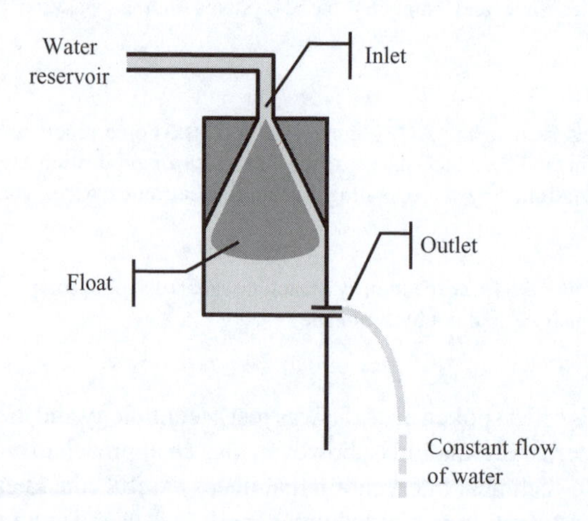

Figure 6.1 The self-regulating valve of Ktesibios

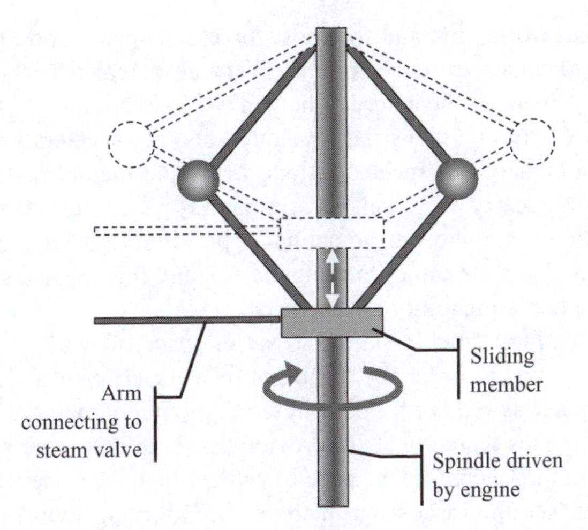

Figure 6.2 Watt's pendulum governor

a member that glided up the spindle. This reduced the steam admitted to the engine, thus reducing the speed. Conversely, if the speed was reduced, the masses moved inward and the member would glide down the spindle, thereby increasing steam flow to the engine. In this way the flying-ball governor automated the task of regulating the steam valve, which otherwise had to be carried out by a person.

Both examples illustrate two fundamental relations between automation and human work. One is that automation ensures a more *precise* performance of a given function, either by the self-regulating flow-valve in the water clock or by the flying-ball governor in the steam engine. The performance, or output, is more precise because a technological artefact on the whole can respond to smaller variations of the input than a human. A second, and somewhat related, feature is that automation improves the *stability* of performance because mechanical devices function at a stable level of effort without any of the short-term fluctuations that are seen in human work. People, on the other hand, are ill-suited to repetitive and monotonous tasks, which they notably do very badly. In both examples, a person could in principle carry out the same functions – and indeed did so initially – but the performance would not be as accurate and smooth as that of the automation, even if full and undivided attention were paid to the task.

6.1.2 Automation as compensation

As these examples show, automation was from the very beginning used to overcome certain shortcomings found in human work, notably that people tended to get distracted or tired, and hence were unable to maintain the required level of precision and stability. The need for automation has thus clearly existed and been recognised for a very long time, even though the practical demands in the beginning were limited. In the 18th century the industrial revolution provided the possibility to amplify many aspects of human work, first and foremost force and speed, and also helped overcome problems of constancy and endurance. Yet it also created a need for automation that dramatically changed the nature of human work. In addition to the demands for precision and stability, the industrial revolution and the acceleration of technology quickly added a demand for speed. As long as people mainly had to interact with other people, the pace of work settled at a natural pace. But when people had to interact with machines, there was no natural upper limit. Indeed, a major advantage of machines was that they could do things faster, and this advantage was not to be sacrificed by the human inability to keep pace.

As new technology quickly started to set the pace of work, humans were left struggling to respond. In order for machines to work efficiently, a high degree of regularity of input was required – both in terms of energy/material and in terms of control. Since humans were unable to provide that, artefacts were soon invented to take over. That in turn increased the pace of work, which led to new demands, hence new automation (see the discussion of the self-reinforcing loop in Section 6.1.3). A significant side effect was that work also became more monotonous. That is one of the unfortunate consequences of automation, and one of the reasons why it is a socio-technical rather than an engineering problem.

Taken together, the three features of precision, stability, and speed meant that humans soon became a bottleneck for system performance. This was actually one of the reasons for the emergence of human factors engineering as a scientific discipline in the late 1940s. In modern terms this is expressed by saying that humans have a capacity limitation or that their capacity (in certain respects) clearly is less than that of machines. One purpose of automation is therefore to overcome the capacity limitations that humans have when they act as control systems, thereby enabling processes to be carried out faster, more efficiently – and hopefully also more safely.

In addition to being seen as a bottleneck, humans were also frequently seen as a source of unwanted variability in the system. This variability might lead not only to production losses, but more seriously to incidents and accidents that could jeopardise the system itself. Consistent with this line of reasoning, the solution was to eliminate humans as a source of variability – as far as possible – by replacing them with automation. While this view no longer is regarded as valid (cf. the discussion in Section 6.1.4), it did have a significant influence on the development and use of automation in the last quarter of the 20th century, and was, paradoxically, one of the reasons for the many problems that this field experienced.

6.1.3 The self-reinforcing loop

While the history of automation is as long as the history of technology and human invention itself, the issue became much more important in the 20th century, especially with the proliferation of digital information technology. One way of understanding that is to see the increasing complexity of technological systems as a self-reinforcing or positive feedback loop [5], cf. Figure 6.3.

The growing potential of technology is an irresistible force which invariably is used to increase system functionality – for instance by reducing production costs,

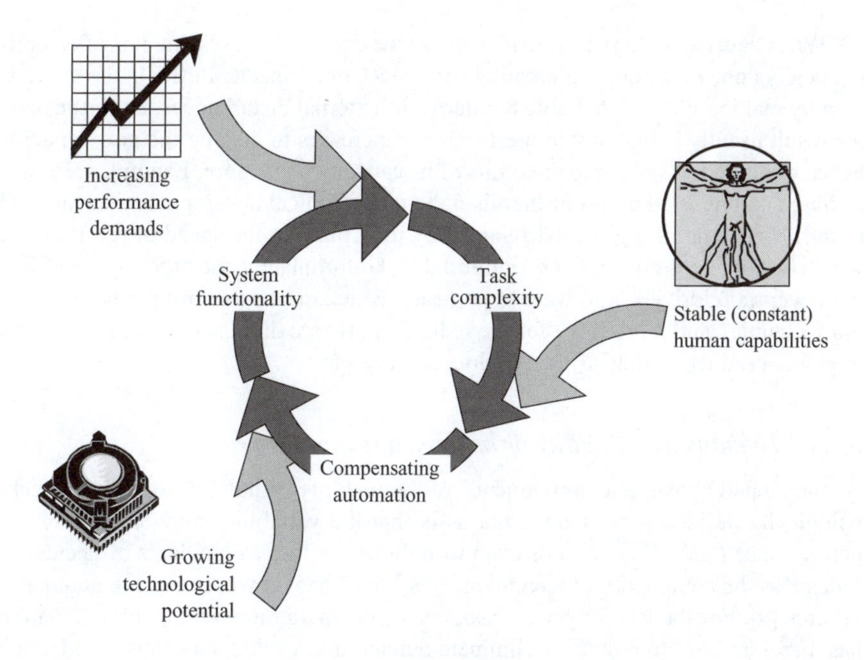

Figure 6.3 The self-reinforcing complexity cycle

improving quality of products, enlarging capacity, and shortening cycle times for production and maintenance. To this are added the effects of a second driving force, namely the growing demands for performance, particularly in terms of speed and reliability. These demands are themselves due to an increasing dependency on technology in all fields of human endeavour. The combination of increased system functionality and growing performance demands leads to more complex tasks and therefore to increased system complexity in general. This may seem to be something of a paradox, especially because technological innovations often purportedly are introduced to make life easier for the users. The paradox is, however, not very deep, and is related to the well-known 'ironies of automation' [6], cf. Section 6.3.2.

The growing complexity of systems and tasks increases the demands for control, in accordance with the Law of Requisite Variety [7]. This leads to a reduction in the tolerance for variability of input, hence to an increase in the demands on the humans who operate or use the systems. Unfortunately, the increased demands on controller performance create a problem because such demands are out of proportion to human capabilities, which in many ways have remained constant since Neolithic times. From the technological point of view humans have therefore become a bottleneck for overall system throughput. The way this problem is overcome, as we have already seen, is to introduce automation either to amplify human capabilities or to replace them – known as the tool and prosthesis functions, respectively [8]. The ability to do so in turn depends on the availability of more sophisticated technology, leading to increased system functionality, thereby closing the loop.

When new technology is used to increase the capacity of a system, it is often optimistically done in the hope of creating some slack or a capacity buffer in the process, thereby making it less vulnerable to internal or external disturbances. Unfortunately, the result usually is that system performance increases to take up the new capacity, hence bringing the system to the limits of its capacity once more. Examples are easy to find, and one need go no further than the technological developments in cars and in traffic systems. If highway driving today took place at the speed of the 1940s, it would be both very safe and very comfortable. Unfortunately, the highways are filled with ever more drivers who want to go faster, which means that driving has become more complex and more risky for the individual. (More dire examples can be found in power production industries, aviation, surgery, etc.)

6.1.4 Humans as a liability or as a resource

As mentioned above, one consequence of the inability of humans to meet the technologically defined performance norms is that the variability may lead to loss of performance (and efficiency) or even to failures – either as incidents or accidents. Indeed, as the complexity of technological systems has grown, so has the number of failures [9]. For the last 50 years or so, one of the main motivations for automation has therefore been to reduce or eliminate human failures that may cause production losses and system accidents. Automation has been used to constrain the role of the operator or preferably to eliminate the need for human operators altogether. Although, from a purely engineering perspective, this may seem to be an appropriate arrangement, there are some sobering lessons to be learned from the study of the effects of automation (e.g. [10]).

An alternative to the view of humans mainly as a source of variability and failures is to view them as a resource that enables the system to achieve its objectives. This view acknowledges that it is impossible to consider every possible contingency during design, and therefore impossible for technological artefacts to take over every aspect of human functioning. Humans are in this way seen as a source of knowledge, innovation and adaptation, rather than as just a limiting factor. This leads to the conclusion that automation should be made more effective by improving the coupling or co-operation between humans and technology, i.e., a decidedly human-oriented view. When this conclusion is realised in practice, two thorny questions are: (1) what should humans do relative to what machines should do, and (2) are humans or machines in charge of the situation? The first question is identical to the issue of function allocation proper, and the second to the issue of responsibility. The two issues are obviously linked, since any allocation of functions implies a distribution of the responsibility as well. The problem is least complicated when function allocation is fixed independently of variations in the system state and working conditions, although it is not easy to solve even then. The responsibility issue quickly becomes complicated when function allocation changes over time, either because adaptation has been used as a deliberate feature of the system, or – more often – because the system design is incomplete or ambiguous.

It is a sobering thought that the motivation for introducing automation generally is technical rather than social or psychological. That is, automation is introduced to answer the needs of the process, rather than to answer the needs of people working with the process. The process needs can be to reduce the number of disturbances or accidents, or simply to improve efficiency. Automation is in both cases allowed to take over from people because it is assumed to do better in some respects. While this may be correct in a narrow sense, an inevitable consequence of introducing automation is that the working conditions of the human operators are affected. Since this will have effects on the functioning of the overall system both short-term and long-term, it is important to consider the automation strategies explicitly rather than implicitly.

6.2 Humans and automation

One way of describing how automation has affected human work is to consider the distribution of functions between humans and machines. This is illustrated in Figure 6.4, which shows the shrinking domain of responsibility from the operators' point of view. The changes brought about by advances in automatic control can be characterised by means of four distinctive types of activity called *tracking, regulating, monitoring*, and *targeting*.

- In *tracking*, the detection (of changes) and correction (execution of control actions) is performed in a closed loop. Any system worth mentioning comprises a number of simultaneous tracking loops, where each is confined to a small number of parameters and controls. Tracking is also done within a limited time horizon,

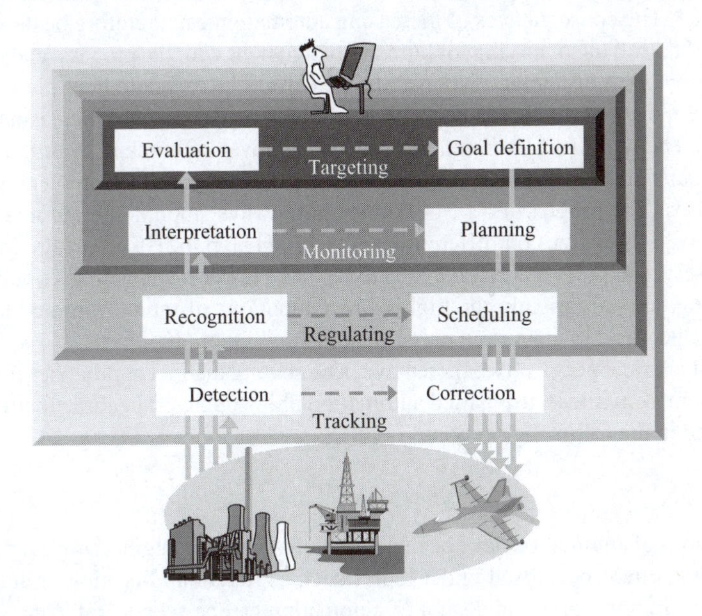

Figure 6.4 The erosion of operator responsibility

typically as part of short-term control of the process. The purpose of tracking is to maintain performance within narrowly specified limits.

- *Regulating* extends over longer periods of time and/or larger parts of the system and mixes open- and closed-loop control. It may, for instance, deal with the transition between system states or the carrying out of a specific set of operations, such as pump rotation. Regulating includes the recognition of the process state and the scheduling of appropriate activities. The carrying out of these activities involves tracking, which thus can be seen as a component of regulating.
- The purpose of *monitoring* is to keep track of how the system behaves, and to generate or select the plans necessary to keep the system within the overall performance envelope. Monitoring, like regulating, combines open- and closed-loop control. The plans may, for instance, be available as procedures of various types. Monitoring comprises interpretation of the state of the system and the selection or specification of proper plans for action. Monitoring thus encompasses both tracking and regulating.
- Monitoring in turn is guided by *targeting*, which sets the overall goals for system performance – in close accordance with, e.g., management policies and regulations. The goals are normally prescribed for a process in advance, but for specific situations such as disturbances, new goals may be defined to match the current conditions. Targeting is decidedly open-loop control.

As this chapter has shown, automation began by taking over the simpler functions on the level of tracking (think of the flying-ball governor), and has slowly expanded to include regulation and monitoring activities. In some cases automation has even covered the aspects of targeting or decision-making, leaving the operators with very little to do. The consequences of increasing automation can therefore be described as a gradual narrowing or eroding of human involvement with the process, and therefore also a narrowing of human responsibilities. (From the technological point of view, the development is obviously one of increasing involvement and increasing responsibility.) The net effect has been the gradual removal of humans from the process, with the unwanted side effect that they are less well prepared to intervene when the need arises. The progression is, of course, not always smooth and automation may exist in various degrees on different levels (e.g., [11], p. 62). The illustration nevertheless serves to point out that humans today have fewer responsibilities and that the remaining tasks are mainly the higher-order functions of monitoring and targeting, and hence depend on cognitive rather than manual functions.

In a historical perspective, there have been several distinct approaches to automation design, sometimes also called automation philosophies. The three main ones are described in the following sections.

6.2.1 *The 'left-over' principle*

Discussions of automation have been part of human factors engineering ever since the late 1940s, often specialised under topics such as function allocation or automation strategies. In retrospect, the design of automation can be seen as referring to one of several different principles, which have developed over the years. Each principle is associated with a view of the nature of human action, although this may be implicit rather than explicit. They are therefore sometimes referred to as automation philosophies or automation strategies, although they have not all been recognised as such at the time. The simplest automation philosophy is that the technological parts of a system are designed to do as much as feasible (usually from an efficiency point of view) while the rest is left for the operators to do. This approach is often called the 'left-over' principle or the residual functions principle. The rationale for it has been expressed as follows:

> The nature of systems engineering and the economics of modern life are such that the engineer tries to mechanize or automate every function that can be .... This method of attack, counter to what has been written by many human factors specialists, does have a considerable logic behind it ... machines can be made to do a great many things faster, more reliably, and with fewer errors than people can .... These considerations and the high cost of human labor make it reasonable to mechanize everything that can be mechanized.
> [12]

Although this line of reasoning initially may seem unpalatable to the human factors community, it does on reflection make good sense. The proviso of the argument is, however, that we should only mechanise or automate functions that can be completely automated, i.e., where it can be guaranteed that automation will always work correctly and not suddenly require the intervention or support of humans. This is a very strong requirement, since such guarantees can only be given for systems where it is possible to anticipate every possible condition and contingency. Such systems are few and far between, and the requirement is therefore often violated [6]. Indeed, if automation were confined to those cases, far fewer problems would be encountered.

Without the proviso, the left-over principle takes a rather cavalier view of humans since it fails to include any explicit assumptions about their capabilities or limitations – other than that the humans in the system hopefully are capable of doing what must be done. Implicitly this means that humans are treated as extremely flexible and powerful machines, which far surpass what technological artefacts can do.

6.2.2 The compensatory principle

A second automation philosophy is the eponymous Fitts' list (named after Paul Fitts, [13]), sometimes also referred to as the compensatory principle. This principle proposes that the capabilities (and limitations) of people and machines be compared on a number of relevant dimensions, and that function allocation is made so that the respective capabilities are used optimally. (The approach is therefore sometimes referred to as the 'Men-Are-Better-At, Machines-Are-Better-At' or 'MABA-MABA' strategy.) In order for this to work it must be assumed that the situation characteristics can be described adequately *a priori*, and that the capabilities of humans (and technology) are more or less constant, so that the variability will be minimal. Humans are furthermore seen as mainly responding to what happens around them, and their actions are the result of processing input information using whatever knowledge they may have – normally described as their mental models.

Table 6.1 The principles of the Fitts list

Attribute	Machine	Operator/human
Speed	Much superior	Comparatively slow, measured in seconds
Power output	Much superior in level and consistency	Comparatively weak, about 1500 W peak, less than 150 W during a working day
Consistency	Ideal for consistent, repetitive actions	Unreliable, subject to learning (habituation) and fatigue
Information capacity	Multi-channel. Information transmission in megabits/sec.	Mainly single channel, low rate <10 bit/sec.
Memory	Ideal for literal reproduction, access restricted and formal	Better for principles and strategies, access versatile and innovative
Reasoning, computation	Deductive, tedious to program. Fast, accurate. Poor error correction	Inductive. Easy to 'programme'. Low, inaccurate. Good error correction.
Sensing	Specialised, narrow range. Good at quantitative assessment. Poor at pattern recognition.	Wide energy ranges, some multi-function capability
Perceiving	Copes poorly with variations in written/spoken material. Susceptible to noise.	Copes well with variation in written/spoken material. Susceptible to noise.

Table 6.1 shows the dimensions or attributes proposed by Paul Fitts as a basis for comparing humans and machines. Even though this list reflects the human factors' perspective of more than 50 years ago, most of the dimensions and characterisations are still highly relevant – the main exception being the attribute of power output. Some of the characterisations need perhaps be adjusted to represent the capabilities of current technology, but despite five decades of development within artificial intelligence, machines still have a long way to go before they can match humans in terms of perception, reasoning, and memory.

The determination of which functions should be assigned to humans and which to machines is, however, not as simple as implied by the categories of the Fitts list. Experience has taught us that it is necessary to consider the nature of the situation, the complexity, and the demands and not just compare the attributes one by one:

> It is commonplace to note the proper apportionment of capabilities in designing the human/computer interface. What humans do best is to integrate disparate data and construct and recognize general patterns. What they do least well is to keep track of large amounts of individual data, particularly when the data are similar in nature and difficult to differentiate. Computers are unsurpassed at responding quickly and predictably under highly structured and constrained rules. But because humans can draw on a vast body of other experience and seemingly unrelated (analogous or 'common-sense') reasoning, they are irreplaceable for making decisions in unstructured and uncertain environments.
>
> ([14], p. 4.)

Furthermore, the function substitution that is part of the compensatory principle disregards the fact that to achieve their goals most systems have a higher order need for co-ordination of functions. Function allocation by substitution is based on a very narrow understanding of the nature of human work and capabilities, and effectively forces a machine-like description to be applied to humans. A substitution implies a rather minimal view of the function in question and blatantly disregards the context and other facets of the situation that are known to be important. Since functions usually depend on each other in ways that are more complex than a mechanical decomposition can account for, a specific assignment of functions will invariably have consequences for the whole system. In this way even apparently small changes will affect the whole. The issue is thus one of overall system design and co-operation between humans and machines, rather than the allocation of functions as if they were independent entities.

6.2.3 *The complementarity principle*

A third automation philosophy, which has been developed during the 1990s, is called the complementarity principle (e.g., [15, 16]). This approach aims to sustain and strengthen human ability to perform efficiently by focusing on the work system in the long term, including how routines and practices may change as a consequence of learning and familiarisation. The main concern is the ability of the overall system to sustain acceptable performance under a variety of conditions rather than a transitory peak in efficiency (or safety). The complementarity principle is concerned with human–machine co-operation rather than human–machine interaction, and acknowledges the significance of the conditions provided by the overall socio-technical system. It is consistent with the view of Cognitive Systems Engineering (CSE), which emphasises the functioning of the joint cognitive system [17]. CSE describes human actions as proactive as well as reactive, driven as much by goals and intentions as by 'input' events. Information is furthermore not only passively received but actively sought, hence significantly influenced by what people assume and expect to happen. A joint cognitive system is characterised by its ability to maintain control of a situation, in spite of disrupting influences from the process itself or from the environment. In relation to automation, CSE has developed an approach called function congruence or function matching, which takes into account the dynamics of the situation, specifically the fact that capabilities and needs may vary over time and depend on the situation [18]. One way of making up for this variability is to ensure an overlap between functions assigned to the various parts of the system, corresponding to having a redundancy in the system. This provides the ability to redistribute functions according to needs, hence in a sense dynamically to choose from a set of possible function allocations.

The main features of the three different automation philosophies are summarised in Table 6.2. This describes the function allocation principle, the purpose of function allocation, i.e., the criteria for successful automation, and the school of thinking (in behavioural science terms) that the philosophy represents. Table 6.2 also indicates the view of humans that is implied by each automation philosophy and thereby also their view of what work is. In the 'left-over' principle few, if any, assumptions are made about humans, except that they are able to handle whatever is left to them.

Table 6.2 Comparison of the three automation philosophies

	Automation principle		
	Residual function ('left-over')	Fitts list	Complementarity/ congruence
Function allocation principle	Leave to humans what cannot be achieved by technology	Avoid excessive demands to humans (juxtaposition)	Sustain and strengthen human ability to perform efficiently
Purpose of function allocation	Ensure efficiency of process by automating whatever is feasible	Ensure efficiency of process by ensuring efficiency of human–machine interaction	Enable the joint system to remain in control of process across a variety of conditions
School of thought	Classical human factors	Human–machine interaction, human information processing	Cognitive systems engineering
View of humans (operators)	None (human as a machine)	Limited capacity information processing system (stable capabilities)	Dynamic, adaptive (cognitive) system, able to learn and reflect

Human work is accordingly not an activity that is considered by itself, but simply a way of accomplishing a function by biological rather than technological means. The strongest expression of this view is found in the tenets of Scientific Management [19]. The compensatory principle does consider the human and makes an attempt to describe the relevant attributes. Although the Fitts list was developed before the onslaught of information processing in psychological theory, the tenor of it corresponds closely to the thinking of humans as information processing systems that has dominated human factors from the 1970s onwards. This also means that work is described in terms of the interaction between humans and machines or, more commonly, between humans and computers, hence as composed of identifiable segments or tasks. Finally, in the complementarity/congruence principle humans are viewed as actively taking part in the system, and as adaptive, resourceful and learning partners without whom the system cannot function. The view of work accordingly changes from one of interaction to one of co-operation, and the analysis starts from the level of the combined or joint system, rather than from the level of human (or machine) functions *per se*.

6.2.4 From function allocation to human–machine co-operation

A common way to introduce automation is to consider the operator's activities in detail and evaluate them individually with regard to whether they can be performed better (which usually means faster and/or more cheaply) by a machine. Although some kind of decomposition is inevitable, the disadvantage of only considering functions one by one – in terms of how they are accomplished rather than in terms of what they achieve – is that it invariably loses the view of the human–machine system as a whole. The level of decomposition is furthermore arbitrary, since it is defined either by the granularity of the chosen model of human information processing, or by a technology-based classification, as epitomised by the Fitts list. To do so fails to recognise important human capabilities, such as being able to filter irrelevant information, scheduling and reallocating activities to meet current constraints, anticipating events, making generalisations and inferences, learning from past experience, establishing and using collaboration, etc. Many of the human qualities are of a heuristic rather than an algorithmic nature, which means that they are difficult to formalise and implement in a machine. Admittedly, the use of heuristics may also sometimes lead to unwanted and unexpected results but probably no more often than algorithmic procedures do. The ability to develop and use heuristics is nevertheless the reason why humans are so efficient and why most systems work.

In a human-centred automation strategy, the analysis and description of system performance should refer to the goals or functional objectives of the system as a whole. One essential characteristic of a well-functioning human–machine system is that it can maintain, or re-establish, an equilibrium despite disturbances from the environment. In the context of process control the human–machine system is seen as a joint cognitive system, which is able to maintain control of the process under a variety of conditions. Since this involves a delicate balance between feedback and feedforward, it is essential that the system be represented in a way that supports this. Although a joint cognitive system obviously can be described as being composed of several subsystems, it is more important to consider how the various parts of the joint system must correspond and co-operate in order for overall control to be maintained. The automation strategy should therefore be one of function congruence or function matching, which takes into account the dynamics of the situation, specifically the fact that capabilities and needs may vary over time and depend on the situation. Function congruence must include a set of rules that can achieve the needed re-distribution, keeping in mind the constraints stemming from limited resources and inter-functional dependencies.

This type of automation strategy reflects some of the more advanced ways of delegating functions between humans and machines. Both Billings [2] and Sheridan [20] have proposed a classification of which the following categories are relevant for the current discussion:

- In *management by delegation* specific tasks or actions may be performed automatically when so ordered by the operators.
- In *management by consent* the machine, rather than the operator, monitors the process, identifies an appropriate action, and executes it. The execution must, however, have the previous consent by the operators.
- In *management by exception* the machine identifies problems in the process and follows this by automatic execution without need for action or approval by operators.

- In *autonomous operation* the machine carries out the necessary tasks without informing the operator about it.

Neither management by exception nor autonomous operation is a desirable strategy from the systemic point of view. In both cases operators are left in the dark, in the sense that they do not know or do not understand what is going on in the system. This means that they are ill-prepared either to work together with the automation or take over from it if and when it fails. Management by delegation and management by consent are better choices, because they maintain the operator in the loop. Since human action is proactive as well as reactive, it is essential to maintain the ability to anticipate. This requires knowledge about what is going on. If the automation takes that away, operators will be forced into a reactive mode, which will be all the more demanding because they have no basis on which to evaluate ongoing events.

6.2.5 Unwanted consequences of automation

The reasons for introducing more technology and/or improving the already existing technology are always noble. A host of empirical studies nevertheless reveal that new systems often have surprising unwanted consequences or that they even fail outright (e.g., [21]). The reason is usually that technological possibilities are used clumsily [10], so that well-meant developments, intended to serve the users, instead lead to increasing complexity and declining working conditions. For example, if users do not understand how new autonomous technologies function, they are bound to wonder 'what is it doing now?', 'what will happen next?', and 'why did it do this?' [22]. Other unwanted, but common, consequences of automation are:

- Workload is not reduced by automation, but only changed or shifted.
- Erroneous human actions are not eliminated, but their nature changes. Furthermore, the elimination of small erroneous actions usually creates opportunities for larger and more critical ones.
- There are wide differences of opinion about the usefulness of automation (e.g., in terms of benefit versus risk). This leads to wide differences in the patterns of utilisation, hence affect the actual outcomes of automation.

System changes, such as automation, are never only simple improvements, but invariably have consequences for which knowledge is required, how it should be brought to bear on different situations, what the roles of people are within the overall system, which strategies they employ, and how people collaborate to accomplish goals. The experience since the beginning of the 1970s has identified some problems that often follow inappropriate employment of new technology in human–machine systems [23], such as:

- *Bewilderment.* Even well-intended system changes may be hard to learn and use. Users may find it difficult to remember how to do infrequent tasks or may only be partially aware of the system's capabilities. This may lead to frustrating breakdowns where it is unclear how to proceed.

- *Input overload.* The technological ability to collect and transmit data has since the 1950s by far exceeded the human capacity to deal with them. The practitioner's problem is rarely to get data, but rather to locate what is informative given current interests and demands. People increasingly find that they do not have the tools to cope with the huge quantities of information to which they are exposed, either at work or at home.
- *Failure modes.* While sophisticated technology may eliminate some failure modes it may also create possibilities for new failures modes, sometimes with larger consequences [24]. A characteristic example of that is the mode error [21].
- *Clumsiness.* Many technological improvements that are intended to help users by reducing workload sometimes have the perverse effect that already easy tasks are made even easier while challenging aspects are made more difficult. This clumsiness arises because designers have inaccurate models of how workload is distributed over time and task states and an incomplete understanding of how humans manage workload to avoid activity bottlenecks.
- *Memory overload.* Information technology systems seem to require that users know more and remember more. In practically every place of work users keep paper notes (often as yellow stickers, cf. [25]) or apply other idiosyncratic solutions to manage the seemingly arbitrary demands a new system introduces.

While most early human–machine systems were designed just to get some specific task done, the reality today encompasses a wide variety of users and tasks within a system of closely coupled devices and activities. It is gradually being acknowledged that the development of human–machine systems has been pushed by technology and driven by engineering purposes, rather than by concerns for human work. Yet in order to produce useful and successful systems the actual impact of technology change on cognition, collaboration and performance must be taken into account.

6.3 Cognitive systems engineering

Cognitive systems engineering emphasises that humans and machines should not be considered as incompatible components or parts that in some way have to be reconciled, but that they rather should be seen as a whole – as a joint cognitive system. The functions and capabilities of the whole are therefore more important than the functions and capabilities of the parts. A cognitive system is formally defined by being able to modify its pattern of behaviour on the basis of past experience to achieve specific anti-entropic ends. Less formally, a joint cognitive system is characterised by its ability to maintain control of a situation. One way of representing the dynamics of system performance is by means of the basic cyclical model shown in Figure 6.5. According to this model, control actions or system interventions are selected based on the construct (or understanding) of the situation to produce a specific effect. If that is achieved, the correctness of the construct is confirmed and it becomes correspondingly easier to select the next action. If the expected effect does not occur, the controlling system must somehow reconcile the difference between the expected and

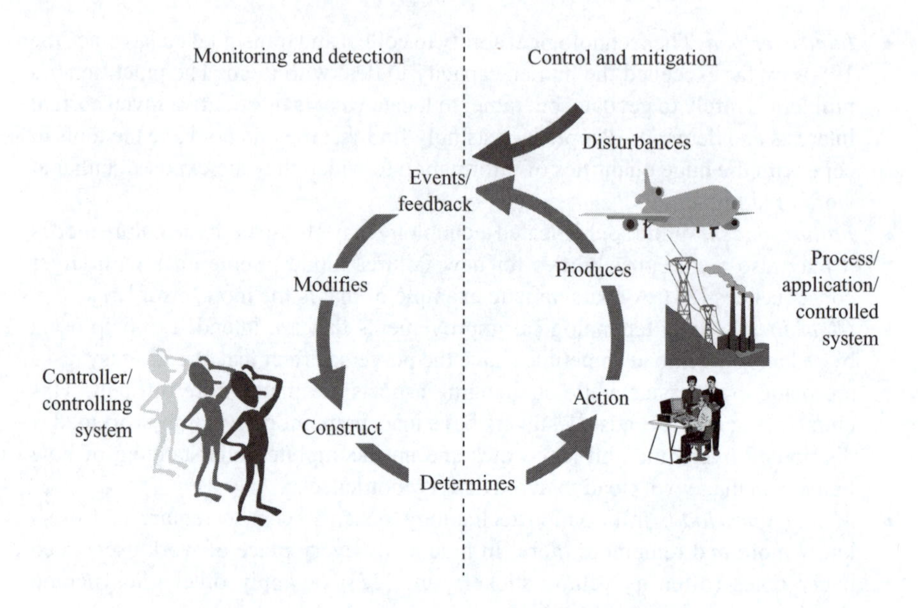

Figure 6.5 The basic cyclical model of cognitive systems engineering

actual outcomes. In the case of humans this means they must revise their understanding of the situation. This is also necessary if an unrelated disturbance interferes, since that will produce effects that were not expected. In both cases more time and effort is required to evaluate the situation and to select the next action. (For a more detailed description of this model see Reference 26).

This model can easily be applied to the description of automation by making a distinction between monitoring and detection functions on the one hand, and control and mitigation functions on the other. Monitoring and detection denote the steps necessary to keep track of how an event develops (parameter values and state transitions), to establish a reference description (construct) and to detect possible significant changes. Control and mitigation denote the actions taken either to regulate the development of the event, for instance to maintain a process within given performance limits, and the actions needed to mitigate unwanted effects and recover from undesirable conditions. The focus on functions rather than structures or components is intentionally consistent with the function congruence principle described above, but differs from the earlier automation philosophies.

6.3.1 *Balancing functions and responsibilities*

For any joint system the two sets of functions can be carried out either by human users or by machines (technology). This provides a convenient way to consider the role of automation in relation to the overall functioning of the joint system (Figure 6.6). For monitoring and detection, some functions can easily be carried out by technology, hence automated. A machine can readily check whether a specific parameter reaches

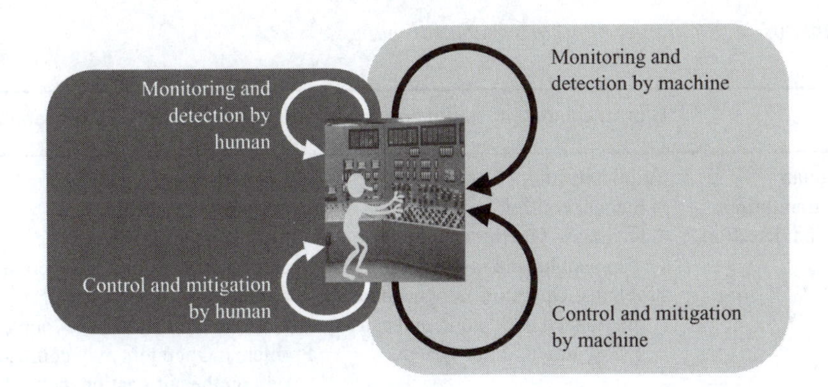

Figure 6.6 Balancing the use of automation

a threshold (e.g., a level or pressure alarm), and it can do it for many parameters and for any length of time. Other functions may better be left to humans, such as recognising patterns of occurrence or detecting dynamic trends and tendencies. Similarly, some functions of control and mitigation may also be taken over by machines. For simple and well-defined conditions mitigation can easily be completely automated, e.g., an automatic fire extinguisher or cruise control in a car. For more complex conditions, the contribution of humans is needed, such as finding ways of recovering from a combination of failures. In general, if automation takes over the detection-correction functions, people lose information about what the system does, and hence lose control.

If, for the sake of simplicity, we consider the extreme situations where the two sets of functions are assigned entirely either to humans or to automation, the result is a definition of four distinct conditions (Table 6.3). Each of these conditions corresponds to a well-known automation scenario, which can be described in terms of the main advantages and disadvantages. In practice it is, however, not reasonable to maintain such a simple distinction. As the small examples above have shown, humans and automation have quite different capabilities with regard to monitoring/detection and control/mitigation. Since these capabilities are dependent on situation characteristics, for instance time demands and work environment, it is ineffectual to allocate functions based on a simple comparison. It would be better to consider a set of distinct and representative scenarios for the joint system, and then for each of these consider the advantages and problems of a specific automation design in accordance with the principles of balancing or function congruence.

As Table 6.3 clearly shows, there is no single or optimal situation that represents human-oriented automation, and the choice of an automation strategy must always be a compromise between efficiency and flexibility. If human functions are replaced by technology and automation, efficiency will (hopefully) increase, but there is a cost in terms of loss of flexibility. Conversely, if flexibility is the more important concern, part of the efficiency may have to be sacrificed. There is no uncomplicated way of designing automation that is applicable across all domains and types of work. As noted in the beginning, automation is introduced to improve precision, improve

Table 6.3 Balancing monitoring-control matrix

	Human control and mitigation	Automated control and mitigation
Human monitoring and detection	Characterisation: Conventional manual control. Advantages: Operators are fully responsible, and in-the-loop. Problems: Operators may become overloaded with work, and cause a slow-down of the system.	Characterisation: Operation by delegation. Advantages: Improved compensatory control. Operators are relieved of part of their work, but are still in-the-loop. Problems: Operators may come to rely on the automation, hence be unable to handle unusual situations.
Automated monitoring and detection	Characterisation: Automation amplifies attention/recognition. Advantages: Reduced monotony. Problems: Operators become passive and have to rely on/trust automation.	Characterisation: Automation take-over. Advantages: Effective for design-based conditions. Problems: Loss of understanding and de-skilling. System becomes effective but brittle.

stability and/or improve speed of production. For some processes, one or more of these criteria are of overriding concern. For other processes different criteria, such as flexibility or adaptability, may be more important. Each case therefore requires a careful consideration of which type of automation is the most appropriate. Human-oriented automation should not in itself be a primary goal, since the joint system always exists for some other purpose. Human-oriented automation should rather be interpreted as the need to consider how the joint system can remain in control, and how this can be enhanced and facilitated via a proper design of the automation involved.

6.3.2 *The ironies of automation*

One of the problems with automation is that it is often difficult to understand how it works as noted by Billings [22], Woods and Sarter [27] and many others. Before the advent of mechanised logic, automation was analogue and therefore relatively straightforward to follow, as illustrated by the self-regulating valve and the flying-ball governor. Today, automation is embedded in logic and far more complex, hence difficult to understand. This is most obviously the case for the end user, as demonstrated by the phenomenon called 'automation surprises' [28]. But it is increasingly also a problem for automation designers, who may be unable to comprehend precisely how functions may be coupled, and hence how they depend on each other [9].

As a simple example, consider attempts to automate the use of room lighting in order to conserve energy. A simple rule for that could be:

IF <lights in room X are off> AND <movement in room X is detected> THEN <turn lights in room X on >
IF <lights in room X are on> AND <no movement is detected in room X for *n* minutes> THEN <turn lights in room X off >

(Even for this simple type of automation, there are problems relating to the sensitivity of detection, i.e., the criterion for whether a movement has taken place. It is not an uncommon experience that a group involved in intense discussions, but with little use of gestures, suddenly find themselves in darkness.)

Not satisfied with this, designers (or the marketing department?) may propose to automate room lighting as a person walks through a house or building. In this case the light should be turned on just before a person enters a room, and be turned off soon after a person has left the room. In order for this to work it is necessary that the automation can determine not just that a person is moving, but also in which direction, at least for rooms that have more than two points of access. This can become very complex, even for a small house and even if only one person is moving at a time. What happens, for instance, if in the middle of going from room B to a non-adjacent room E, the person realises that s/he has forgotten something in room B and therefore starts to walk back again? It is easy to see that the automation of such a relatively simple function may become very complex, and that it may be difficult for both designers and users to have a clear understanding of how it works. If this example seems too trivial, try to consider the automation required for an adaptive gearbox or intelligent cruise control, or for landing an airplane. Or what about the attempts to design automatic filtering of spam email or undesirable web contents?

Bainbridge [6] has described the difficulties that designers have in understanding how automation works as an irony of automation. The noble ambition of automation has generally been to replace human functions by technological artefacts. From that perspective it is something of a paradox that automated systems still are human–machine systems and that humans are needed as much as ever. Bainbridge argued that 'the increased interest in human factors among engineers reflects the irony that the more advanced a control system is, so the more crucial may be the contribution of the human operator' ([6], p. 775).

The first irony is that designer errors can be a major source of operating problems. In other words, the mistakes that designers make can create serious problems for operators. It can be difficult to understand how automation works because it is so complex, but the situation is obviously made worse if there are logical flaws in the automation. This may happen because the design fails to take into account all possible conditions because something is forgotten, because of coding mistakes and oversights, etc.

The second irony is that the designer, who tries to eliminate the operator, still leaves the operator to do the task that the designer cannot imagine how to automate. This problem is more severe since it obviously is unreasonable to expect that an operator in an acute situation can handle a problem that a designer, working under more serene conditions at his desk, is unable to solve. Indeed, automation is usually designed by a team of people with all possible resources at their disposal, whereas the operator often is alone and has limited access to support.

6.3.3 Automation as a socio-technical problem

The discussion of balancing/function congruence and the 'ironies of automation' should make it obvious that automation is a socio-technical rather than an engineering problem. This means that automation cannot be seen as an isolated problem of substituting one way of carrying out a function with another. The classical automation philosophies, the residual functions principle and the Fitts list, both imply that functional substitution is possible – something known in human factors as the substitution myth [28]. The basic tenets of the substitution myth are, (1) that people can be replaced by technology without any side effects and (2) that technology can be upgraded without any side effects. (Notice that the issue is not whether the side effects are adverse or beneficial, but whether they are taken into account.)

The first tenet means that there are no side effects of letting technology carry out a function that was hitherto done by humans. In other words, people can simply let go of part of their work, without the rest being the least affected. It is obviously very desirable that people are relieved of tedious and monotonous work, or that potential human risks are reduced. But it is incorrect to believe that doing so does not affect the work situation as a whole. New technology usually means new roles for people in the system, changes to what is normal and exceptional, and changes in manifestations of failures and failure pathways. There are therefore significant consequences both in the short term on the daily routines at the individual or team level, and in the long term on the level of organisations and society, such as structural changes to employment patterns.

The second tenet means that a function already carried out by technology can be replaced by an updated version, which typically means increased capacity but also increased complexity, without any side effects. This tenet is maintained despite the fact that practically every instance of technological upgrading constitutes evidence to the contrary. To illustrate that, just consider what happens when a new model of photocopier, or printer, is installed.

An excellent example of how the substitution principle fails can be found in the domain of traffic safety. Simply put, the assumption is that traffic safety can be improved by providing cars with better brakes, in particular with ABS. The argument is presumably that if the braking capability of a vehicle is improved then there will be fewer collisions, hence increased safety. The unspoken assumption is, of course, that drivers will continue to drive as they did before the change. The fact of the matter is that drivers do notice the change and that this affects their way of driving, specifically in the sense that they feel safer and therefore drive faster or with less separation from the car in front.

That automation is a socio-technical rather than an engineering problem means that automation design requires knowledge and expertise from engineering as well as from human factors. Automation is essentially the design of human work, and for this to succeed it is necessary that humans are not seen simply as sophisticated machines [29]. This is even more important because automation for most systems is bound to remain incomplete, as pointed out by the second irony. As long as automation cannot be made completely autonomous, but requires that some kind of human–machine co-operation take place, the design must address the joint system as a whole rather than focus on specific and particular functions. Even if automation autonomy may be achieved for some system states such as normal operation, human–machine co-operation may be required for others (start-up, shut-down, disturbances or emergencies), for repair and maintenance, or for upgrading.

The solution to these problems does not lie in any specific scientific discipline or method. This chapter has not even attempted to provide a complete coverage, since that would have to include control theory, cybernetics, and parts of cognitive science – and possibly also risk analysis, total quality management, social psychology, organisational theory, and industrial economics. It is, of course, possible to design automation without being an expert in, or having access to expertise from, all of these fields. Indeed, few if any systems have included this broad a scope. Designing automation neither is, nor should be, an academic exercise with unlimited time and resources. It is rather the ability to develop an acceptable solution to a practical problem given a (usually) considerable number of practical constraints. But what this chapter has argued is that no solution is acceptable unless it acknowledges that automation design is also the design of human work, and that it therefore must look at the system as a whole to understand the short- and long-term dynamics that determine whether the final result is a success or a failure.

6.4 Exercises

- Where do we find the self-regulating flow valve that is used every day?
- Why can a person not perform continuous regulation as well as a mechanical artefact? (Use the water clock and the steam valve as examples.)
- Suggest some examples of other devices (mechanisms) that serve to produce a more steady (less variable) performance of a machine.
- Find some examples where the speed of human responses is a limiting factor. How can automation be used to overcome that? Are there any potential negative consequences to consider?
- Discuss the nature of the self-reinforcing loop. Find some examples of the self-reinforcing loop in different industrial domains.
- Moore's law represents a major driving technological force. Find other examples.
- Describe some ways in which humans are a resource for a system.
- Give some examples of how technology/automation is used on the levels of tracking, regulating, monitoring and targeting, respectively. How did the transition take place? Are there some domains where the development is further ahead than others? Why is that the case?
- Find three clear examples of each of the three main automation philosophies.
- Which dimensions should be added to the Fitts list today? How should the characterisations of the attributes be changed? Is it meaningful to continue to use the Fitts list?

- Find examples of how technological improvement has resulted in increased performance rather than increased safety. (This is sometimes known as the phenomenon of risk homeostasis.)
- Discuss the details of the assumptions about humans, systems, and failures in the three automation philosophies. What are the practical consequences of each assumption?
- Find examples that illustrate both the first and the second irony of automation.
- Specify the logic for an automated system to control the lighting in a normal house.
- Study Table 6.3, and find examples of systems for each of the four main cells.

6.5 References

1 RAOUF, A.: 'Effects of automation on occupational safety and health', in KARWOWSKI, W., PARAEI, H. R., and WILHELM, M. R. (Eds): 'Ergonomics of hybrid automated systems 1' (Elsevier Science Ltd, USA, 1988)

2 BILLINGS, C. E.: 'Human-centered aircraft automation: a concept and guidelines'. NASA Technical Memorandum 103885, August, 1991

3 PARASURAMAN, R., and RILEY, V.: 'Humans and automation: use, misuse, disuse, abuse', *Human Factors*, 1997, **39** (2), 230–253

4 JAMES, P., and THORPE, N.: 'Ancient inventions' (Ballantine Books, New York, 1994)

5 MARUYAMA, M.: 'The second cybernetics: deviation-amplifying mutual processes', *American Scientist*, 1963, **55**, 164–179

6 BAINBRIDGE, L.: 'Ironies of automation', *Automatica*, 1983, **19** (6), 775–779

7 ASHBY, W. R.: 'An introduction to cybernetics' (Methuen & Co, London, 1956)

8 REASON, J. T.: 'Cognitive aids in process environments: prostheses or tools?' in HOLLNAGEL, E., MANCINI, G., and WOODS, D. D. (Eds): 'Cognitive engineering in complex dynamic worlds' (Academic Press, London, 1988)

9 PERROW, C.: 'Normal accidents', (Basic Books, New York, 1984)

10 WIENER, E. L.: 'Fallible humans and vulnerable systems: lessons learned from aviation', in WISE, J. A., and DEBONS, A. (Eds): 'Information systems: failure analysis' (NATO ASI Series vol. F32) (Springer-Verlag, Berlin, 1987)

11 SHERIDAN, T. B.: 'Humans and automation: system design and research issues' (Wiley, New York, 2002)

12 CHAPANIS, A.: 'Human factors in systems engineering', in DE GREENE, K. B. (Ed.): 'Systems psychology' (McGraw-Hill, New York, 1970) pp. 51–78.

13 FITTS, P. M. (Ed.).: 'Human engineering for an effective air navigation and traffic-control system' (Ohio State University Research Foundation, Columbus, Ohio, 1951)

14 ROCHLIN, G.: '"High-reliability" organizations and technical change: some ethical problems and dilemmas', *IEEE Technology and Society Magazine*, September 1986

15 GROTE, G., WEIK, S., WÄFLER, T., and ZÖLCH, M.: 'Complementary allocation of functions in automated work systems', in ANZAI, Y., OGAWA, K., and MORI, H. (Eds): 'Symbiosis of human and artifact' (Elsevier, Amsterdam, 1995)

16 WÄFLER, T., GROTE, G., WINDISCHER, A., and RYSER, C.: 'KOMPASS: a method for complementary system design', in HOLLNAGEL, E. (Ed.): 'Handbook of cognitive task design' (Erlbaum, Mahwah, NJ, 2003)

17 HOLLNAGEL, E., and WOODS, D. D.: 'Cognitive systems engineering: new wine in new bottles', *International Journal of Man-Machine Studies*, 1983, **18**, 583–600

18 HOLLNAGEL, E.: 'From function allocation to function congruence', in DEKKER, S., and HOLLNAGEL, E. (Eds): 'Coping with computers in the cockpit' (Ashgate, Aldershot, UK, 1999)

19 TAYLOR, F. W.: 'The principles of scientific management' (Harper, New York, 1911)

20 SHERIDAN, T. B.: 'Telerobotics, automation, and human supervisory control' (MIT Press, Cambridge, MA, 1992)

21 NORMAN, D. A.: 'The problem of automation: inappropriate feedback and interaction not overautomation', ICS Report 8904 (Institute for Cognitive Science, University of California-San Diego, La Jolla, CA, 1989)

22 BILLINGS, C. E.: 'Aviation automation, the search for a human centered approach' (Erlbaum, Hillsdale, NJ, 1996)

23 WINOGRAD, T., and WOODS, D. D.: 'The challenge of human-centered design', in FLANAGAN, J., HUANG, T., JONES, P., and KASIF, S. (Eds): 'Human-centered systems: information, interactivity, and intelligence' (National Science Foundation, Washington DC, 1997)

24 WOODS, D. D., JOHANNESEN, L. J., COOK, R. I., and SARTER, N. B.: 'Behind human error: cognitive systems, computers, and hindsight' (Crew Systems Ergonomic Information and Analysis Center, Dayton, OH, 1994)

25 TIMOTHY, D., and LUCAS, D.: 'Procuring new railway control systems: a regulatory perspective.' Proceedings of people in control: second international conference on *Human interfaces in control rooms, cockpits and command centres*, 19–21 June, 2001, Manchester, UK

26 HOLLNAGEL, E.: 'Time and time again', *Theoretical Issues in Ergonomics Science*, 2002, **3** (2), 143–158

27 WOODS, D. D., and SARTER, N. B.: 'Learning from automation surprises and "going sour" accidents: progress on human-centered automation.' Report ERGO-CSEL-98-02 (Institute for Ergonomics, the Ohio State University, Columbus, OH, 1998)

28 WOODS, D. D.: 'Automation: apparent simplicity, real complexity. Proceedings of the *First automation technology and human performance conference*, Washington, DC, April 7–8, 1994

29 HOLLNAGEL, E. (Ed.): 'Handbook of cognitive task design' (Lawrence Erlbaum Associates, Hillsdale, NJ, 2003)

Further Reading

BAINBRIDGE, L.: 'Ironies of automation', *Automatica*, 1983, **19** (6), pp. 775–779.

This article presented a penetrating analysis of automation from the human factors perspective, and dealt a severe blow to engineering optimism. The paradoxes pointed out by Bainbridge still remain unsolved, and have if anything gained more importance.

SHERIDAN, T. B.: 'Humans and automation: system design and research issues' (Wiley, New York, 2002).

A comprehensive and excellent overview of all the main automation issues. It describes the background for humans and automation with examples from many domains; the design of human-automation systems; and the generic research issues. It is destined to become a classic text in this discipline.

WIENER, E. L.: 'Fallible humans and vulnerable systems: lessons learned from aviation' in WISE, J. A., and DEBONS, A. (Eds): 'Information systems: failure analysis' (NATO ASI Series vol. F32) (Springer-Verlag, Berlin, 1987).

Earl Wiener's studies of automation problems in aviation are classic material, and may be found in several edited books. This chapter discusses the roles of humans and machines and the importance of information management as a crucial element in the design of human-automation systems.

Chapter 7
To engineer is to err

Sidney Dekker

Engineers make many assumptions about human error and about their ability to design against it. This chapter tries to unpack some of those assumptions. Does 'human error' exist as a uniquely sub-standard category of human performance? Are humans the most unreliable components in an engineered human–machine assembly? Once we embrace the idea that errors are consequences, not causes, can we still distinguish between mechanical failure and human error? In fact, engineers themselves are prone to err too – not only with respect to the assumptions they make about operators, but because the very activity of engineering is about reconciling irreconcilable constraints. The optimal, perfect engineered solution does not exist because by then it has already violated one or more of the original requirements. The chapter also discusses two popular ways of restraining human unreliability: procedures and automation. It tries to shed some light on why these 'solutions' to human error do not always work the way engineers thought they would.

7.1 Humans degrade basically safe systems. Or do they?

The most basic assumption that engineers often bring to their work is that human error exists. Human error is 'out there': it can be measured, counted, and it can be designed or proceduralised against – at least to some extent. Among engineers, the traditional idea has often been that human error degrades basically safe systems. Engineers do their best to build safe systems: to build in redundancies, double-checks and safety margins. All would go well (i.e. all can be predicted or calculated to go well) until the human element is introduced into that system. Humans are the most unreliable components in an engineered constellation. They are the least predictable: their performance degrades in surprising and hard to anticipate ways. It is difficult to even say when errors will occur. Or how often. 'Errors' arise as a seemingly random by-product of having people touch a basically safe, well-engineered system.

The common engineering reflex (which is assumed to create greater safety) is to keep the human away from the engineered system as much as possible (by automation), and to limit the bandwidth of allowable human action where it is still necessary (through procedures). Neither of these 'solutions' work unequivocally well.

Human factors, as we know it today, got its inspiration from these basic ideas about human error. It then showed something different: an alternative way of looking at human error. As a result, there are basically two ways of looking at human error today. We can see human error as a cause of failure, or we can see human error as a symptom of failure. These two views have recently been contrasted as *the old view* of human error versus *the new view* – fundamentally irreconcilable perspectives on the human contribution to system success and failure. In the old view of human error:

- Human error is the cause of accidents.
- The system in which people work is basically safe; success is intrinsic. The chief threat to safety comes from the inherent unreliability of people.
- Progress on safety can be made by protecting the system from unreliable humans through selection, proceduralisation, automation, training and discipline.

In the new view of human error:

- Human error is a symptom of trouble deeper inside the system.
- Safety is not inherent in systems. The systems themselves are contradictions between multiple goals that people must pursue simultaneously. People have to create safety.
- Human error is systematically connected to features of people's tools, tasks and operating environment. Progress on safety comes from understanding and influencing these connections.

The groundwork for the new view of human error was laid at the beginning of human factors. Fitts and Jones described back in 1947 how features of World War II airplane cockpits systematically influenced the way in which pilots made errors [1]. For example, pilots confused the flap and gear handles because these typically looked and felt the same and were co-located. Or they mixed up the locations of throttle, mixture and propeller controls because these kept changing across different cockpits. Human error was the starting point for Fitts' and Jones' studies – not the conclusion. The label 'pilot error' was deemed unsatisfactory, and used as a pointer to hunt for deeper, more systemic conditions that led to consistent trouble. The idea these studies convey to us is that mistakes actually make sense once we understand features of the engineered world that surrounds people. Human errors are systematically connected to features of people's tools and tasks. The insight, at the time as it is now, was profound: the world is not unchangeable; systems are not static, not simply given. We can re-tool, re-build, re-design, and thus influence the way in which people perform. This, indeed, is the historical imperative of human factors – understanding why people do what they do so we can change the world in which they work and shape their assessments and actions accordingly. Human factors is about helping engineers build systems that are error-resistant (i.e. do not invite errors) and error-tolerant (i.e. allow for recovery when errors do occur).

But what is an error, really? There are serious problems with assuming that 'human error' exists as such, that it is a uniquely identifiable category of sub-standard human performance, that it can be seen, counted, shared and designed against. For what do engineers refer to when they say 'error'? In safety and engineering debates there are at least three ways of using the label 'error':

- Error as the *cause* of failure. For example: this accident was due to operator error.
- Error as the *failure itself*. For example: the operator's selection of that mode was an error.
- Error as a *process*, or, more specifically, as a departure from some kind of standard. For example: the operators failed to follow procedures. Depending on what you use as standard, you will come to different conclusions about what is an error.

This lexical confusion, this inability to sort out what is cause and what is consequence, is actually an old and well-documented problem in human factors and specifically in error classifications [2]. Research over the past decade has tried to be more specific in its use of the label 'human error'. Reason, for example, contends that human error is inextricably linked with human intention. He asserts that the term error can only be meaningfully applied to planned actions that fail to achieve their desired consequences without some unforeseeable intervention. Reason identified the basic types of human error as either slips and lapses or as mistakes. Specifically, slips and lapses are defined as errors that result from some failure in the execution or storage stage of an action sequence, regardless of whether or not the plan that guided the action was adequate to achieve its objective. In this context, slips are considered as potentially observable behaviour whereas lapses are regarded as unobservable errors. In contrast, Reason defines mistakes as the result of judgmental or inferential processes involved in the selection of an objective or in the specification of the means to achieve it. This differentiation between slips, lapses and mistakes was a significant contribution to the understanding of human error.

Reason's error type definitions have limitations when it comes to their practical application. When analysing erroneous behaviour it is possible that both slips and mistakes can lead to the same action although they are both the results of different cognitive processes. This can have different implications for the design and assessment of human–computer interfaces. To understand why a human error occurred, the cognitive processes that produced the error must also be understood, and for that the situation, or context, in which the error occurred has to be understood as well. Broadly speaking, slips or lapses can be regarded as action errors, whereas mistakes are related more to situation assessment, and people's planning based on such assessment (see Table 7.1).

Much of Reason's insight is based on, and inspired by Jens Rasmussen's proposal [3], where he makes a distinction between skill-based, rule-based and knowledge-based errors. It is known as the SRK framework of human performance which is shown in Table 7.2.

In Rasmussen's proposal, the three levels of performance in the SRK framework correspond to decreasing levels of familiarity with a task or the task context; and increasing levels of cognition. Based on the SRK performance levels, Reason argues

Table 7.1 Behavioural errors and cognitive processes (from [2], p. 13)

Behavioural error type	Erroneous cognitive process
Mistakes	Planning/situation assessment
Lapses	Memory storage
Slips	Execution

Table 7.2 Skill–rule–knowledge framework (from [3])

Performance level	Cognitive characteristics
Skill-based	Automatic, unconscious, parallel activities
Rule-based	Recognising situations and following associated procedures
Knowledge-based	Conscious problem solving

Table 7.3 Human performance and behavioural errors (from [2])

Performance level	Behavioural error type
Skill-based	Slips and lapses
Rule-based	Rule-based mistakes
Knowledge-based	Knowledge-based mistakes

that a key distinction between the error types is whether an operator is engaged in *problem solving* at the time an error occurs. This distinction allowed Reason to identify three distinct error types which are shown in Table 7.3.

Jens Rasmussen's SRK framework, though influential, may not be as canonical as some literature suggests. Especially, Rasmussen's so-called 'skilful' behaviour is often thought to actually occur across all three levels, if indeed there are different levels: think, for example, about decision making skills, or pattern recognition skills.

7.2 Human error or mechanical failure?

Both Rasmussen and Reason, and a host of other influential authors in human factors, have embraced and propagated the view of error-as-consequence. Errors are consequences, or symptoms, of deeper causes: problems and failures deeper inside the system in which human performance is embedded. Yet, adopting the view that human error is the symptom of failure still embraces error as something that is 'out there', and can be measured or captured independently. This may actually be difficult. One corollary to the engineering assumption that human error exists is that there is a basic distinction between human error and mechanical failure. Engineers want to make this distinction because it can help them identify and deal with their potential areas of responsibility. Did the problem arise because something broke that the engineer built? Or did the system work as designed, only to be degraded by random human unreliability?

The distinction between human error and mechanical failure is often false and can be deeply misleading. Consider the following case of a 'human error' – failing to arm the ground spoilers before landing in a commercial aircraft. Passenger aircraft have 'spoilers' (panels that come up from the wing upon landing) to help brake the aircraft during its landing roll-out. To make these spoilers come out, pilots have to manually 'arm' them by pulling a lever in the cockpit. Many aircraft have landed without the spoilers being armed, some cases even resulting in runway overruns. Each of these events gets classified as 'human error' – after all, the human pilots forgot something in a system that is functioning perfectly otherwise. But deeper probing reveals a system that is not functioning perfectly at all. Spoilers typically have to be armed after the landing gear has come out and is safely locked into place. The reason is that landing gears have compression switches that communicate to the aircraft when it is on the ground (how else would an aircraft know that?). When the gear compresses, the logic tells the aircraft that it has landed. And then the spoilers come out (if they are armed, that is). Gear compression, however, can also occur *while* the gear is coming out, because of air pressure from the enormous slip stream around a flying aircraft, especially if landing gear folds open *into* the wind (which many do). This would create a case where the aircraft thinks it is on the ground, but it is not. If the spoilers were already armed at that time, they would come out too – which would be catastrophic if still airborne. To prevent this from happening, all these aircraft carry procedures that say the spoilers may only be armed when the gear is fully down and locked. It is safe to do so, because the gear is then orthogonal to the slipstream, with no more risk of compression. But the older an aircraft gets, the longer a gear takes to come out and lock into place. The hydraulic system no longer works as well, for example. In some aircraft, it can take up to half a minute. By that time, the gear extension has begun to seriously intrude into other cockpit tasks that need to happen by then – selecting wing flaps for landing, capturing and tracking the electronic glide slope towards the runway, and so forth. These are items that come *after* the 'arm spoilers' item on a typical before-landing checklist. If the gear is still doing its thing, while the world has already pushed you further down the checklist, not arming the spoilers is a slip that is only too easy to make. Combine this with a system that, in many aircraft, never warns pilots that their spoilers are not armed and a spoiler handle that sits over to one dark side of the centre cockpit console, obscured for one pilot by power levers, and whose difference between armed and not-armed may be all of one inch, and the question becomes: is this mechanical failure or human error?

'Human error', if there were such a thing, is not a question of an individual single-point failure to notice or process that can simply be counted and added up – not in the spoiler story and probably not in any story of breakdowns in flight safety. Practice that goes wrong spreads out over time and in space, touching all the areas that usually make people successful. It extends deeply into the engineered, organised, social and operational world in which people carry out their work. Were we to trace 'the cause' of failure, the causal network would fan out immediately, like cracks in a window, with only us determining when to stop looking because the evidence will not do it for us. Labelling certain assessments or actions in the swirl of human and social and technical activity as causal, or as 'errors', is entirely arbitrary and ultimately meaningless.

Indeed, real stories of failure often resist the identification of a clear cause. Even if we want to find the 'eureka part', the central error, the 'nucleus of cause', we may only be fooling ourselves. What 'caused' the Challenger Space Shuttle disaster? Was it mechanical failure? Was it human error? Violations? Was it managerial wrongdoing? All of those, and thus perhaps none. Certainly none in isolation. As Diane Vaughan ([4], p. xiv) concludes:

> No extraordinary actions by individuals explain what happened: no intentional managerial wrongdoing, no rule violations, no conspiracy. (These are) mistakes embedded in the banality of organisational life and facilitated by environments of scarcity and competition, uncertain technology, incrementalism, patterns of information, routinisation and organisational and interorganisational structures.

Error, then, may not really exist 'out there'. 'Errors' that occur (engineering miscalculations about the redundancy of O-rings, in this case) are systematically produced by the normal organisational processes and factors that make up everyday engineering life (including: 'scarcity and competition, uncertain technology, incrementalism, patterns of information, routinisation and organisational and interorganisational structures'). More than anything, 'error' may be a product of our hindsight, a product of our strong desire to find a cause, a product of our strong desire to imagine ourselves in control over the systems we help engineer. There is this desire to label something as *the* cause, because then we can identify something as *the* solution. This is fundamental to engineering: uncertainty must be closed. If the cause is human error, then the solution can be more automation or more procedures (or disciplining, training, removal, etc. but those will not be considered here). But neither works well. Here is why.

7.3 Why don't they just follow the procedures?

Procedures are a way of feedforward control. Operators can more or less do what they want once a system is fielded. Procedures constrain what operators can do and align their actions with what the designers of the system saw as rational when they conceived the system. But people do not always follow procedures. This observation is easy to make while watching people at work. Engineers may see this as an even greater testimony to the fundamental unreliability of people and an even greater motivation for automating. The problem, however, is that the links between procedures and work, and between procedure-following and safety, are not as straightforward as engineers may sometimes think. Of course, there is a persistent notion that not following procedures can lead to unsafe situations. For example, a study carried out for an aircraft manufacturer identified 'pilot deviation from basic operational procedure' as the primary factor in almost 100 accidents [5]. This study, as well as many engineers, think that:

- Procedures represent the best thought-out, and thus the safest way to carry out a job.
- Procedure-following is mostly simple IF-THEN rule-based mental activity: IF this situation occurs, THEN this algorithm (e.g. checklist) applies.
- Safety results from people following procedures.
- For progress on safety, people need to know the procedures and follow them.

But procedures are not the same as work. Real operational work takes place in a context of limited resources and multiple goals and pressures. Work-to-rule strikes show how it can be impossible to follow the rules and get the job done at the same time. Aviation line maintenance is emblematic: A 'job perception gap' exists where supervisors are convinced that safety and success result from mechanics following procedures – a sign-off means that applicable procedures were followed. But mechanics may encounter problems for which the right tools or parts are not at hand; the aircraft may be parked far away from base. Or there may be too little time: aircraft with a considerable number of problems may have to be 'turned around' for the next flight within half an hour. Mechanics, consequently, see success as the result of their evolved skills at adapting, inventing, compromising and improvising in the face of local pressures and challenges on the line – a sign-off means the job was accomplished in spite of resource limitations, organisational dilemmas and pressures [6]. Those most adept are valued for their productive capacity even by higher organisational levels. Unacknowledged by those levels, though, are the vast informal work systems that develop so mechanics can get work done, advance their skills at improvising, impart them to one another and condense them in unofficial, self-made documentation [7]. Seen from the outside, a defining characteristic of such informal work systems would be routine non-conformity. But from the inside, the same behaviour is a mark of expertise, fuelled by professional and inter-peer pride. And of course, informal work systems emerge and thrive in the first place because procedures are inadequate to cope with local challenges and surprises, and because the procedures' conception of work collides with the scarcity, pressure and multiple goals of real work. In fact, procedure-following can be antithetical to safety. In the 1949 US Mann Gulch disaster, firefighters who perished were the ones sticking to the organisational mandate to carry their tools everywhere [8]. In this case, as in others [9], operators faced the choice between following the procedure or surviving.

This, then, is the tension. Procedures are an investment in safety – but not always. Procedures are thought to be required to achieve safe practice, yet they are not always necessary, nor likely ever sufficient for creating safety. Procedures spell out how to do the job safely, yet following all the procedures can lead to an inability to get the job done. Engineers assume that order and stability in operational systems are achieved rationally, mechanistically, and that control is implemented vertically (e.g. through task analyses that produce prescriptions of work-to-be-carried-out). But this does not necessarily apply. People at work must interpret procedures with respect to a collection of actions and circumstances that the procedures themselves can never fully specify [10]. In other words, procedures are not the work itself. Work, especially that in complex, dynamic workplaces, often requires subtle, local judgments with regard to timing of subtasks, relevance, importance, prioritisation and so forth. For example, there is no technical reason why a before-landing checklist in a commercial aircraft could not be automated. The kinds of items on such a checklist (e.g. hydraulic pumps OFF, gear down, flaps selected) are mostly mechanical and could be activated on the basis of pre-determined logic without having to rely on, or constantly remind, a human to do so. Yet no before-landing checklist is fully automated today. The reason is that approaches for landing differ – they can differ in terms of timing, workload, priorities and so forth. Indeed, the reason is that the checklist is not the job itself. The checklist is a resource for action; it is one way for people to help structure activities across roughly similar yet subtly different situations.

Circumstances change, or are not as was foreseen by those who designed the procedures. Safety, then, is not the result of rote rule following; it is the result of people's insight into the features of situations that demand certain actions, and people being skillful at finding and using a variety of resources (including written guidance) to accomplish their goals. This suggests a different idea of procedures and safety:

- Procedures are resources for action. Procedures do not specify all circumstances to which they apply. Procedures cannot dictate their own application. Procedures can, in themselves, not guarantee safety.
- Applying procedures successfully across situations can be a substantive and skillful cognitive activity.
- Safety results from people being skilful at judging when and how (and when not) to adapt procedures to local circumstances.
- For progress on safety, organisations must monitor and understand the reasons behind the gap between procedures and practice. Additionally, organisations must develop ways that support people's skill at judging when and how to adapt.

Pre-specified guidance is inadequate in the face of novelty and uncertainty. But adapting procedures to fit circumstances better is a substantive cognitive activity. Take for instance the crash of a large passenger aircraft near Halifax, Nova Scotia in 1998. After an uneventful departure, a burning smell was detected and, not much later, smoke was reported inside the cockpit. Reference 9 characterises the two pilots as respective embodiments of the models of procedures and safety: the co-pilot preferred a rapid descent and suggested dumping fuel early so that the aircraft would not be too heavy to land. But the captain told the co-pilot, who was flying the plane, not to descend too fast, and insisted they cover applicable procedures (checklists) for dealing with smoke and fire. The captain delayed a decision on dumping fuel. With the fire developing, the aircraft became uncontrollable and crashed into the sea, taking all 229 lives on board with it.

The example illustrates a fundamental double bind for those who encounter surprise and have to apply procedures in practice [11]:

- If rote rule following persists in the face of cues that suggest procedures should be adapted, this may lead to unsafe outcomes. People can get blamed for their inflexibility; their application of rules without sensitivity to context.
- If adaptations to unanticipated conditions are attempted without complete knowledge of circumstance or certainty of outcome, unsafe results may occur too. In this case, people get blamed for their deviations, their non-adherence.

In other words, people can fail to adapt, or attempt adaptations that may fail. In the Halifax crash, rote rule following became a de-synchronised and increasingly irrelevant activity, de-coupled from how events and breakdowns were really unfolding and multiplying throughout the aircraft. But there was uncertainty about the very need for adaptations (how badly ailing was the aircraft, really?) as well as uncertainty about the effect and safety of adapting: how much time would the crew have to change their plans? Could they skip fuel dumping and still attempt a landing? Potential adaptations, and the ability to project their potential for success, were not necessarily supported by specific training or overall professional indoctrination. Many industries, after all, tend to emphasise model 1): stick with procedures and you will most likely be safe. Tightening procedural adherence, through threats of punishment or other supervisory interventions, does not remove the double bind. In fact, it may tighten the double bind, making it more difficult for people to develop judgment of how and when to adapt. Increasing the pressure to comply increases the probability of failures to adapt, compelling people to adopt a more conservative response criterion. People will require more evidence of the need to adapt, which takes time, and time may be scarce in cases that call for adaptation (as in the accident described above).

The gap between procedures and practice is not constant. After the creation of new work (e.g. through the introduction of new technology), considerable time can go by before applied practice stabilises – likely at a distance from the rules as written for the system 'on-the-shelf'. Social science has characterised this migration from tightly coupled rules to more loosely coupled practice as 'fine-tuning' [12] or 'practical drift' [13]. Through this shift, applied practice becomes the pragmatic imperative; it settles into a system as normative. Deviance (from the original rules) becomes normalised; non-conformity becomes routine [4]. The literature has identified important ingredients in this process, which can help engineers understand the nature of the gap between procedures and practice:

- Rules that are overdesigned (written for tightly coupled situations, for the 'worst-case') do not match actual work most of the time. In real work, there is time to recover, opportunity to reschedule and get the job done better or more smartly [12]. This mismatch creates an inherently unstable situation that generates pressure for change [13].

- Emphasis on local efficiency or cost-effectiveness pushes operational people to achieve or prioritise one goal or a limited set of goals (e.g. customer service, punctuality, capacity utilisation). Such goals are typically easily measurable (e.g. customer satisfaction, on-time performance), whereas it is much more difficult to measure how much is borrowed from safety.
- Past success is taken as guarantee of future safety. Each operational success achieved at incremental distances from the formal, original rules, can establish a new norm. From here a subsequent departure is once again only a small incremental step [4]. From the outside, such fine-tuning constitutes incremental experimentation in uncontrolled settings [12] – on the inside, incremental non-conformity is not recognised as such.
- Departures from the routine become routine. Seen from the inside of people's own work, violations become compliant behaviour. They are compliant with the emerging, local ways to accommodate multiple goals important to the organisation (maximising capacity utilisation but doing so safely; meeting technical requirements but also deadlines). They are compliant, also, with a complex of peer pressures and professional expectations in which unofficial action yields better, quicker ways to do the job; in which unofficial action is a sign of competence and expertise; where unofficial action can override or outsmart hierarchical control and compensate for higher-level organisational deficiencies or ignorance.

There is always a tension between feedforward guidance (in the form of procedures) and local practice. Sticking to procedures can lead to ineffective, unproductive or unsafe local actions, whereas adapting local practice in the face of pragmatic demands can miss global system goals and other constraints or vulnerabilities that operate on the situation in question. Helping people solve this fundamental trade-off is not a matter of pushing the criterion one way or the other. Discouraging people's attempts at adaptation can increase the number of failures to adapt in situations where adaptation was necessary. Allowing procedural leeway without investing in people's skills at adapting, on the other hand, can increase the number of failed attempts at adaptation.

7.4 Automation as a solution to the human error problem?

The idea that engineers may bring to automation is that automation supplants human work; that it replaces human work. By replacing human work, automation takes away the human actions that could go wrong, it takes away the error potential. Automation, in this sense is thought to bring quantifiable benefits. People will have a lower workload, there will be better, more economical and accurate system performance, there will be fewer errors, fewer training requirements. While automation sometimes lives up to those quantifiable promises, and indeed delivers new capabilities, it introduces new complexities as well. Operators may report higher workloads at certain times; they may feel that training for the automated system should be more, not less. And indeed, they still make errors. Automation does not just replace human work.

Automation creates new human work. And by creating new human work, automation creates new error opportunities: pathways to breakdown and failure that did not exist before. Automation thus induces qualitative shifts in the way work is done; in the way work can be done successfully (and unsuccessfully).

Indeed, automation is never just a replacement of human work. Automation is not a mere division of labour across people and engineered systems. Instead:

- Automation transforms human roles (operators become supervisor, monitor, manager) and creates new human work (typing, searching databases, remembering modes). By creating new work and roles, automation presents new error opportunities (e.g. mode errors) and pathways to system breakdown (e.g. fighting with the autopilot).
- Automation introduces new or amplifies existing human weaknesses (skill erosion with lack of practice; vigilance decrements with lack of change and events).
- Automation changes relationships between people (the one who better understands or is quicker on the engineered system in effect has control over the system) and between people and their monitored process.

Indeed, the real effects of automation are not (just) quantitative (less or more of this or that). Its real effects are qualitative. Automation creates qualitative changes in the distribution of workload over operational phases; it shifts where and how errors will occur; it transforms roles and relationships; it redirects, recodes and re-represents information flows across the human–machine ensemble.

Human error does not go away with more automation. Instead, different problems occur. Coordination breakdowns in automated systems, for example, are common. Operators think they told the automation to do one thing, and it is in fact doing something else (for example, because it was in a different mode). The symptom of this breakdown is the automation surprise – the realisation on the part of operators that the automation has misinterpreted their instructions ('What is it doing now? Why is it doing that?'). Sometimes the discovery of this divergence comes too late to avert an accident.

At the heart of the automation surprise lies a tension, an imbalance. Today's (and tomorrow's) automated systems are high in authority and autonomy: engineers ensure that these systems can carry out long sequences of actions without human interference. Yet these systems can be low on observability. They may look simple, or neat at the outside, especially when compared to the old, hardwired system. Their actual complexity is hidden – obscured behind a few displays or mode annunciations that really say little about the behaviour of the system or about the future of the monitored process. Automated systems can also be hard to direct, with people forced to weed serially through a multitude of interrelated display pages or manipulate a large number of functions through only a few buttons (or touch-screens that function as controls) and set different modes with single knobs that do double duty. Systems that are highly autonomous should be high on observability and directability if they want to keep their human operators in the loop. Representations of automation behaviour that highlight process changes and events (that are future-oriented and pattern-based for quick perceptual pick-up of deviances) are an important step in the right direction.

Automation does not remove human error just as it does not (totally) remove human beings. There are supervisors or system monitors somewhere. By displacing human work, and by changing it, automation changes the error potential of the remaining human–machine ensemble. The error potential will not go away.

7.5 Unknowable futures and irreconcilable constraints

New technology changes, in unpredictable ways, the tasks it was designed to support or replace. Engineers build a system with a particular future in mind, yet that future will be changed by the very introduction of that system. The cycle will never end, and it can become very discouraging. New technology creates new tasks, which may call for new technology, and so forth. New designs that engineers come up with represent a hypothesis about what would be useful for operators in a field of practice, and as hypotheses, engineers' ideas are, or should be, open to empirical scrutiny. Indeed, engineers should take an experimental stance – see their designs as an intervention in a field of ongoing activity that will create effects, both anticipated and unanticipated. An experimental stance means that designers need to

- recognise that design concepts represent hypotheses or beliefs about the relationship between technology and human work;
- subject these beliefs to empirical jeopardy by a search for disconfirming and confirming evidence;
- recognise that these beliefs about what would be useful are tentative and open to revision as engineers learn more about the mutual shaping that goes on between their products and operators in a field of practice.

This experimental stance for engineering is needed, not because engineers should mimic traditional research roles, but because it will make a difference in developing useful systems. Assessing these effects is also a way of (re-)calibrating the engineer – was the design idea based on an accurate understanding of people and technology, and the coordinative and role transformations that occur? Was it grounded in what we already know about automation and human performance?

Many will say that accurate assessment of such effects is impossible without fielding a functional system. But this puts every engineer in Newell's catch. By the time the engineer can actually measure the effects, so much capital (political, organisational, financial, personal) has been invested that changing anything on the basis of lessons learned is impossible. One direction out of the dilemma is to 'front-load' human factors – to do much of the hard analytical work before a prototype has even been designed, for example through approaches like 'contextual design' [14], 'contextual inquiry' [15] or the 'envisioned world problem' [16] that have gained some popularity over the last decade.

Despite such novel approaches to anticipating the effects of new technology on human error, engineers will never get it entirely right. Engineering is always erring. Not because engineers want to err, or do not want to get things entirely right, but because getting it entirely right is a logical impossibility. Engineering design is about the reconciliation of basically irreconcilable constraints. Creating an optimal design is impossible in every principle; compromise is fundamentally inevitable:

> The requirements for design conflict cannot be reconciled. All designs are in some degree failures, either because they flout one or another of the requirements or because they are compromises, and compromise implies a degree of failure . . .. It is quite impossible for any design to be 'the logical outcome of the requirements' simply because, the requirements being in conflict, their logical outcome is an impossibility. [17]

Engineering design is about trade-offs and compromise. In making trade-offs, engineers have to rely on knowledge that is often incomplete, uncertain and ambiguous. Of course, engineers want to follow rules, get the numbers right and close uncertainty. But this may not always be possible. Many may think (even engineers themselves) that engineering is about precision, rationality and precise knowledge. This idea is sponsored by the fact that their resulting technology is most often successful (accidents are pretty rare and problems are obscured by responsible operators who want to get technology to work despite its shortcomings), and by the formal language and specialised skills associated with engineering that obscure the process from public understanding. When technical systems fail, however, outsiders quickly find an engineering world characterised by ambiguity, lack of testing, deviation from design specifications and operating standards, and ad-hoc rule making. This messy situation, when revealed to the public, automatically becomes the explanation for the accident. But much of this is the bias inherent in accident investigation. The engineering process behind a 'non-accident' is never subjected to the same level of scrutiny, but if it was, a similarly messy interior would undoubtedly be found. After the accident, it all may look like it was 'an accident waiting to happen'. In reality, it was more or less normal operations, normal technology, normal engineering decision making [4].

It is not that engineering work is not rule-following. The rules that are followed, however, are 'practical rules': operating standards consisting of numerous ad-hoc judgments and assumptions that are grounded in evolving engineering practice. Such rules are experience-driven. Rules follow evolving practice; the practical rules that engineers and technicians apply in their work often emerge from the practical success they obtained earlier. Learning proceeds through iteration; in all engineering organisations learning is routine-based, and dependent on history. Practical rules that follow from evolving practice also mean that absolute certainty is impossible to obtain. It is impossible to test conclusively that error opportunities do not exist in the system to be fielded. To engineer is to err: the potential for error and for inviting error is inherent in the very nature of the activity.

7.6 References

1 FITTS, P. M., and JONES, R. E.: 'Analysis of factors contributing to 460 "pilot error" experiences in operating aircraft controls', Memorandum Report TSEAA-694-12 (Aero Medical Laboratory, Air Material Command, Wright-Patterson Air Force Base, Dayton, Ohio, July 1 1947)

2 REASON, J.: 'Human error' (Cambridge University Press, Cambridge, 1990)
3 RASMUSSEN, J.: 'Skills, rules, knowledge: signals, signs and symbols and other distinctions in human performance models', *IEEE Transactions on Systems, Man and Cybernetics*, 1983, SMC **13** (3), pp. 257–267
4 VAUGHAN, D.: 'The Challenger launch decision: risky technology, culture and deviance at NASA' (University of Chicago Press, Chicago, IL, 1996)
5 LAUTMAN, L., and GALLIMORE, P. L.: 'Control of the crew caused accident: results of a 12-operator survey', *Boeing Airliner*, April–June 1987, pp. 1–6
6 VAN AVERMAETE, J. A. G., and HAKKELING-MESLAND, M. Y.: 'Maintenance human factors from a European research perspective: results from the ADAMS project and related research initiatives' (National Aerospace Laboratory NLR, Amsterdam, 2001)
7 MC DONALD, N., CORRIGAN, S., and WARD, M.: 'Well-intentioned people in dysfunctional systems' (Keynote speech presented at *5th Workshop on human error, safety and systems development*, Newcastle, Australia, 2002)
8 WEICK, K. E.: 'The collapse of sensemaking in organisations', *Administrative Science Quarterly*, 1993, **38**, pp. 628–652
9 CARLEY, W. M.: 'Swissair pilots differed on how to avoid crash', *The Wall Street Journal,* January 21, 1999
10 SUCHMAN, L. A.: 'Plans and situated actions' (Cambridge University Press, Cambridge, 1987)
11 WOODS, D. D., and SHATTUCK, L. G.: 'Distant supervision – local action given the potential for surprise', *Cognition Technology and Work*, 2000, **2** (4), pp. 242–245
12 STARBUCK, W. H., and MILLIKEN, F. J.: 'Challenger: fine-tuning the odds until something breaks', *Journal of Management Studies*, 1988, **25** (4), pp. 319–340
13 SNOOK, S. A.: 'Friendly fire' (Princeton University Press, Princeton, NJ, 2000)
14 BEYER, H., and HOLTZBLATT, K.: 'Contextual design: defining customer-centered systems' (Academic Press, San Diego, CA, 1998)
15 DEKKER, S. W. A., NYCE, J. M., and HOFFMAN, R.: 'From contextual inquiry to designable futures: what do we need to get there?', *IEEE Intelligent Systems*, 2003, **2**, pp. 2–5
16 WOODS, D. D., and DEKKER, S. W. A.: 'Anticipating the effects of technology change: a new era of dynamics for Human Factors', *Theoretical Issues in Ergonomics Science*, 2001, **1** (3), pp. 272–282
17 PETROSKI, H.: 'To engineer is human: the role of failure in successful design' (St Martin's, New York, 1985) pp. 218–219

Chapter 8
Qualitative and quantitative evaluation of human error in risk assessment

David Embrey

8.1 Introduction

Following the occurrence of major disasters in the petrochemical industry (Piper Alpha, Bhopal, Texas City), nuclear power (Three Mile Island, Chernobyl), marine transport (*Herald of Free Enterprise*, *Exxon Valdez*) and rail transport (Ladbroke Grove, Southall, Potters Bar) there is an increasing requirement by regulatory authorities for organisations to conduct formal safety assessments of systems. As part of these assessments, risk analysts are now required to perform evaluations of human reliability in addition to the analyses of hardware systems that are the primary focus of a typical risk assessment.

Historically, the emphasis in human reliability analysis (HRA) has been on techniques for the derivation of human error probabilities (HEPs) for use in systems analysis techniques such as fault tree analysis (see Section 8.2). However, HRA should be an integrated process that includes a systematic and rigorous qualitative analysis to identify the nature of the errors that can arise prior to any attempt at quantification. This qualitative analysis (sometimes referred to as Human Error Identification, HEI) must ensure that no significant failures are omitted from the analysis.

It is widely recognised that there are considerable uncertainties in the quantitative data available for inclusion in HRA. However, as long as the qualitative error identification process is sufficiently comprehensive, valuable insights will emerge with regard to the sources of risk, and where resources should be most cost effectively applied in minimising these risks.

In subsequent sections, an overall framework will be introduced which will integrate the qualitative and quantitative aspects of HRA, illustrated, where appropriate, with case studies. The issue of the availability of data sources to use with the quantitative aspects of HRA will also be discussed. Although a large number of tools and techniques have been developed for use in HRA, this chapter will focus on those that have been applied in practical assessments. Where appropriate, references will be provided to the research literature on other techniques.

8.2 Human reliability analysis in risk assessment

8.2.1 Introduction to the risk analysis process

Since HRA is usually applied in the context of technical safety and risk assessment, sometimes referred to as Formal Safety Analysis (FSA), we will first provide a brief overview of this process. The overall purpose of technical safety analysis and risk assessment is the identification and management of risks so that they are reduced to an acceptable level. The stages of the process can be summed up as follows:

- Identification of the hazards. These are aspects of a system likely to cause harm to people (e.g. high temperatures, pressures, toxic substances, voltages, high velocities) or financial loss.
- Evaluation of scenarios or credible incidents. These are events or sequences of events that could release the hazards.
- Evaluation of consequences. This is concerned with the different ways in which the hazard could exert its effects or influences on people, company assets or the environment once released.
- Evaluation of the probability or frequency with which the hazard is likely to be released (e.g. once every 10,000 operations of the system, once every 10 years).
- Evaluation of the risk. The product of the severity of the consequences and the frequency of its occurrence is the risk (alternatively the product of the severity of consequences, the frequency of exposure and the probability of the incident leading to the release of the consequences).
- Assessment of whether the risk is acceptable, using risk criteria, or bands of acceptability.
- Modification of the system if the risk is deemed to be unacceptable.

These stages are represented in the flow diagram shown in Figure 8.1. Risk assessment is often used to perform cost effectiveness evaluations, to decide which of a possible set of interventions will achieve a required level of risk for the lowest cost. For analyses of this type, it is necessary to be able to assess both the severity and the probability of a particular set of consequences occurring as a function of a number of mitigation options, each of which may have different associated costs. In the context of aircraft accidents, for example, it may be possible to demonstrate that a particular type of accident is due to a certain type of skill deficiency. This could be remedied by extending the training period of pilots, but this will have implications for both costs and operational availability. In order to assess whether or not the training option is viable, it would be necessary to evaluate the probability of the accident type as a function of the degree of training.

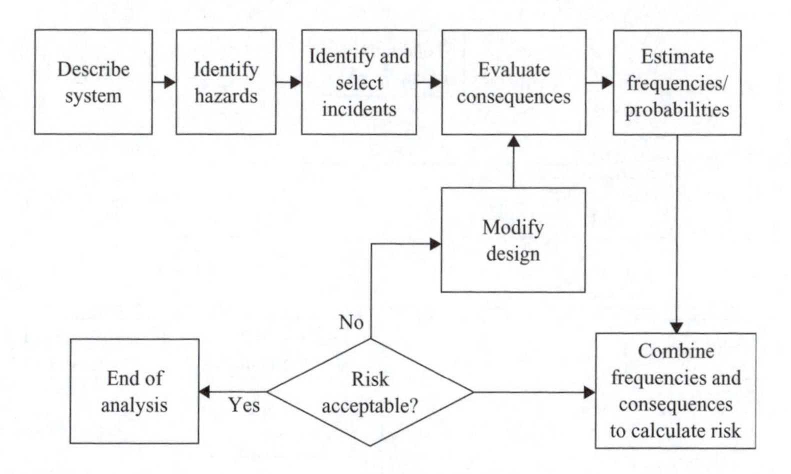

Figure 8.1 Overall risk analysis process

The potential costs of accidents arising from this cause will then have to be compared with the increased costs of training and reduced operational availability. In order to assess the risk and perform the cost-effectiveness calculations, the different ways in which the failures occur will need to be modelled, and appropriate costs and probabilities inserted in this model. This will require the use of a range of tools and techniques, which will be discussed in the following sections of this document.

8.2.2 Modelling tools used in risk analysis

In technical risk assessments and safety analyses, typical failures of safety critical systems are modelled in the form of representations such as event trees and fault trees. The failure probabilities of various hardware components in the system are then combined together using the logic represented by these models to give the probability of the failure. Where there are a number of systems that could contribute to the mitigation of an accident sequence, the probabilities of failure of each of the individual systems (which have been evaluated using fault tree analysis) are then combined using an event tree. This process can be used for both hardware and human failure probabilities and will be illustrated later.

The structure of a typical fault tree is illustrated in Figure 8.2. The event being analysed, the 'Total failure of car brakes' is called the 'top event'. It can be seen that if appropriate failure probabilities are inserted in such a tree for each of the items, then the overall failure probability of the top event can be calculated from the logic of the fault tree model. Probabilities are multiplied at AND gates, added at OR gates, and the result propagated up the tree. It will be apparent that some of the failure probabilities in the event tree could easily arise from human error. For example, although the wear-out of the brake linings is a property of the hardware, it could arise from a failure to carry out routine servicing correctly (a 'latent' failure,

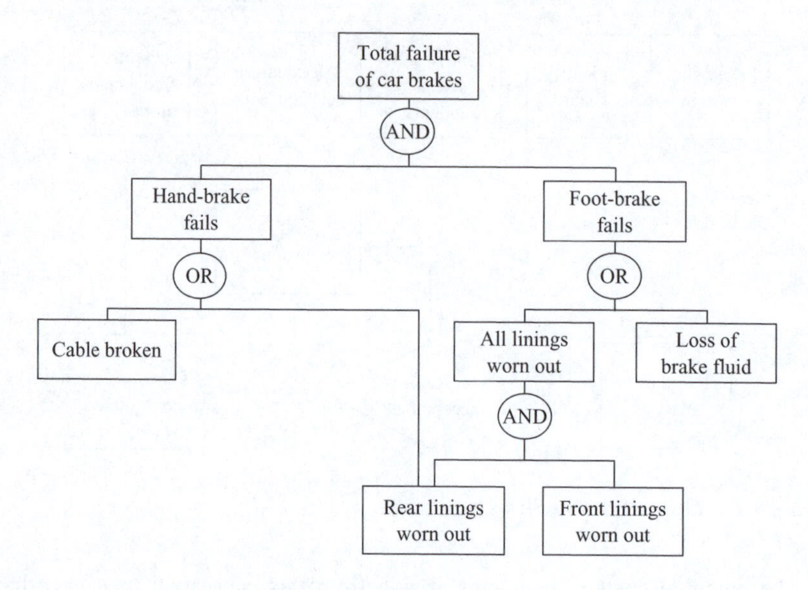

Figure 8.2 Fault tree for brake failure

where the effects are delayed). The same considerations apply to loss of brake fluid or the probability of the cable being broken.

Both of these latter failures could also arise from failures in maintenance activities. Another way in which human activities could be involved in an apparently hardware-based system would be in the safety-related aspects of the design. For example, the linings may not have been designed for the types of braking encountered in the use of the vehicle. It might also be argued that some form of redundant system would have been appropriate so that the loss of brake fluid did not fail the foot-brake. In addition, certain failures, in this case the failure of the rear brake linings, can affect more than one branch of the fault tree and may therefore have a greater impact than failures that only make one contribution to the top event. Failure probabilities for fault trees of this type are usually derived from databases of failure probabilities for components, which are obtained by observing failure rates *in situ*, or as a result of testing programmes.

Failure rates can be expressed in terms of time, such as a mean time between failures, or as a failure probability per demand. These measures can be converted to one another if the failure distribution is known. It is important to emphasise that these probabilities are conditional probabilities, since they are dependent on the conditions under which the data was collected. Although the issue of the context within which a failure occurs is not insignificant for hardware, it is particularly important in the case of human failure. For both humans and hardware components, such as pumps, failure probabilities arise from the interaction between the person (or component) and the environment. However, whilst the failure probability of a pump can be largely predicted by its basic design and its level of use, human error probabilities are influenced by a much wider range of contextual factors, such as the quality of the training, the design of the equipment and the level of distractions. These are sometimes referred to as Performance Shaping Factors (PSFs). The term 'Performance Shaping' originated in the context of the early days of psychology where various types of conditioning were used to shape the performance of simple tasks performed by animals under laboratory conditions. However, this context has little relevance to the performance of skilled people in technical systems, and hence the term Performance Influencing Factors (PIFs) will be used to refer to the direct and indirect factors that influence the likelihood that a task will be performed successfully. Williams [1] uses the term 'Error Producing Conditions' with a similar meaning.

The event tree is used to model situations where a number of events need to occur (or be prevented) in sequence for an undesirable outcome to arise or be averted. Typically, these events can be either initiating events (hardware/software or human), which start the accident sequence, or possible preventative actions (again hardware/software or human), which may prevent the sequence proceeding to the final undesirable consequence. Depending on which failure occurs and whether it can be recovered by subsequent actions, a range of paths through the event tree is possible. The overall probability of the undesirable consequence, therefore, has to take into account all of these paths.

The event tree shown in Figure 8.3 shows the different types of event that could give rise to a signal passed at danger (SPAD) in a railway situation. This is based on the MARS approach (Model for Assessing and Reducing SPADs) developed by Embrey *et al.* [2]. The probability of each failure is evaluated by multiplying the probabilities along each of the routes that could be traversed. If S stands for successes at each node of the tree, and F for the corresponding failures, the probabilities of the

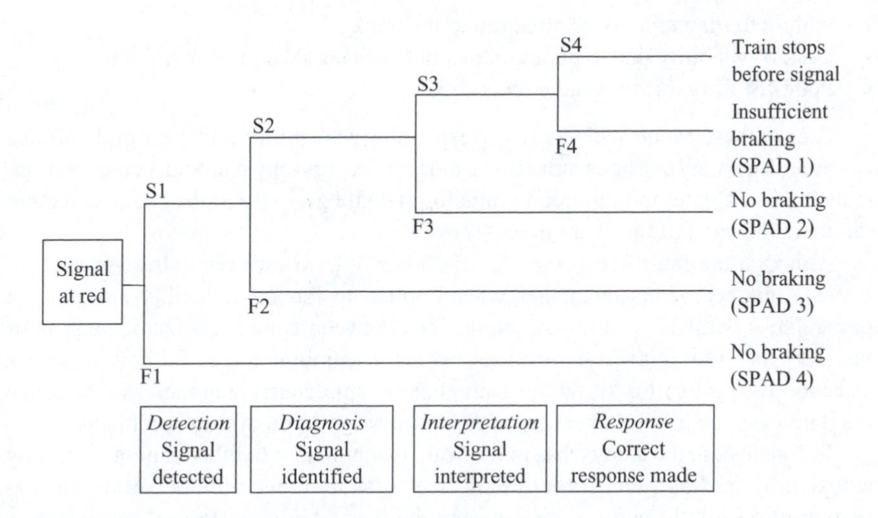

Figure 8.3 Event tree for signals passed at danger (SPADs)

different types of SPAD are given by:

$$P(SPAD\ 1) = S1 \times S2 \times S3 \times F4$$

$$P(SPAD\ 2) = S1 \times S2 \times F3$$

$$P(SPAD\ 3) = S1 \times F2$$

$$P(SPAD\ 4) = F1$$

where P(X) = probability of SPAD type X occurring.

The overall combined probability of a SPAD arising from those modelled in the event tree is therefore:

$$P(SPAD\ 1) + P(SPAD\ 2) + P(SPAD\ 3) + P(SPAD\ 4)$$

Although mathematically equivalent to a fault tree, the event tree has certain useful features when used for modelling human performance. In particular, it is possible to take into account modifications in the probabilities of the event tree as a result of antecedent events. For example, if one event in the sequence failed, but was subsequently recovered, the time taken to recover could lead to less time to perform subsequent operations, raising the probability of failure for these operations. In the case of the SPAD event tree shown above, a signal aspect (red or green) could initially be misdiagnosed and then recovered. However, the time lost in correcting the initial failure could give the driver less time to brake. Therefore, the probabilities in the event tree are not necessarily independent, as is usually assumed to be the case in a fault tree. The structure of the event tree often makes it easier to understand the nature of these dependencies. In practice, the probabilities in event trees are evaluated by fault trees that contain lower level events for which probabilities are more readily obtainable than at the event tree level. For example, the first failure event in the SPAD event tree in Figure 8.3 'Failure to detect signal' could be decomposed into the events:

- Failure to maintain visual attention to the trackside.
- Lack of visibility due to poor contrast between signal and background.
- Poor visibility due to weather conditions.

Each of these events will have a corresponding probability, and these probabilities can be added if they can be assumed to be independent and any one or any combination, (called an 'OR gate' in fault tree terminology) could give rise to the signal detection failure at the level of the event tree.

Although the fault tree is the tool of choice in hardware reliability assessment, it has a number of disadvantages when applied to human reliability analysis. In particular, it is difficult to model interactions between contextual factors, e.g. poor visibility and a high level of workload, which in combination could have a greater effect on the probability of failure than either event occurring alone. The influence diagram (see Section 8.7.6) can be used to overcome some of these problems.

In conclusion, the analyst has to be aware that human reliability cannot be blindly assessed by feeding numbers into fault tree software tools without careful consideration of the subtle interactions between these probabilities. In particular, human error probabilities are frequently subject to 'Common Causes', i.e. factors operating globally across a number of tasks or task elements that may negate any assumptions of independence. For example, the probabilities of failures of two independent checks of an aviation maintenance re-assembly task may be naively multiplied together by the analyst on the assumption that both must fail for the re-assembly to be performed incorrectly. In reality, if both checkers are old friends and have a high opinion of each other's capabilities, their checks may in fact be far from independent, and hence the actual failure probability may be that of the worst checker. Modelling these failure routes and interactions, is often the most difficult but most useful aspect of a human reliability assessment. Once a realistic model of the ways in which a system can fail has been constructed, a significant portion of the benefits of HRA has been achieved. Unfortunately HRA is often carried out as an engineering number crunching exercise without the development of comprehensive and complete qualitative models. Although a selection of techniques for generating the numerical data to populate the failure models will be discussed in the next section it cannot be emphasised too strongly that applying these techniques to inadequate failure models can produce completely incorrect results. Developing accurate models of human failures in systems requires considerable experience, and hence it is recommended that professional advice be sought if the analyst lacks experience in these areas.

8.3 A systematic human interaction reliability assessment methodology (HIRAM)

The HIRAM framework is designed to be used either as a stand-alone methodology to provide an evaluation of the human sources of risk, or in conjunction with hardware-orientated analyses to provide an overall system safety assessment. The overall structure of the framework is set out in Figure 8.4. Figure 8.5 provides a flow diagram showing how this human-orientated analysis is integrated with the engineering risk analysis process described in Figure 8.1.

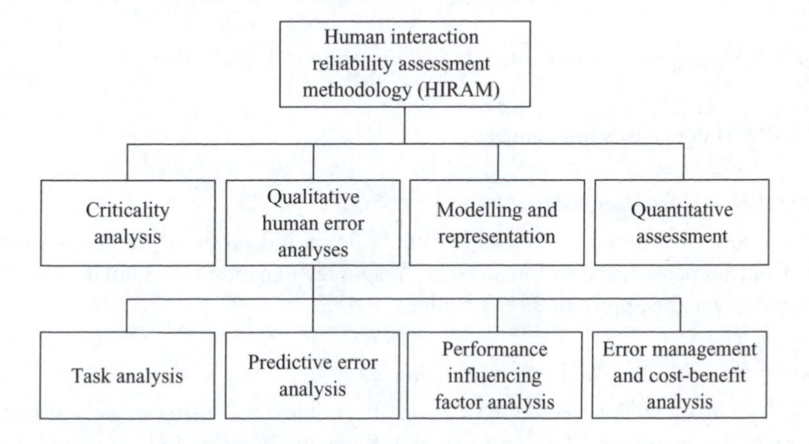

Figure 8.4 Human interaction reliability assessment methodology (HIRAM)

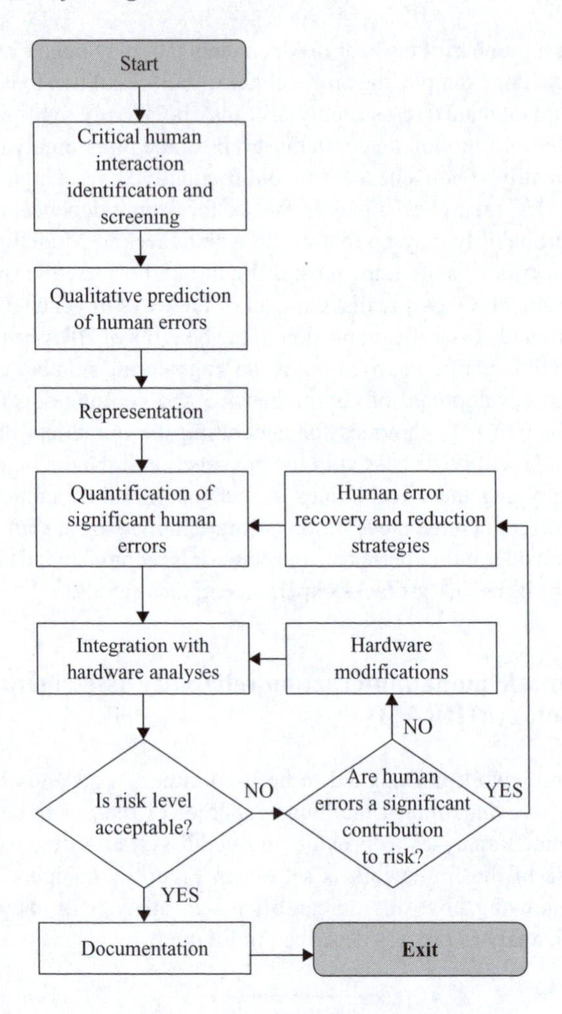

Figure 8.5 Integration of human and engineering risk analyses

HIRAM comprises four stages.

Criticality analysis (stage 1)

The purpose of criticality analysis is to provide an indication of the main areas at risk from human-caused failures so that resources expended on human reliability assessment can be appropriately prioritised.

Qualitative analysis of human errors (stage 2)

This stage involves the prediction of errors that could arise, using models of human error and the analysis of the Performance Influencing Factors, and the nature of the human interactions involved (e.g. actions, checking, communication). Only if human errors with significant consequences (loss of life or major damage) are identified will the subsequent stages of the process be performed. This stage, therefore, includes a consequence and error reduction analysis.

Representation (stage 3)

This stage involves representing the structure of the tasks in which errors with severe consequences could occur, in a manner that allows the probabilities of these consequences to be generated. The usual forms of representation are event trees and fault trees, as discussed earlier.

Quantification (stage 4)

The quantification process involves assigning numerical probabilities or frequencies to the errors (or error recovery opportunities) that have been identified during the preceding stages. Following the quantification process, error probabilities are combined with hardware analyses to provide an overall measure of risk. If this level of risk is unacceptable, then changes will be made in the human or hardware systems to reduce it (see Figure 8.5). This will involve cost-effectiveness considerations relating to alternative strategies.

In the following sections, each of the stages of HIRAM will be considered in detail.

8.4 Criticality analysis

8.4.1 Developing the task inventory

The first stage of criticality analysis is the development of a *task inventory*. A task inventory is a high level description of the various human activities in the system of interest. Often, the task of inventory is classified in terms of the main engineering areas in the system. For example in a refinery, activities associated with process units such as furnaces, distillation columns and cat crackers, may be the natural way to organise the task inventory. In the railways, tasks might be organised into groups such as train operations, signalling, train maintenance, and trackside activities.

Often there are discrete stages of operation that can be identified such as start-up, routine operation and shutdown. In the marine and aviation environments there are phases of operation such as starting and completing a voyage or flight sector that will have associated human activities. Because of the wide range of systems that HIRAM might be applied to, it is difficult to stipulate rules for organising the task inventory. However, it must be comprehensive to ensure that no significant activity (in terms of risk potential) is omitted.

Initially, the activities will be considered at quite a high level of detail, e.g. ‘carry out maintenance on system X’. The criticality screening criteria, discussed below, are then applied at this level to eliminate areas that do not require detailed application of human reliability techniques. If the criteria cannot be applied at this coarse level of detail, it will be necessary to break down the task further into subtasks and re-apply the screening criteria to each of the subtasks. Hierarchical Task Analysis (HTA) is a useful tool for this purpose (as illustrated in Section 8.5.3).

8.4.2 Screening the tasks

Once all of the human activities have been grouped into meaningful categories it is then necessary to apply the screening criteria to each of the human activities (tasks or subtasks). Some of the dimensions that are used to evaluate criticality are as follows:

- Likelihood of the task or subtask failing.
- Severity of consequences (safety, financial, quality, environmental impact, company reputation are common measures).
- Frequency of performance of the task (exposure).
- Likelihood of recovery from the error (recovery means some intervention that occurs to avert the consequences actually occurring).

8.4.2.1 Likelihood of task failure

At first sight it seems paradoxical that an evaluation of the failure likelihood of the tasks is required during the criticality analysis. It can be reasonably argued that the likelihood of the subtask failing is unlikely to be known at this stage. However, since the criticality analysis is primarily directed at rank ordering the tasks in terms of their priorities for analysis only relative likelihoods of failure, rather than absolute probabilities are required. Thus, a scale composed of qualitative statements of likelihoods of failure, as shown in Table 8.1 would be adequate for this purpose. In the table some credible ranges for the failure frequencies in typical safety critical systems are shown, but these are not strictly necessary for use in the criticality analysis, where a task classification of high, medium and low is usually sufficient. When the analyst is assigning tasks to these categories, a process of comparative judgements will be carried out, which are considerably more reliable than absolute judgements of probability. A similar approach of using high, medium and low assignments can also be used for the other dimensions described above.

Table 8.1 Mapping of qualitative scales on to probabilities

Verbal description	Expected frequency of failure
High (H)	1 in 100
Medium (M)	1 in 1000
Low (L)	1 in 10,000

8.4.2.2 Severity of consequences

As discussed above, the severity of consequences measure could focus on safety, finance, quality, environmental impact, company reputation (or even some combination of these factors), depending on the nature of the criticality analysis. In many risk analyses these factors are mapped onto a common dimension, which is usually monetary cost.

The process involves identifying and describing human interactions with the system that will have major impact on risk. A human interaction can in some cases comprise a single operation, e.g. operating a flap or detecting a temperature increase. Usually, however, a human interaction will consist of a task directed at achieving a particular system objective, for example, responding correctly in an emergency. Human interactions are obviously not confined to operational situations. They may also be involved in maintenance and system changes. Errors in these operations can give rise to latent failures, which are not manifested immediately, but which may produce failures in the future when a particular combination of operational conditions occurs.

8.4.3 Developing a screening index

It is useful to develop a numerical measure of task criticality for screening purposes. If such a measure is developed for each task in the set identified by the task inventory, tasks can then be rank ordered and placed into bands such as high, medium and low criticality. This provides a means of prioritising which tasks should be analysed first, and which should receive the most analytical resources.

- Approximate Error Likelihood (AEL). Normally, only assessments at the level of high (H), medium (M) or low (L) are possible at the screening stage. If nothing is known about the Performance Influencing Factors likely to affect failures in the situation being assessed, then this factor can be omitted from the screening analysis.
- The potential severity of the consequences if the hazard were released, referred to as the Outcome Severity Index (OSI).
- The likelihood that barriers (hardware or software) or recovery actions (actions taken by people after the accident sequence has commenced which prevent the consequences occurring) are effective (Recovery Likelihood, RL).
- The exposure to the task, e.g. the frequency (F) with which it is performed.

If these factors are evaluated on appropriate scales, an index of task criticality can be generated as follows:

$$\text{Task Criticality Index (TCI)} = \text{AEL} \times \text{OSI} \times \text{RL} \times \text{F}$$

Each task can then be assessed on this basis to produce a ranking of criticality, which is essentially equivalent to a scale of risk potential. Only those tasks above a predetermined level of the TCI (a function of time and resources) will be subjected to a detailed analysis.

8.5 Qualitative human error analysis

As discussed in Section 8.2, a comprehensive modelling of the types of failure likely to occur in human interactions in systems is probably the most important aspect of assessing and reducing the human contribution to risk. Qualitative human error analysis is a set of tools designed to support the modelling activity. The qualitative analysis performed in HIRAM involves the following techniques:

- Task analysis
- Predictive human error analysis
- Consequence analysis
- Performance influencing factor analysis
- Error reduction analysis.

These techniques will be illustrated later with reference to a simple example, the loading of a chlorine tanker.

8.5.1 Task analysis

Task analysis is a very general term that encompasses a wide variety of techniques (see [3] and Chapter 5 of this book for an overview of available methods). In this context, the objective of task analysis is to provide a systematic and comprehensive description of the task structure and to give insights into how errors can arise. The structure produced by task analysis is combined with the results of the PIF analysis as part of the error prediction process.

The particular type of task analysis used in this example is Hierarchical Task Analysis (HTA). This has the advantage that it has been applied extensively in a number of safety critical industries. HTA breaks down the overall objective of a task by successively redescribing it in increasing detail, to whatever level of description is required by the analysis. At each of the levels, a 'plan' is produced that describes how the steps or functions at that level are to be executed. More comprehensive descriptions of HTA are provided in Chapter 5 of this book. The HTA form of task analysis will be illustrated in the case study provided in Section 8.5.3.

8.5.2 Predictive human error analysis

Predictive Human Error Analysis (PHEA) is a process whereby specific errors associated with tasks or task steps are postulated. The process also considers how these predicted errors might be recovered before they have negative consequences. The technique has some affinities with Failure Mode Effects and Criticality Analysis (FMECA) used in engineering risk analyses. The basic aim is to consider all credible failures so that their consequences and possible mitigations can be explored. Failures that have severe consequences and low probability of recovery will normally be incorporated into the modelling of the overall system risks. The screening criticality index can be used to decide which of the identified failure types will be explored further.

Table 8.2 Error classification

Action		*Retrieval*	
A1	Action too long/short	R1	Information not obtained
A2	Action mistimed	R2	Wrong information obtained
A3	Action in wrong direction	R3	Information retrieval incomplete
A4	Action too little/too much	*Communication*	
A5	Misalign	T1	Message not transmitted
A6	Right action on wrong object	T2	Message information transmitted
A7	Wrong action on right object	T3	Message transmission incomplete
A8	Action omitted	*Selection*	
A9	Action incomplete	S1	Selection omitted
A10	Wrong action on wrong object	S2	Wrong selection made
Checking		*Plan*	
C1	Checking omitted	P1	Plan preconditions ignored
C2	Check incomplete	P2	Incorrect plan executed
C3	Right check on wrong object		
C4	Wrong check on right object		
C5	Check mistimed		
C6	Wrong check on wrong object		

The inputs to the process are the task structure and plans, as defined by the task analysis, and the results of the PIF analysis.

The types of failures considered in PHEA are set out in Table 8.2 and are described in detail below. It should be emphasised that the categorisation of errors described is generic, and may need to be modified for specific industries.

8.5.2.1 Action errors

Action errors are errors associated with one or more actions that change the state of the system, e.g. steps such as close fill valve on equalisation line, slowly open the vent valve, etc.

8.5.2.2 Checking errors

These are errors such as failing to perform a required check, which will usually involve a data acquisition process such as verifying a level or state by visual inspection, rather than an action.

8.5.2.3 Retrieval errors

These are concerned with retrieving information from memory (e.g. the time required for a reactor to fill), or from a visual display or a procedure.

8.5.2.4 Communication or transmission errors

These errors are concerned with the transfer of information between people, either directly or via written documents such as permit systems. These errors are particularly pertinent in situations where a number of people in a team have to co-ordinate their activities.

8.5.2.5 Selection errors

These are errors that occur in situations where the operator has to make an explicit choice between alternatives. These may be physical objects (e.g. valves, information displays) or courses of action. An example would be where a control room operator goes to the wrong compressor, following an alarm indicating a blockage (see Section 8.5.3 below).

8.5.2.6 Planning errors

If the procedures were not regularly updated, were incorrect, or if training were inadequate, P1 errors could occur. P2 errors would often arise as a result of misdiagnosing a situation, or if the entry conditions for executing a sequence of operations were ambiguous or difficult to assess and therefore the wrong procedure was selected. It is important to note that if a planning error occurs, then this implies that a detailed analysis needs to be conducted of the alternative course of action that could arise.

The basic stages of a PHEA are follows:

- Error identification.
- Recovery analysis.
- Consequence analysis.
- Error reduction analysis.

Each of these stages will be illustrated in the case study described in the next section.

8.5.3 Case study illustrating qualitative analysis methods in HIRAM

To illustrate the qualitative approaches to modelling we will introduce a simple example. This is drawn from the chemical processing industry and involves the switchover between two filters associated with the lube oil flowing through a compressor (see Figure 8.6). When the filter currently in use becomes blocked, the resulting high differential pressure produces an alarm in the control room. An operator is then required to go to the compressor and change from the on-line filter to the off-line filter. Prior to this switchover, the off-line filter has to be filled with oil (primed) to ensure that no air bubbles get into the oil during the changeover. If this happens the compressor could cavitate, causing severe and expensive damage.

8.5.3.1 Perform task analysis

The HTA for this task is shown in Figure 8.7. The main subtasks are to identify the off-line filter, prime it, change the filter and renew the dirty off-line filter. The lines drawn below subtasks 1 and 4 indicate that these tasks will not be decomposed further for the purpose of these analyses.

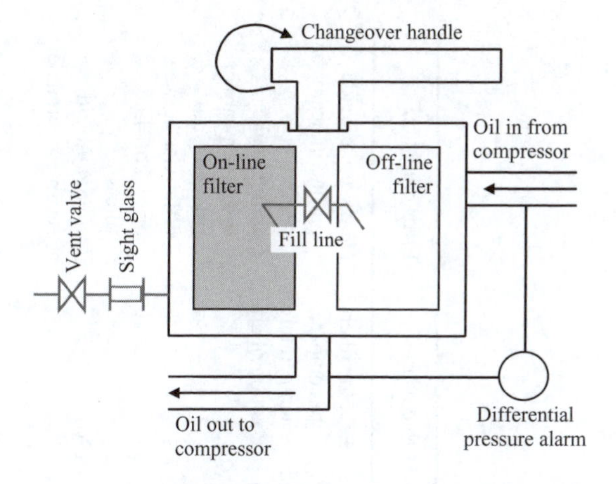

Figure 8.6 Compressor filter changeover task

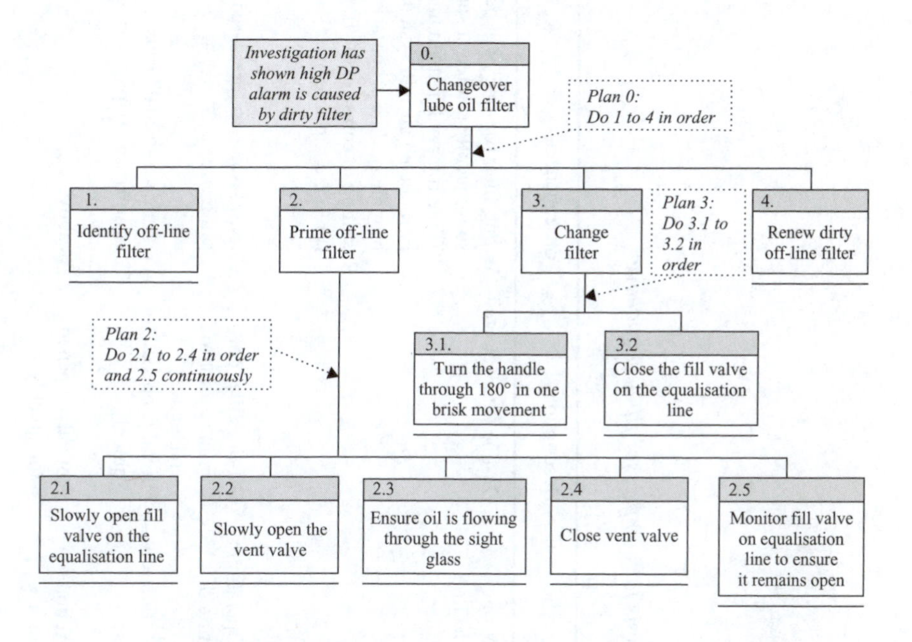

Figure 8.7 HTA for compressor filter changeover

8.5.3.2 Perform Predictive Human Error Analysis (PHEA)

The PHEA analysis for the HTA in Figure 8.7 is provided in Table 8.3. Example analyses are only provided for certain subtasks. Table 8.3 lists the subtasks and task steps from the HTA in the first column. Each of these task steps is considered from the point of view of whether any of the errors in Table 8.2 are credible. For subtask 1 (identify the off-line filter), the PHEA error type 'S2: Wrong selection made' is

Table 8.3 PHEA analysis for compressor filter changeover

Task step	Error type	Description	Consequences	Recovery	Error reduction
1. Identify off-line filter	S2 Wrong selection made	On-line filter selected	Wrong filter (on-line) will be subsequently primed at subtask 2. Off-line filter will remain un-primed introducing air into system when brought on-line at subtask 3.1. Resulting in **compressor trip** (pressure trip safeguard).	Feel the filters to compare temperatures. Position of arm.	*Hardware*: provide on-line indicator on pump *Training*: make operator aware of risk and temp. difference between on and off-line filters
2. Prime off-line filter	A9 Operation omitted	Off-line filter not primed prior to changeover	**Compressor trip** (pressure trip safeguard)	No	*Training*: job aid (warning, reminder)
2.1 Slowly open fill valve on the equalisation line	A5 Action too fast	Fill valve opened too quickly	Pressure surge could cause a **compressor trip** (pressure trip safeguard)	No	*Training*: job aid
2.2 Slowly open vent valve	A5 Action too fast	Vent valve opened too quickly	As above	No	*Training*: job aid

2.3 Ensure oil is flowing through the sight glass	C1 Check omitted	Flow through sight glass not verified	Possible blockages that could prevent priming will not be detected. Air will be introduced into system, on changeover, **tripping the compressor**.	No	*Training*: job aid
	A1 Action too short	Not enough time allowed to purge air bubbles from the system	Air will be introduced into system, on changeover, **tripping the compressor**.	No	
2.4 Close vent valve	A9 Operation omitted	Vent valve left open	Filter will be by-passed when brought on-line. Lube oil quality will continually degrade and **potentially damage the compressor**.	Regular oil sample checks	*Training*: job aid
2.5 Ensure fill valve on equalisation line remains open	C1 Check omitted	If fill valve has been previously closed this erroneous condition will remain undetected (recovery step)	Potential pressure surge during changeover will not be dampened out resulting in **tripping the compressor**.	No	*Training*: job aid

postulated. The error in this context will be that the on-line filter is selected for priming instead of the off-line.

8.5.3.3 Perform consequence analysis

The consequence analysis is carried out in the fourth column. It can be seen that a number of consequences, both immediate and delayed, can arise from the error. The objective of consequence analysis is to evaluate the safety (or quality) consequences to the system of any human errors that may occur. Consequence analysis obviously impacts on the overall risk assessment within which the human reliability analysis is embedded. In order to address this issue, it is necessary to consider the nature of the consequences of human error in more detail.

At least three types of consequences are possible if a human error occurs in a task sequence:

- The overall objective of the task is not achieved.
- Some other negative consequence occurs in addition to the task not achieving its intended objective.
- The task achieves its intended objective but some other negative consequence occurs (either immediate or latent), which may be associated with a system unrelated to the primary task.

Generally, risk assessment has focused on the first type of error, since the main interest in human reliability has been in the context of human actions that were required as part of an emergency response. However, a comprehensive consequence analysis has to also consider other types, since both of these outcomes could constitute sources of risk to the individual or the plant.

One example of a particularly hazardous type of consequence in the second category is where, because of misdiagnosis, the person performs some alternative task other than that required by the system. For example, irregular engine running may be interpreted as being the result of a blockage in a fuel line, which would lead to an incorrect mitigation strategy if the problem were actually a result of a bearing failure.

8.5.3.4 Perform recovery analysis

Once errors have been identified, the analyst then decides if they are likely to be recovered before a significant consequence occurs. Consideration of the structure of the task (e.g. whether or not there is immediate feedback if an error occurs) will usually indicate if recovery is likely. In general, this column refers to self-recovery by the individual who initiated the error or a member of the operating team. Recovery by hardware systems such as interlocks could also be included at this point if required.

8.5.3.5 Error reduction analysis

The final column addresses the issue of error reduction or prevention. Methods to enhance the probability of recovery can also be considered at this stage. In order to develop appropriate error reduction strategies, it is necessary to identify the primary PIFs that appear to be driving the likelihood of the identified errors. This process will be discussed later in this chapter in the context of quantification techniques.

8.6 Representation of the failure model

If the results of the qualitative analysis are to be used as a starting-point for quantification, they need to be represented in an appropriate form. As discussed, the form of representation can be a fault tree, as shown in Figure 8.2, or an event tree (Figure 8.3). The event tree has traditionally been used to model simple tasks at the level of individual task steps, for example in the THERP (Technique for Human Error Rate Prediction) method for human reliability assessment [4] (see Section 8.7.3). It is most appropriate for sequences of task steps where few side effects are likely to occur as a result of errors, or when the likelihood of error at each step of the sequence is independent of previous steps.

If we consider the compressor filter case study described previously, the PHEA analysis shown in Table 8.3 indicates that the following outcomes (hazards) are possible for an unplanned compressor trip:

- loss of production;
- possible knock-on effects for downstream process;
- compressor life reduced;
- filter not changed;
- oil not filtered will eventually damage compressor.

The overall fault tree model of the errors that could give rise to a compressor trip is shown in Figure 8.8. (For the purposes of illustration, technical causes of these failures are not included in the representation.)

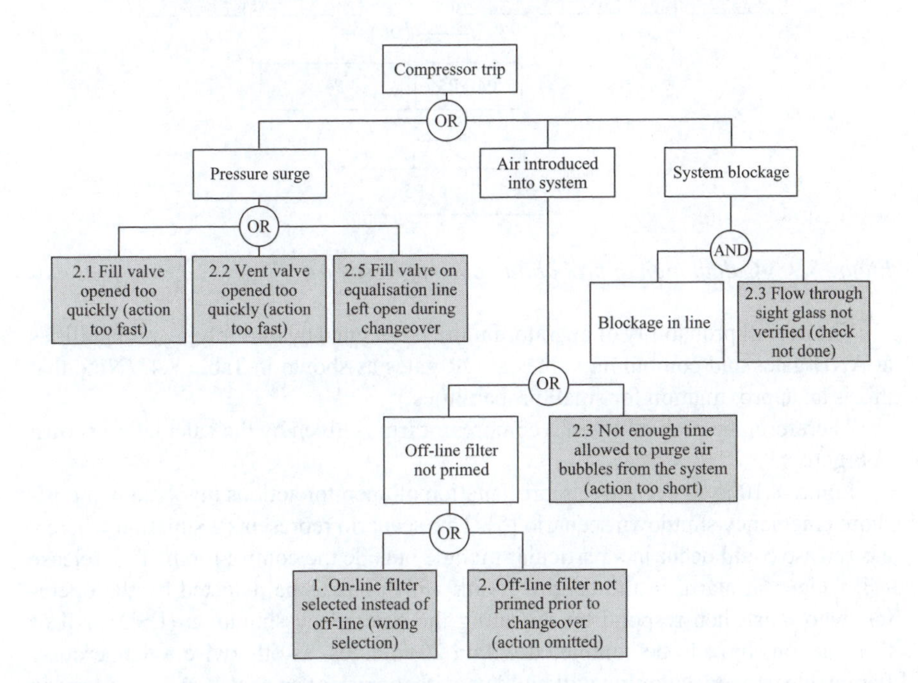

Figure 8.8 Fault tree for compressor trip

Table 8.4 Probabilities for each failure mode

Failure modes	Probabilities
2.1 Fill valve opened too quickly (action too fast)	P1
2.2 Vent valve opened too quickly (action too fast)	P2
2.5 Fill valve on equalisation line left open (action omitted)	P3
1. On-line filter selected instead of off-line (right action, wrong object)	P4
2. Off-line filter not primed prior to changeover (action omitted)	P5
2.3 Not enough time allowed to purge air bubbles from the system (action too short)	P6
Blockage in line	P7
2.3 Flow through sight glass not verified (check omitted)	P8

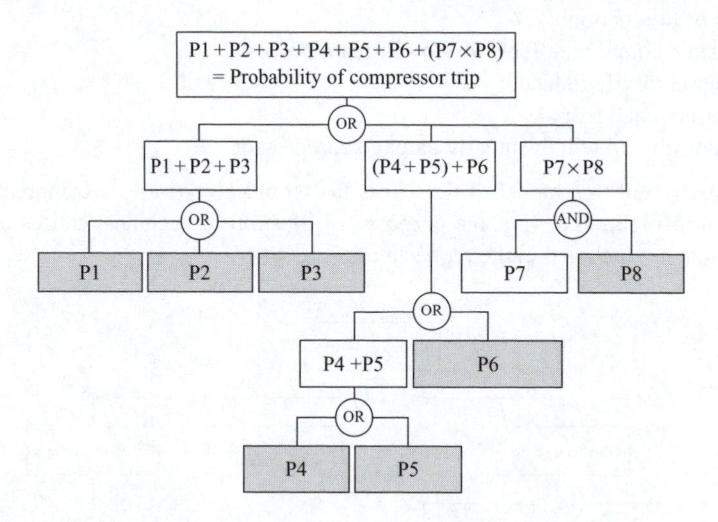

Figure 8.9 Calculation of probability of compressor trip

The overall probability of operator failure is obtained by multiplying probabilities at AND gates and combining values at OR gates as shown in Table 8.4. (Note that this is an approximation for small probabilities.)

Therefore, the probability of a compressor trip is given by the calculation shown in Figure 8.9.

Figure 8.10 is an event tree representation of operator actions involved in an offshore emergency shutdown scenario [5]. This scenario represents a situation where a gas release could occur in a particular module outside the control room. This release will trigger an alarm in the control room, which has to be detected by the operator, who must then respond by operating the emergency shutdown (ESD). All of these actions have to be completed within 20 minutes, as otherwise a dangerously flammable gas concentration will build up with the risk of an explosion. The scenario

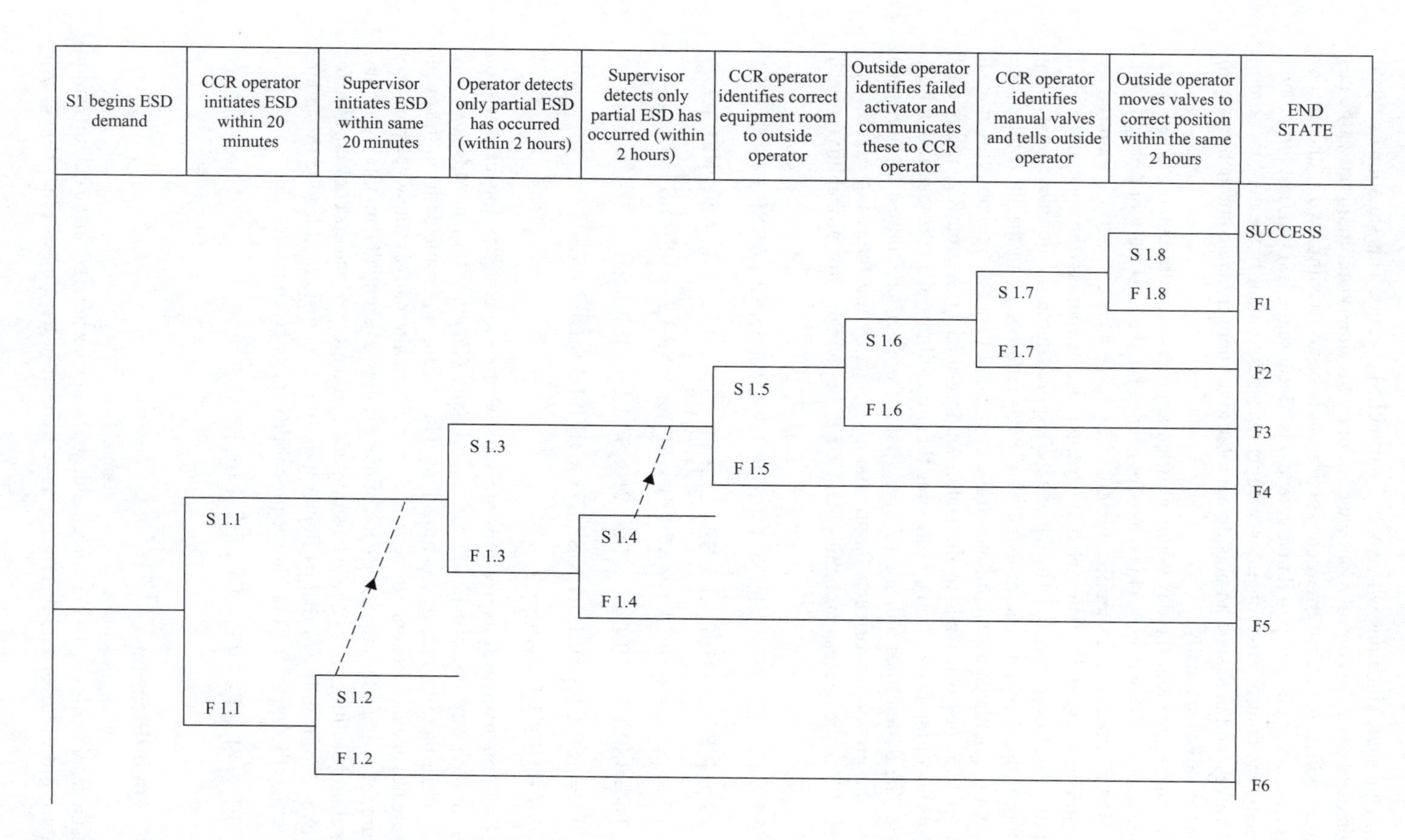

Figure 8.10 Operator action tree for ESD failure scenario [5]

also includes the situations where only a partial ESD occurs. This situation could be detected either by the control room operator or by his supervisor. In this case, the gas flow rate will be lower, but emergency action has to be taken within two hours in order to prevent the gas build-up to flammable levels. Because the ESD has not functioned correctly, the control room operator or supervisor has to identify the malfunctioning ESD valves, and then send an outside operator to the appropriate equipment room to close the valves manually.

It can be seen that the development of an appropriate representation even for a simple scenario is not trivial. This type of event tree is called an Operator Action Event Tree (OAET) because it specifically addresses the sequence of actions required by some initiating event. Each branch in the tree represents success (the upper branch, S) or failure (the lower branch, F) to achieve the required human actions described along the top of the diagram. The probability of each failure state to the far right of the diagram is calculated as the product of the error and/of success probabilities at each node of branches that leads to the state. As described in Section 8.2, the overall probability of failure is given by summing the probabilities of all the resulting failure states. The dotted lines indicate recovery paths from earlier failures. In numerical terms, the probability of each failure state is given by the following expressions (where SP is the success probability and HEP the human error probability at each node):

$$F1 = [SP\ 1.1 + HEP\ 1.1 \times SP\ 1.2] \times SP\ 1.3 \times SP\ 1.5 \times SP\ 1.6 \times SP\ 1.7 \times HEP\ 1.8$$

$$F2 = [SP\ 1.1 + HEP\ 1.1 \times SP\ 1.2] \times SP\ 1.3 \times SP\ 1.5 \times SP\ 1.6 \times HEP\ 1.7$$

$$F3 = [SP\ 1.1 + HEP1.1 \times SP\ 1.2] \times SP\ 1.3 \times SP\ 1.5 \times HEP\ 1.6$$

$$F4 = [SP\ 1.1 + HEP\ 1.1 \times SP\ 1.2] \times SP\ 1.3 \times HEP\ 1.5$$

$$F5 = [SP\ 1.1 + HEP\ 1.1 \times SP\ 1.2] \times HEP\ 1.3 \times HEP\ 1.4$$

$$F6 = HEP\ 1.1 \times HEP\ 1.2$$

These mathematical expressions are somewhat more complex for this situation than for the event tree considered earlier in Figure 8.3, since they need to take into account the fact that recovery is modelled. This means that some paths in the event tree are traversed twice, once directly and once via the alternative route when recovery occurs following a failure. Since each of these routes is essentially an 'OR' combination of probabilities, they have to be summed together to produce the correct value for the overall probability of that particular train of events that could lead to the failure mode (i.e. F1, F2, etc.). Total failure probability PT is given by:

$$PT = F1 + F2 + F3 + F4 + F5 + F6$$

8.7 Quantification

Most research effort in the human reliability domain has focused on the quantification of error probabilities, and therefore a large number of techniques exist. However, a relatively small number of these techniques have actually been applied in practical risk assessments. For this reason, in this section only four techniques will be described in detail. More extensive reviews are available from other sources, e.g. References 5, 6 and 7. Following a brief description of each technique, a case study will be provided to illustrate the application of the technique in practice. As emphasised earlier, quantification has to be preceded by a rigorous qualitative analysis in order to ensure that all errors with significant consequences are identified. If the qualitative analysis is incomplete, then quantification will be inaccurate. It is also important to be aware of the limitations of the accuracy of the data generally available for human reliability quantification.

The various techniques that are available represent different balances in the extent to which they attempt to model the full richness of the large number of factors that are likely to impact on human performance in real situations, or adopt a simplified approach that enables them to be used more easily by analysts without a specialised knowledge of human factors principles.

8.7.1 The quantification process

All quantification techniques follow the same four basic stages:

- modelling the task;
- representing the failure model;
- deriving error probabilities for task steps (if the task has been broken down into subtasks or task steps during the modelling process);
- combining task element probabilities to give overall task failure probabilities.

8.7.1.1 Modelling the task

This involves analysing the task of interest and identifying which aspects should be quantified. In some cases, the analyst will be interested in the probability of a discrete human action, e.g. 'what is the likelihood that the pilot will respond with the appropriate action within ten seconds of a ground proximity alarm?'

In other cases, the interest will be in quantifying a complete task, e.g. 'what is the probability that the pilot will change the mission plan following the receipt of a deteriorating weather message?' In this case, quantification can be carried out at the global level of the whole task, or the task can be broken down into task elements, each of which is quantified (the decomposition approach). The overall probability of success or failure for the whole task is then derived by combining the individual task elements in accordance with the qualitative model that has been developed.

Quantification at a global task level is essentially the same process as with a single discrete operation. A single probability is assigned without explicit reference to the internal structure of the task. There are arguments for and against both the global and the decomposition approach. The advantages of the decomposition approach are as follows:

- It can utilise any databases of task element probabilities that may be available.
- Recovery from errors in individual task steps can be modelled.
- Consequences to other systems arising from failures in individual task steps (e.g. the results of alternative actions as opposed to simply omitted actions) can be modelled and included in the assessment.
- Effects of dependencies between task steps can be modelled.

The main argument for the global approach is that human activities are essentially goal-directed, and this cannot be captured by a simple decomposition of a task into its elements. If an intention is correct (on the basis of an appropriate diagnosis of a situation), then errors of omission are unlikely, because feedback will constantly provide a comparison between the expected and actual results of the task. From this perspective, the focus would be on the reliability of the cognitive rather than the action elements of the task.

On the whole, most quantification exercises have employed the decomposition approach, partly because most engineers (the most common clients for the results of such analyses) are more comfortable with the analysis and synthesis approach, which is very similar to that used in formal safety analyses.

8.7.1.2 Developing the failure model

If the decomposition approach is used, it is necessary to represent the way in which the various task elements (e.g. task steps or subtasks) and other possible failures are combined to give the failure probability of the task as a whole. Generally, the most common form of representation is the event tree. This is the basis for the Technique for Human Error Prediction (THERP), which will be described in Section 8.7.3. Fault trees are usually used when discrete human error probabilities are combined with hardware failure probabilities.

8.7.1.3 Deriving error probabilities for tasks or task elements

Error probabilities that are used in decomposition approaches are all derived in basically the same manner. Some explicit or implicit form of task classification is used to derive categories of tasks in the domain addressed by the technique. For example, typical THERP categories are selections of switches from control panels, walk-around inspections, responding to alarms and operating valves.

A basic human error probability (BHEP) is then assigned to tasks (or task steps depending on the level of decomposition of the analysis) in each category or subcategory. It usually represents the error likelihood under 'average' conditions. This probability is then modified by evaluating the performance influencing factors in the specific situation being considered, to adjust the baseline probability to the specific characteristics of the situation being assessed. Thus, a baseline probability of, say, 10^{-3} for the probability of correctly operating a control under normal conditions may be degraded to 10^{-1} under the effects of high stress, and when wearing cumbersome protective clothing.

8.7.1.4 Combining task element probabilities to give overall task failure probabilities

During the final stage of the decomposition approach, the task element probabilities in the event tree are combined together to give the overall task failure probability. At this stage, various corrections for dependencies between task elements may be applied.

8.7.2 Overview of quantitative human error probability assessment techniques

One of the problems with quantifying human error probabilities is that the effort required to produce realistic estimates is frequently underestimated. The typical user of human error data is an engineer wishing to populate the fault trees produced during a safety assessment. Engineering analysts would usually prefer to insert generic data into the fault trees from a standard database in a similar manner to the way in which hardware failure rates are used for engineering components such as pumps or valves. However, it has to be recognised that human performance is influenced by a very wide range of factors including the characteristics of the individual, the task, the equipment being used and the physical environment. Many of the tools that have been developed for quantifying human error have attempted to simplify the consideration of these factors in order to make the task of human error quantification more accessible to the non-specialist (i.e. engineering) user. However, although such an approach will certainly simplify the quantification process, the greater the attempt made to model the whole range of factors that influence performance in the situation being assessed, the more likely it is to produce credible and valid results.

Another aspect of human reliability assessment is its role in gaining a comprehensive understanding of the range of factors that influence error probability. This is an input to the decision making process with regard to which intervention will produce the greatest reduction in error probability and risk. Simple techniques designed to minimise the amount of effort required to generate a numerical error probability may lead to an inadequate understanding of what needs to be addressed in order to decide whether the risk is tolerable, and if not, what needs to be done. In the techniques that will be described below, it will be apparent that some techniques require a greater level of resource to apply than others, both in terms of time and the level of understanding of the user. When choosing a technique the trade-offs discussed above in terms of the potential accuracy of the results and the amount of effort invested in the analysis should always be borne in mind.

Human reliability quantification techniques can be divided into the following three broad categories.

8.7.2.1 Database techniques

These techniques utilise a comprehensive database of human error probabilities that are organised according to the typical tasks in the domain to which the approach is most commonly applied. The Technique for Human Error Rate Prediction (THERP), which will be described in detail below, is typical of these techniques. The tasks included in this database are primarily drawn from nuclear power plants and similar situations such as chemical plants.

8.7.2.2 Performance Influencing Factors (PIFs) based approaches

These techniques use a generic or situation specific model of the PIFs that is assumed to influence the probability of human error. In the case of the HEART technique, for example, a set of basic generic task categories is provided, each of which has an associated base error probability. These base probabilities are then adjusted by means of a set of provided PIFs (called Error Producing Conditions in HEART), to reflect the degree to which they are present in the situation being assessed.

The other main PIF based approach is the Success Likelihood Index Methodology (SLIM). The main difference between these two techniques is that SLIM does not provide an explicit set of generic baseline tasks or PIFs. This is because it is assumed that the PIFs that influence performance for specific tasks need to be individually identified.

8.7.2.3 Combined Accident Investigation/Human Reliability Quantification approaches

Given that one of the fundamental problems with all human reliability quantification approaches is the difficulty of obtaining data to populate and calibrate the techniques, there is a compelling argument for developing techniques that are able to analyse human-caused incidents retrospectively and then to use these analyses to develop quantitative data bases that can subsequently be used to support prediction. This approach is likely to produce predictions that are much more defensible than those based on data from laboratory studies, since they provide the possibility of allowing the predictions of the quantification methods to be verified. The Influence Diagram Evaluation and Assessment System (IDEAS) will be used to illustrate this approach.

8.7.3 Database approaches: technique for human error rate prediction (THERP)

This technique is the longest established of all the human reliability quantification methods. Dr. A. D. Swain developed it in the late 1960s, originally in the context of military applications. It was subsequently developed further in the nuclear power industry. A comprehensive description of the method and the database used in its application is contained in Reference 4. Later developments are described in Reference 8. The THERP approach is probably the most widely applied quantification technique. This is due to the fact that it provides its own database and uses methods such as event trees that are readily familiar to the engineering risk analyst.

The most extensive application of THERP has been in nuclear power, but it has also been used in the military, chemical processing, transport and other industries.

The technical basis of the THERP technique is identical to the event tree methodology. The basic level of analysis in THERP is the task, which is made up of elementary steps such as closing valves, operating switches and checking. THERP predominantly addresses action errors in well-structured tasks that can be broken down to the level of the data contained in the THERP handbook [4]. Cognitive errors such as misdiagnosis are evaluated by means of a time-reliability curve, which relates the time allowed for a diagnosis to the probability of misdiagnosis.

8.7.3.1 Stages in applying the technique

Problem definition

This is achieved through plant visits and discussions with risk analysts. In the usual application of THERP, the scenarios of interest are defined by the hardware orientated risk analyst, who would specify critical tasks (such as performing emergency actions) in scenarios such as major fires. Thus, the analysis is usually driven by the needs of the hardware assessment to consider specific human errors in pre-defined, potentially high-risk scenarios. This is in contrast to the qualitative error prediction methodology described in Section 8.5, where all interactions by the operator with critical systems are considered from the point of view of their risk potential.

Qualitative error prediction

The first stage of quantitative prediction is a task analysis. THERP is usually applied at the level of specific tasks and the steps within these tasks. The form of task analysis used therefore focuses on the operations that would be the lowest level of a Hierarchical Task Analysis such as that shown in Figure 8.7. The main types of error considered are as follows:

- Errors of omission (omit step or entire task).
- Errors of commission.
- Selection error:
  - selects wrong control;
  - mis-positions control;
  - issues wrong command.
- Sequence error (action carried out in wrong order).
- Time error (too early/too late).
- Quantitative error (too little/too much).

The analyst also records opportunities to recover errors, and various Performance Shaping Factors (called Performance Influencing Factors in this document) which will subsequently be needed as part of the quantification process.

Representation

Having identified the errors that could occur in the execution of the task, these are then represented in the form of an event tree (Figure 8.11). This event tree is taken from

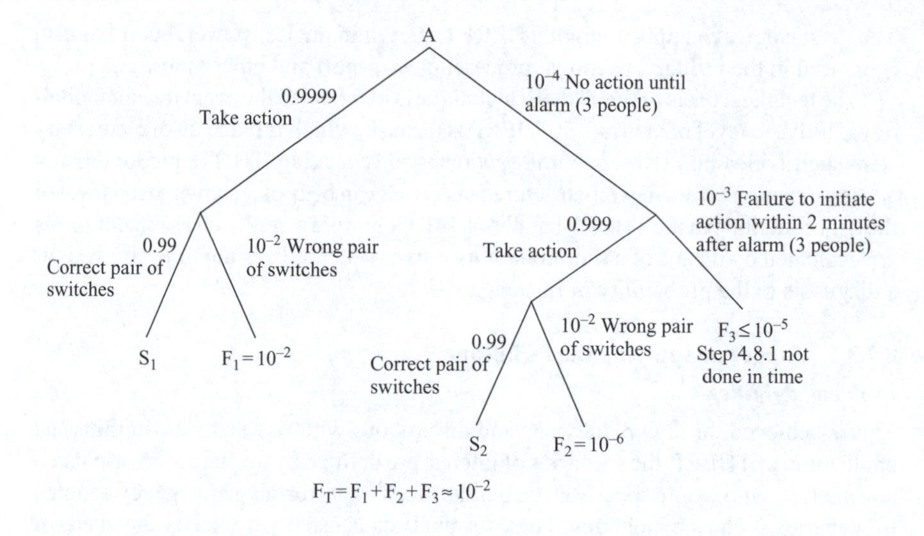

Figure 8.11 THERP event tree [4]

Reference 4. Branches of the tree to the left represent success and those to the right indicate the corresponding failures. Note that event trees are represented vertically in THERP but the calculations are identical to those used in the conventional horizontal representation. Although the event tree in Figure 8.11 is quite simple, complex tasks can generate very elaborate event trees. Error recovery is represented by a dotted line as in the event tree shown in Figure 8.10.

Quantification

Quantification is carried out in the THERP event tree as follows:

- Define the errors in the event tree for which data are required. In Figure 8.11, these errors are:
  - No action taken until alarm (action omitted).
  - Failure to initiate action within two minutes of alarm (action too late).
  - Wrong pair of switches chosen (selection error).
- Select appropriate data tables in [4]. This handbook contains a large number of tables giving error probabilities for operations commonly found in control rooms, e.g. selecting a switch from a number of similar switches. Because the handbook was originally written for the nuclear industry, the data reflect the types of operations frequently found in that industry. The source of these data is not defined in detail by the authors, although it appears to be partly based on the American Institute for Research human error database [9] together with plant data extrapolated and modified by the authors' experience.
- Modify the basic data according to guidelines provided in the handbook, to reflect differences in the assumed 'nominal' conditions and the specific conditions for the task being evaluated. The major factor that is taken into account is the level of perceived stress of the operator when performing the task.

- Modify the value obtained from the previous stage to reflect possible dependencies between error probabilities assigned to individual steps in the task being evaluated. A dependence model is provided that allows for levels of dependence from complete dependence to independence to be modelled. Dependence could occur if one error affected the probability of subsequent errors, for example, if the total time available to perform the task was reduced.
- Combine the modified probabilities to give the overall error probabilities for the task. The combination rules for obtaining the overall error probabilities follow the same addition and multiplication processes as for standard event trees (see last section).

Integration with hardware analysis
The error probabilities obtained from the quantification procedure are incorporated in the overall system fault trees and event trees.

Error reduction strategies
If the error probability calculated by the above procedures leads to an unacceptable overall system failure probability, then the analyst will re-examine the event trees to determine if any PIFs can be modified or task structures changed to reduce the error probabilities to an acceptable level.

8.7.3.2 Advantages of THERP

The technique has been extensively applied, particularly to areas such as nuclear power plant safety analyses, and uses an approach that will be very familiar to engineers and safety analysts who use Probabilistic Safety Analysis (PSA) or Quantitative Risk Analysis (QRA). The technique provides a comprehensive database of human error probabilities, together with sets of PIFs (called Performance Shaping Factors (PSFs) in THERP) to assist the analyst in modifying the base data to take into account the specific task conditions of the task being assessed.

8.7.3.3 Disadvantages of THERP

The central criticism of THERP is that it takes a very simplified view of human performance in terms of its assumption that complex tasks can simply be synthesised from linear combinations of simpler task elements. The technique is also limited in the extent to which it can take into account PIFs/PSFs that are not embedded in its data tables. It is usually argued that the basic human error probabilities in its database represent average values when all the PSFs are at the middle of their range (between best and worst case values). For certain PSFs deemed to be significant in the domain for which the database was first developed (nuclear power generation), e.g. stress, wearing protective clothing, explicit extrapolations from the base values are provided. However, it is easy to find situations where the predominant drivers of human error probabilities are not included in the tables. There are also problems in assigning the tasks that are to be assessed to the basic task types provided in the database, a problem also found with HEART (see Section 8.7.4).

The provenance of the data contained in the database is unclear, although its developers have mentioned the AIR database as one of its sources, together with field data and expert judgement-based extrapolations from these sources.

8.7.4 Human error assessment and reduction technique (HEART)

HEART is documented comprehensively in References 1 and 5. It is an example of a model-based human reliability quantification technique which is based on the idea of selecting a set of Performance Influencing Factors (called Error Producing Conditions (EPCs) in HEART). These PIFs are the main drivers of error probability in the task or scenario being assessed. The task of interest is first assigned to a generic set of task categories that are shown in Table 8.5. Each of these generic tasks has an associated baseline error probability that is assumed to apply under 'average' operational conditions, i.e. average levels of training, equipment design, procedures etc.

The analyst then selects which of a set of supplied PIFs are relevant to the task being assessed. These PIFs (EPCs) are set out in Tables 8.6 and 8.7. The EPCs are

Table 8.5 HEART generic task categories [1]

	Generic task category	Proposed nominal human unreliability (5th–95th percentile bounds)
A	Totally unfamiliar, performed at high speed with no real idea of likely consequences	0.55 (0.35–0.97)
B	Shift or restore system to a new or original state on a single attempt without supervision or procedures	0.26 (0.14–0.42)
C	Complex task requiring high degree of comprehension and skill	0.16 (0.12–0.28)
D	Fairly simple task performed rapidly or given scant attention	0.09 (0.06–0.13)
E	Routine, highly practised, rapid task involving relatively low level of skill	0.02 (0.007–0.045)
F	Restore or shift a system to original or new state following procedures, with some checking	0.003 (0.0008–0.007)
G	Completely familiar, well-designed, highly practised, routine task occurring several times per hour, performed to highest possible standards by highly motivated, highly trained and experienced person, totally aware of implications of failure, with time to correct potential error, but without the benefit of significant job aids	0.0004 (0.00008–0.009)
H	Respond correctly to system command even when there is an augmented or automated supervisory system providing accurate interpretation of system stage	0.00002 (0.000006–0.0009)

Table 8.6 Primary HEART error producing conditions

	Error-producing condition	Maximum predicted amount by which error probability changes from best to worst case conditions
1	Unfamiliarity with a situation that is potentially important but only occurs infrequently, or is novel	×17
2	A shortage of time available for error detection and correction	×11
3	A low signal-to-noise ratio	×10
4	A means of suppressing or overriding information or features that are too easily accessible	×9
5	No means of conveying spatial or functional information to operators in a form that they can readily assimilate	×8
6	A mismatch between an operator's model of the world and that imagined by the designer	×8
7	No obvious means of reversing an unintended action	×8
8	A channel capacity overload, particularly one caused by simultaneous presentation of non-redundant information	×6
9	A need to unlearn a technique and apply one that requires the application of an opposing philosophy	×6
10	The need to transfer specific knowledge from task to task without loss	×5.5
11	Ambiguity in the required performance standards	×5
12	A mismatch between perceived and real risk	×4
13	Poor, ambiguous or ill-matched system feedback	×4
14	No clear, direct and timely confirmation of an intended action from the portion of the system over which control is to be exerted	×4
15	Operator inexperience (e.g. a newly qualified tradesman, but not an 'expert')	×3
16	An impoverished quality of information conveyed by procedures and person to person interaction	×3
17	Little or no independent checking or testing of output	×3

based on experimental studies of the effects of these factors on human performance. The sources of these data are given in Reference 1. The PIFs are assumed to have differing levels of influence on degrading the baseline probability implicitly assigned to the task. The analyst then has to make an assessment of the proportion of the EPC that exists in the task being assessed. For example, if the assessor decides that the degrading effect of time pressure would be at its maximum, and if it were the only factor affecting the error probability, it would increase this probability by a factor

Table 8.7 Secondary HEART error producing conditions

	Error-producing condition	Maximum predicted amount by which error probability changes from best to worst case conditions
18	A conflict between immediate and long-term objectives	×2.5
19	No diversity of information input for veracity checks	×2.5
20	A mismatch between the educational-achievement level of an individual and the requirements of the task	×2
21	An incentive to use other more dangerous procedures	×2
22	Little opportunity to exercise mind and body outside the immediate confines of the job	×1.8
23	Unreliable instrumentation (enough that it is noticed)	×1.6
24	A need for absolute judgements that are beyond the capabilities or experience of an operator	×1.6
25	Unclear allocation of function and responsibility	×1.6
26	No obvious way to keep track or progress during an activity	×1.4

of 11 (from Table 8.6). Thus a task with a baseline probability of 0.02 (generic task category E) would have an actual failure probability of 0.22 as a result of the negative effects of extreme time pressure. This process is normally performed using all the EPCs that are deemed to apply to the situation being assessed. The primary factors to be considered in the analysis are 1–17. Because of their comparatively small effect on error probability, factors 18–26 are only to be used after initial screening using factors 1–17.

8.7.4.1 HEART case study

The application of HEART is illustrated by the following case study. A chemical plant operator has to perform a bypass isolation using the standard operating procedures.

Assess task conditions

The operational conditions are as follows:

- The operator is inexperienced.
- The task requires the operator to close valves in the opposite manner to which he is accustomed for his usual tasks.
- He is unaware of plant hazards.
- There is a staffing reduction planned for the plant and hence the operator's morale is likely to be low.
- There is a 'production before safety' culture.

Table 8.8 HEART calculations

Task type = F Nominal human error probability = 0.003 (3×10^{-3})			
Error producing conditions	HEART effect	Assessed proportion of effect	Assessed effect
Inexperience	×3	0.4	$((3 - 1) \times 0.4) + 1 = 1.8$
Opposite technique	×6	1.0	$((6 - 1) \times 1.0) + 1 = 6.0$
Risk perception	×4	0.8	$((4 - 1) \times 0.8) + 1 = 3.4$
Conflict of objectives	×2.5	0.8	$((2.5 - 1) \times 0.8) + 1 = 2.24$
Low morale	×1.2	0.6	$((1.2 - 1) \times 0.6) + 1 = 1.12$
Total assessed EPC effect ('degrading factor') = $(1.8 \times 6.0 \times 3.4 \times 2.2 \times 1.12) = 90.5$			
Assessed human error probability = $(0.003) \times 90.5 = 0.27$			

Assign task to appropriate generic category

The initial classification of the task is that it is a category F generic task because it involves 'following procedures to restore or shift a system to its original or new state, with some checking' (see Table 8.5). The baseline error probability for this task type is 0.003.

Assess proportion of EPC effect and calculate human error probability

The analyst decides on which EPCs are present in the task and evaluates the proportion of the effect that is likely to be present, i.e. the extent to which the EPC is at the worst or best end of its range in terms of its negative influence on the error probability. These assessments are set out in Table 8.8. Based on the description of the task, the analyst estimates that about 40 per cent of the full effect of inexperience is likely to be present (i.e. the operator is not completely inexperienced). However, the valve closure is in a completely opposite direction to that normally carried out by the operator, and hence the full negative effect of 'opposite technique' is likely to occur. A value of 1.0, i.e. 100 per cent of the full effect is therefore assigned. The other proportions are assigned using a similar approach. The calculation of the assessed effect is based on the extent to which the effect of the EPCs deviates from the ideal value of 1. A value of 1 would mean that when multiplied by the baseline generic task error probability, it would have no negative effect at all. The greater the product of the size of the effect and the assessed proportion of this effect that is present (e.g. 3×0.4 for inexperience), the larger the negative effect will be in terms of increasing the error probability. The calculations take this into account in the fourth column of the table.

To calculate the overall error probability, taking into account the EPCs in the situation, the various assessed effects are first multiplied together, to give a total assessed 'degrading factor' of approximately 90. The generic task baseline error probability is then multiplied by this factor to give an assessed error probability of 0.27.

Table 8.9 Percentage contribution of EPCs

Error producing conditions	Percentage contribution to increase in error probability from baseline
Opposite technique	41
Risk mis-perception	24
Conflict of objectives	16
Inexperience	12
Low morale	8

Perform sensitivity analysis

Sensitivity analysis is performed to identify which factors have the greatest effect on the error probability, to indicate the most cost effective error reduction strategies.

The relative contribution to the error probability of each of the EPCs can be evaluated by dividing each of the EPC-assessed effects by the total. This produces Table 8.9, which shows the percentage contribution made to the total degradation factor by each of the constituent EPCs.

This analysis suggests that redesigning the task to eliminate the 'opposite technique' problem will produce the greatest impact in reducing the error probability. This would be factored into the calculations by changing the assessed effect of this factor from 6 (maximum effect) to 1 (minimum effect). The total assessed effect of all the EPCs (degrading factor) would then be about 15 and the evaluated error probability would be $15 \times 0.003 = 0.045$. The choice of interventions also needs to consider the relative costs of alternative solutions.

8.7.4.2 Advantages of HEART

Since its initial development HEART has proven to be an extremely popular technique, especially within the engineering community. The technique is easy to understand and use by non-specialists, and the EPCs and their multipliers are based on experimental data on human performance. One of its primary attractions is that the technique contains everything that is required to perform human reliability assessments. In particular, no external databases are required. To this extent the technique will be highly attractive to non-specialist users, in that no special training is required to apply it, and it produces results that are auditable and plausible. Its ease of use make it an attractive proposition when, for example, an engineer performing a safety analysis requires some human error probabilities to insert into a fault tree.

8.7.4.3 Disadvantages of HEART

The very ease of use of HEART by non-specialists is both one of its major strengths and also the source of possible weaknesses. In order to achieve the ease of use by non-specialists, which is such an attractive feature of the approach, the methodology greatly simplifies the complex issues of how human performance is influenced by many factors that are specific to a particular situation. It does this by providing a library of factors influencing human performance. These appear to have been selected largely from experimental studies suitable as sources of the 'multipliers' used to modify the baseline probabilities in HEART. The problems associated with HEART are summarised below.

Problems with the generic task categories

Since the baseline probability associated with a generic task is the starting point for all subsequent extrapolations using the EPCs, the initial assignment of tasks to categories is critical. Although the task categories in Table 8.5 present a plausible range of tasks, assessors often find difficulties in assigning the task being analysed to one of the categories. This is partly because of the fact that certain EPCs/PIFs are implicitly built into the descriptions of the generic task descriptions. If the EPCs/PIFs in the task being analysed are different from these factors, this creates a classification problem for the assessor. For example, how would a task be categorised that is similar to category A in that it is performed at high speed but where the consequences of failure are well known?

Limitations of a fixed set of EPCs/PIFs

There will be problems if HEART is applied to situations where human error is driven by factors other than the EPCs/PIFs provided because it is not possible to anticipate all the factors that influence human performance in all situations. There is often a temptation to apply the HEART factors rather than to perform a more detailed analysis, which might identify other PIFs that are more important drivers of performance. For example, in many driving situations, the level of distractions, fatigue or multitasking being performed by a driver are recognised as being major influences on error probability. However, these are not considered by HEART.

The EPCs focus on individual performance, whereas in many situations team performance is critical. There is also limited attention to the factors affecting cognitive performance, giving rise to mistakes and paying little attention to factors influencing violations and circumventions, which are arguably one of the most insidious forms of human-caused failures in safety critical systems.

A problem also exists with the multipliers that are assigned to each of the EPCs. Although these are based on experimental studies, it has to be recognised that such studies rarely use tasks that are as complex as those encountered in the real world. Experimental studies also usually evaluate the effect of one variable being changed at a time, whereas human performance is likely to involve interactions between variables. Assessing the proportion of effect that is present is also likely to be subject to a wide range of variability between analysts unless clear anchor points are provided to define the various points on the range between best and worst cases for an EPC.

Overall conclusions on HEART

HEART is a useful technique, which allows a quick evaluation by non-specialists of error probabilities for simple tasks (other than violations and mistakes) for which the database of supplied EPCs is relevant. However, for critical applications it is recommended that suitably qualified human factors specialists carry out a more comprehensive identification and evaluation of the context-specific factors influencing performance for safety critical tasks. As discussed in subsequent sections, we believe that the development of credible absolute human error probabilities requires actual data from the situation being evaluated or from similar situations where credible extrapolations can be made.

8.7.5 The success likelihood index method (SLIM)

The SLIM technique is described in detail in References 5 and 10. The technique was originally developed with the support of the US Nuclear Regulatory Commission but, as with THERP, it has subsequently been used in the chemical, transport and other industries. The technique is applied to tasks at any level of detail. Thus, in terms of the HTA in Figure 8.7, errors could be quantified at the level of whole tasks, subtasks, task steps of even individual errors associated with task steps. This flexibility makes it particularly useful in the context of task analysis methods such as HTA.

The basic premise of the SLIM technique is that the probability of error is a function of the PIFs in the situation. An extremely large number of PIFs could potentially impact on the likelihood of error. Normally the PIFs which are considered in SLIM analyses are the direct influences on errors such as levels of training, quality of procedures, distraction level, degree of feedback from the task, level of motivation, etc. However, in principle, there is no reason why higher-level influences such as management policies should not also be incorporated in SLIM analyses. The Influence Diagram Evaluation and Assessment System (IDEAS) (see Section 8.7.6) is a generalisation of the SLIM approach to address the hierarchy of direct and indirect causes of errors that need to be considered to provide a more comprehensive evaluation of the underlying causes of errors. SLIM can in fact be regarded as a special case of IDEAS.

There are many similarities between the HEART and SLIM approaches in that they both make use of the concept of PIFs/EPCs to modify baseline error probabilities. However, there are also substantial differences between the approaches, as will be described in subsequent sections.

In the SLIM procedure, tasks are numerically rated on the PIFs that influence the probability of error and these ratings are combined for each task to give an index called the Success Likelihood Index (SLI). This index is then converted to a probability by means of a general relationship between the SLI and error probability that is developed using tasks with known probabilities and SLIs. These are known as calibration tasks.

8.7.5.1 SLIM case study

SLIM can therefore be used for the global quantification of tasks. Task elements quantified by SLIM may also be combined together using event trees similar to those used in THERP, or by using fault trees such as those illustrated in Figure 8.8.

Form groups of homogenous operations

SLIM is based on the idea of assessing groups of tasks that are influenced by similar PIFs. The first stage is therefore to group together similar operations or tasks. Using the compressor pump case study (introduced in Section 8.5.3) we can see that failure modes 2.1 (fill valve opened too quickly) and 2.2 (vent valve opened too quickly) are similar types of operation, involving the operation of valves. For the purposes of this exercise we shall assume that the error rate data is available for three other tasks that are affected by similar PIFs. In order to derive absolute error probabilities for our failure modes at least two (and preferably three) or more tasks with known error probabilities are required. These are ideally derived from frequency data on observed failure rates in the actual task context. If these data are unavailable, other sources such as existing databases like those associated with THERP or judgement-based techniques will have to be used.

Probability of failure of valve task A $= 5 \times 10^{-3}$
Probability of failure of valve task B $= 1 \times 10^{-2}$
Probability of failure of valve task C $= 7 \times 10^{-3}$

Models have been developed that specify appropriate sets of PIFs for specific types of tasks, e.g. chemical plant maintenance, cockpit operations, train driving, and healthcare. If such models are not available for the task being assessed, the analyst develops a suitable set of PIFs, based on comprehensive task and PHEA analyses, inputs from subject matter experts and appropriate human factors data sources. In this example, it is assumed for illustrative purposes that the main PIFs that determine the likelihood of error are time stress, level of experience, level of distractions, and accessibility of the items being manipulated.

Rate each operation on each PIF

A numerical rating (1–9 in this example) is made for each operation or task being assessed on each PIF. This represents the actual conditions being assessed, e.g. a high time stress situation would be rated as 9 on the time stress scale, and low distractions as 1 on the distractions scale.

For example, the operations might be rated as shown in Table 8.10.

These ratings can be interpreted as follows. In the case of the time stress PIF, all the tasks involve a low level of time stress, apart from 'Valve task B', where stress is high. The operators are very experienced in carrying out all the tasks with the exception of 'Valve task B' where the operator's experience is average. The level of distractions is high for 'Valve task B', but otherwise average, with the exception of 'Valve task A' where there are very few distractions. As we would expect, the PIF ratings for each of the filter changeover tasks are very similar, with the exception of 'Accessibility'. This is because the fill valve is much harder to reach than the vent valve.

Assign weights if appropriate and rescale ratings

Based on the analyst's experience, or upon error theory, it is possible to assign weights to the various PIFs to represent the relative influence that each PIF has on all the tasks

Table 8.10 PIF Ratings for example tasks and errors

Tasks	Performance influencing factor ratings (R)			
	Time stress	Experience	Distractions	Accessibility
Valve task A	1	8	2	6
Valve task B	7	5	7	6
Valve task C	1	6	5	2
Fill valve opened too quickly	1	8	4	2
Vent valve opened too quickly	1	8	4	8

in the set being evaluated. In this example it is assumed that in general the level of experience has the least influence on these types of errors, and time stress the most influence. The weights are generic, in that it is assumed, for example, that variations in time stress will always have four times the influence of variations in experience for all the tasks in the set.

It should be noted that the weights and ratings are independent. The ratings measure the specific quality for each the tasks on all the PIFs being considered. The weights multiply each of the ratings on each task to give a measure of where each task lies on the scale from 1 (worst case) to 9 (best case). The analyst should only assign different weights to the PIFs if he or she has real knowledge or evidence that the relative influence of the factors actually differs. As a starting assumption all the weights are assumed to be equal.

This means that the weights are distributed equally across all the PIFs in the set being assessed. For example, if there were four PIFs, the weights would be 0.25 for each PIF. If there were three PIFs, the weights would be 0.33 for each PIF, and so on.

In the SLIM calculations, it is assumed that as a PIF rating increases, it will lead to an increasing success likelihood. This is obviously true in the cases of PIFs such as experience and accessibility in the current example. However, other PIFs, such as time stress and distractions in the current example, actually lead to a decrease in the success likelihood as they increase. In order to take these reversed scales into account, the PIF ratings for these scales have to be subtracted from the end of the scale range (9 in this case), before being entered into the calculation of the success likelihood index. This adjustment produces rescaled rating for the reverse scaled PIFs as shown in Table 8.11.

Calculate the success likelihood indices

The SLI is given by the following expression:

$$SLI_j = \Sigma R_{ij} W_i$$

Table 8.11 Rescaled PIF Ratings, Weights & SLI calculations for example tasks & errors

Tasks	Rescaled Performance influencing factor ratings (RS)				SLIs Σ (RS × W)
	Time Stress	Experience	Distractions	Accessibility	
Valve task A	8	8	7	6	7.3
Valve task B	2	5	2	6	3.1
Valve task C	8	6	4	2	5.4
Fill valve opened too quickly	8	8	5	2	5.9
Vent valve opened too quickly	8	8	5	8	7.1
PIF weights (W)	0.4	0.1	0.3	0.2	

where SLI_j = SLI for task j, W_i = relative importance weight for the ith PIF (weights sum to 1) and R_{ij} = rating of the jth task on the ith PIF.

The SLI for each task is the weighted sum of the ratings for each task on each PIF.

Convert the success likelihood indices to probabilities

The SLIs represent a measure of the likelihood that the operations will succeed or fail, relative to one another. In order to convert the SLI scale to a probability scale, it is necessary to calibrate it. If a reasonably large number of operations in the set being evaluated have known probabilities (for example, as a result of incident data having been collected over a long period of time), then it is possible to perform a regression analysis that will find the line of best fit between the SLI values and their corresponding error probabilities. The resulting regression equation can then be used to calculate the error probabilities for the other operations in the group by substituting the SLIs into the regression equation.

If, as is often the case, there are insufficient data to allow the calculation of an empirical relationship between the SLIs and error probabilities, then a mathematical relationship has to be assumed. In the original version of the SLIM technique, a logarithmic relationship was assumed, based on data collected on maintenance tasks by Pontecorvo [11]. However, in more recent applications of SLIM, a linear relationship of the form HEP = A SLI + B has been assumed (where A and B are constants, HEP is the error probability and SLI is the SLI value measured in the situation where the error probability was estimated), since this is the simplest form of relationship in the absence of empirical data. In order to calculate the constants A and B in the equation, at least two tasks with known SLIs and error probabilities must be available in the set of tasks being evaluated.

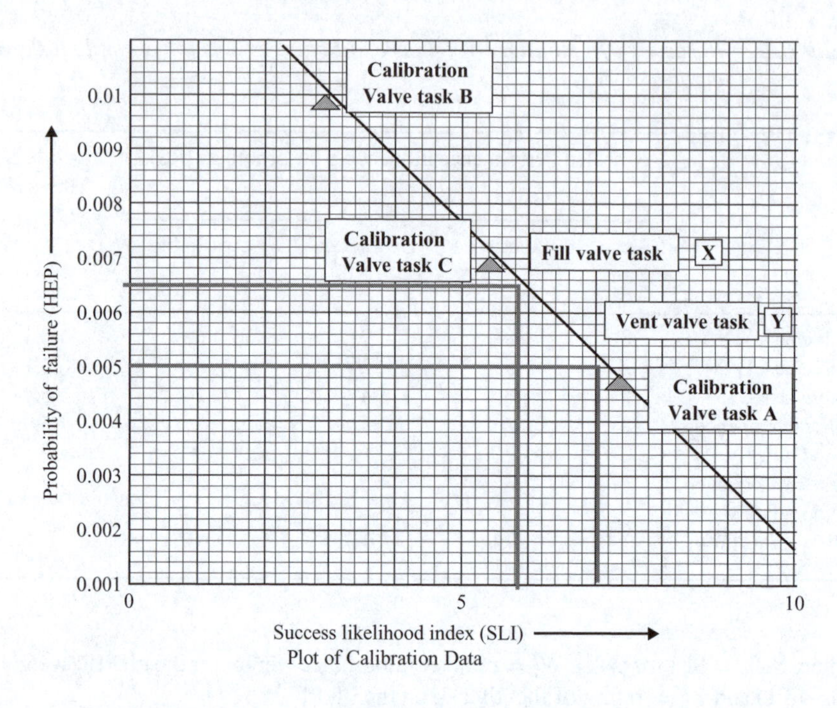

Figure 8.12 *Figure 8.12 Calibration graph developed from calibration tasks A, B & C & used for converting Fill valve & Vent valve SLI values to failure probabilities*

In the example under discussion, error rate data was available for valve tasks A, B and C from Table 8.11.

Probability of failure of valve task A = 5×10^{-3} SLI = 7.3
Probability of failure of valve task B = 1×10^{-2} SLI = 3.1
Probability of failure of valve task C = 7×10^{-3} SLI = 5.4

These data can be plotted on a calibration graph (see Figure 8.12). A linear relationship is assumed and a line of best-fit is drawn between the three data points.

In Figure 8.12 the SLIs for the two unknown tasks (X) and (Y), are plotted on the calibration graph. This allows conversion from the SLIs of 5.9 and 7.1 to give probabilities of approximately:

Probability of fill valve being opened too quickly = 7×10^{-3}
Probability of vent valve being opened too quickly = 5×10^{-3}

Note that once the calibration has been established it can be used to convert SLIs to probabilities for all tasks that are influenced by the same set of PIFs with the same weights.

Table 8.12 Cost benefit analysis

PIF	Weight	Current actual rating	Revised ratings following cost effectiveness analysis	Cost of change
Time stress	0.4	1	1	Medium
Experience	0.1	8	8	High
Distractions	0.3	4	1	Low
Accessibility	0.2	2	2	High

Perform sensitivity analysis

The nature of the SLIM technique renders it very suitable for 'what if' analyses to investigate the effects of changing some of the PIF values on the resulting error probabilities. For example, the error 'Fill valve opened too quickly' has a relatively high probability (7×10^{-3}); therefore an organisation might examine ways they could reduce this probability. Table 8.12 shows how potential improvements could be assessed in terms of their importance, the amount of scope for improvement and the cost of alterations.

For this task, the analyst felt that time stress was already too low to improve, and that the technicians already had considerable experience. Because improving accessibility would be expensive it was decided that modifications to the overall task to reduce distractions was the most cost-effective error reduction strategy, as this factor had a high weight and a low cost of change. The effects of reducing distractions can be investigated by assigning a rating of 1 to this PIF from its current rating of 4. This has the effect of improving the SLI for 'Fill valve opened too quickly' from 5.9 to 8.9. If the new SLI value is converted to a probability using Figure 8.12, the revised probability for 'Fill valve opened too quickly' (with reduced distractions) is 2×10^{-3}.

This represents a considerable improvement on the original error probability of 5×10^{-3}. The analyst would then need to decide whether the gain was significant enough to justify the cost of the alteration. This type of analysis provides the analyst with the ability to compare the possible impact of various interventions, and is therefore a useful tool for cost-benefit analysis.

8.7.5.2 Advantages of SLIM

The SLIM technique is a highly flexible method that allows a precise modelling of the PIFs (or causal context) that are likely to influence performance in a particular situation. Unlike HEART and THERP, it places no constraints on the analyst in terms of the factors that are assumed to influence error probability in the task being assessed. The analyst is also able to take into account the differential weights or levels of influence that each PIF may have in a particular situation. The technique allows the effects of changes in the quality of the PIFs and also assumptions about their relative influence to be evaluated as part of a cost-effectiveness analysis designed to achieve the greatest improvements in human reliability at minimum costs.

8.7.5.3 Disadvantages of SLIM

Since the analyst needs to construct a model of the factors influencing performance in the situation being assessed, some degree of human factors knowledge will be necessary to use the technique effectively. The technique is therefore likely to be less favoured by engineering users than by human factors specialists. The technique also requires that calibration data are available, preferably from the domain in which the technique is being applied, although expert judgement can be used for this purpose if no hard data are available.

8.7.6 The influence diagram evaluation and assessment system (IDEAS)

The IDEAS approach is based on the use of influence diagrams to provide a comprehensive modelling of the direct, indirect and organisational factors that impact on human error in a particular context. Influence diagrams (IDs) were first developed for modelling and evaluating complex decision problems involving a range of stakeholders. They provide a method for comprehensively modelling the conditioning factors (i.e. PIFs) that influence the probabilities of events in real systems. In the 1980s, some researchers applied IDs for evaluating the probabilities of human errors in nuclear power emergencies [12], and for modelling more general major accident scenarios [10]. However, the original Bayesian mathematical basis of IDs required that difficult expert judgements of the influence of combinations of PIFs on error probabilities needed to be performed (see [10] for examples of these evaluations and calculations), in moderately complex assessments. In recent years, a simplified approach to the numerical aspects of IDs has been developed, which considerably reduces the complexity of the evaluations required and makes the approach more accessible to non-specialists. This approach emphasises the value of the ID as a means for identifying and structuring the factors that influence human and system failures in incidents that have already occurred, and then generalising these causal models to allow predictions of the likelihood of similar types of failure occurring in the future.

8.7.6.1 Overview and examples

Figure 8.13 illustrates how an ID was used to model the factors conditioning the probability of an error in a medical context. This example concerns the injection of incorrect medications into the spinal cord of patients. This type of drug administration is referred to as the intrathecal route, and over the past ten years there have been five accidents in the UK alone, where substances intended for intramuscular injection have instead been injected into the spine, with catastrophic results [13]. The model was developed as part of the investigation of a number of incidents indicating that similar types of factors appeared to influence the likelihood of the event occurring. This model was then revised for use in prediction.

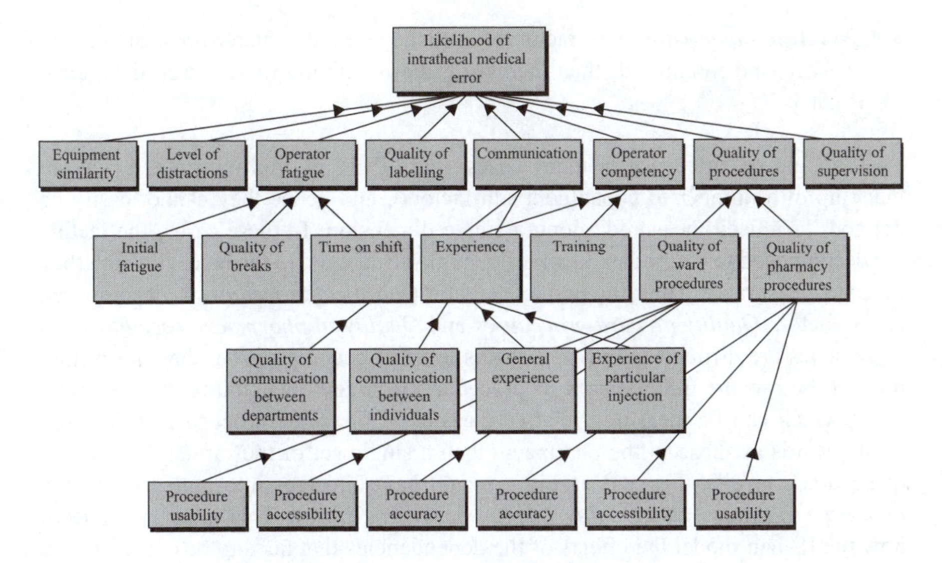

Figure 8.13 Influence diagram for a medical accident scenario

The box at the top of the ID is the outcome box. This describes the likelihood (or frequency) of the event for which the predictive analysis is required, given the state of the nodes below it. The boxes or nodes at the next level are called the direct influences. The state of any box or node in the diagram can be influenced by the state of the nodes below it, and in turn can influence any node above it. PIFs influence the likelihood of the outcome (in this case an intrathecal medical error) box via the other nodes in the network. In the case of the incident under consideration, a number of analyses have suggested that the occurrence of this type of incident is influenced directly by the following factors:

- *Equipment similarity*. The extent to which equipment (e.g. syringes containing different drugs) is readily confusable.
- *Level of distractions*. If a number of procedures are being carried out simultaneously, or there are distracting events in the area where the drugs are being administered, then the likelihood of the error will be increased.
- *Operator fatigue*. In this context, the 'operator' is likely to be a medical specialist such as a doctor or nurse, who often experience excessive levels of fatigue.
- *Quality of labelling*. For example, the extent to which the route of injection is clearly indicated in the labelling.
- *Communication*. In incidents of this type, there are typically a number of communication failures, e.g. between the pharmacy and the ward, and between members of the ward team.
- *Operator competency*. This relates to the knowledge and skills relevant to the task possessed by the individual or the team performing it.

- *Quality of supervision*. This factor refers to the extent to which work is effectively checked and monitored, thus increasing the likelihood of recovery if an error is made.

It should be noted that in many cases, these PIFs are multi-attributed (i.e. are made up of a number of constituent sub-factors), and hence they cannot easily be defined, or indeed measured, along a single dimension. In these cases, the factors are decomposed to their constituent influential sub-factors, to the level at which they become measurable. Thus, the first level factor *Quality of procedures* is decomposed to the factors *Quality of ward procedures* and *Quality of pharmacy procedures*. As these states are difficult to measure at this level of decomposition, they are further decomposed to the three factors of procedure accuracy, availability and usability, all of which can be measured relatively easily. If the same procedures were used on the wards as those in the pharmacy, then a single set of the influencing factors of accuracy, availability and usability would have linkages into both the *Quality of ward procedures* and *Quality of pharmacy procedures*. This is an example of how the ID can model the effects of the dependencies that need to be considered in order to obtain accurate results in reliability analyses. In summary, the ID approach allows factors to be decomposed to a level at which they can be more easily assessed.

The decision regarding the depth of decomposition is a pragmatic one depending on the application. In general, the greater the decomposition the better specified and accurate is the model. This may be particularly important if it is to be used for quantitative applications.

The ID structure is developed by combining information about factors influencing the probability of failures from a variety of sources. Although the above example has focused on data from incident investigations, sources such as research findings, the analysis of incident reports and near misses, and inputs from subject matter experts are also important. Normally, an initial ID is constructed off-line, based on research findings, incident reports and other formal sources of evidence. This ID is then developed further in an interactive group setting, which includes subject matter experts with direct working knowledge of the domain of interest. This group is called a consensus group, and is designed to ensure that all available sources of evidence are combined within a highly structured process. The consensus group is led by a facilitator, whose role is to manage the group interaction effectively, to ensure that the views of all the participants are captured. A software tool called IDEX allows the ID structure to be developed interactively in a group setting, and also allows the assessment of the quality of the factors in the particular context being assessed on numerical scales similar to those used in SLIM.

Once a comprehensive model of the factors influencing human performance has been built up from these sources, quantification can then be performed using a calculation process which is essentially an extension of the SLIM approach to model more complex interrelationships between factors influencing performance that are found in real world situations (see Section 8.7.6.2). However, there is a much greater emphasis in IDEAS on achieving a reasonably complete qualitative model of the significant

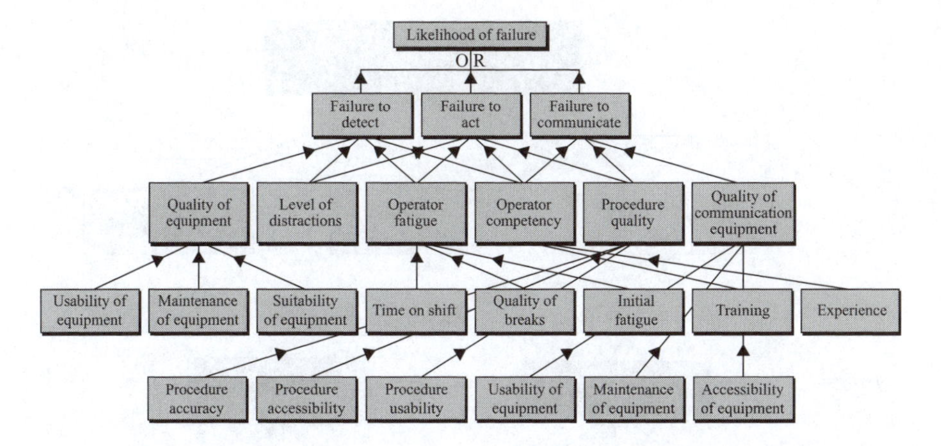

Figure 8.14 Generic ID human failure model showing use of logical combination rules

influencing factors, based on the idea that this is the primary input to ensuring that any risk management decisions and strategies are comprehensive. The quantification aspect of IDEAS is used primarily for choosing between alternative error prevention strategies, and hence the emphasis is on relative rather than absolute quantification. Nevertheless, the outputs from IDEAS can still be used as inputs to applications such as safety cases.

The preceding example has illustrated an ID for a very specific type of medical scenario. It is also possible to develop IDs for broad classes of situations, and to combine together a number of discrete sub-models, using Boolean AND and OR gates, to produce more complex representations. In Figure 8.14 a generic human error model is illustrated that represents a number of ways in which failures could arise, and then combines these together using an OR gate.

This type of model can be used to evaluate the relative contributions that different types of errors (slips, mistakes or violations) are likely to make to the situations being assessed, and to provide the basis for evaluating the effects of different mitigation strategies, e.g. is it more cost effective to improve the training or the equipment usability?

8.7.6.2 Using the ID for human reliability assessment

The calculations that are used to generate human reliability estimates in IDEAS are identical to those used in SLIM, except that they are performed at each level in the hierarchy of influences, to generate a derived rating at each node in the network. These ratings are then combined to generate an SLI in a similar manner to SLIM. The SLI is then input into a calibration model to generate probability estimates. This approach is illustrated in the following example.

The first stage of the MARS model for railway signals passed at danger shown in Figure 8.3 (Section 8.2.2) is 'Detect Signal'. One of the main factors influencing

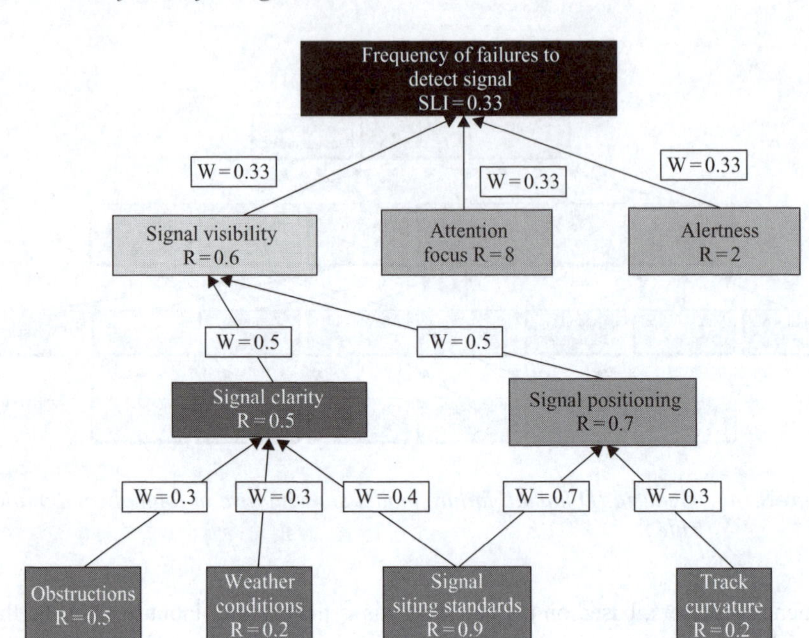

Figure 8.15 Example ID illustrating calculations to produce derived ratings

signal detection is signal visibility. A simplified ID for this situation is shown in Figure 8.15. This diagram shows how evaluating the factors at the base of the ID, i.e. obstructions, weather conditions, signal sighting standards and track curvature, generates a derived rating for signal visibility. These ratings are then combined with weights for the corresponding factors in the same manner as for SLIM, to develop derived ratings for signal clarity and signal positioning at the next level of the ID. The final derived ID for signal visibility of $R = 0.6$ is then derived from these ratings. If similar calculations had been performed for the other top-level factors of 'attention focus' and 'alertness', an overall SLI of 0.33 would have been created. In order to convert this to a probability, calibration data would be required as shown in the SLIM example in the previous section.

It can be seen that even for quite complex IDs, as long as the bottom level factors can be assessed and the structure of the model is correct, then a value can be derived that is subsequently quantified to produce a probability value. Sensitivity analyses are performed in a similar manner to SLIM, except that changes to the bottom level of the ID may have implications for a number of levels, depending on the interconnections between different factors at different levels in the network. For more complex networks, it is somewhat time consuming to perform these calculations by hand. The IDEX (Influence Diagram Expert) software tool [14] allows the IDs to be generated interactively with subject matter experts, which can then be assessed by evaluating the bottom level PIFs. IDEX then automatically performs the appropriate calculations. This makes sensitivity and cost-benefit analyses very easy to perform.

8.7.6.3 Advantages of IDEAS

The IDEAS technique provides the most comprehensive and versatile approach to modelling the factors that influence human reliability in a particular context, such as medicine, train operations or process plants. The process also allows a specific predictive model to be used that is tailored to a particular context where the assessment has been carried out. The use of this comprehensive model of the factors affecting error likelihood means that any prediction based on this model is likely to be more accurate than one in which significant factors have been omitted. Although there are obviously resources required to construct the domain-specific models required for particular applications, this effort can be tailored to the degree of comprehensiveness required. Over a period of time, analysts working in particular areas such as medicine and aviation are likely to build up core models that can be modified with relatively little effort for specific applications.

The ability of the model to address underlying causes such as management and organisational influences on the likelihood of error makes it a useful tool in studies where these factors need to be addressed, e.g. the evaluation of safety management systems.

The close linkage between the use of the technique as an incident investigation tool and its application for predicting failures means that there is the potential for gathering data to populate the model and for verifying its predictions.

8.7.6.4 Disadvantages of IDEAS

The main disadvantage of IDEAS is the effort required to construct the ID. This will require work on the part of the analysts to find out which factors need to be considered in the particular application of interest. Engineering users may well feel that they would prefer to use an existing generic model, rather than being required to identify the specific factors that may influence error in a particular situation. With regard to the quantification aspects of IDEAS, problems with regard to obtaining calibration data exist that are similar to those encountered in SLIM. However, the use of the same model for incident investigation and prediction may mean that calibration data can be generated.

8.8 Choice of techniques

In the preceding sections, a number of human reliability quantification techniques have been reviewed from the point of view of their strengths and weaknesses. The question that arises is which techniques should be deployed for a particular human reliability assessment application? In Table 8.13 and the subsequent sections, the techniques discussed in this Chapter have been compared using a number of criteria. The evaluations will be useful to assist the analyst in the choice of a technique for a particular assessment.

It may seem surprising that the criterion of accuracy, which may appear to be the most relevant dimension, is omitted from the table. This is because, in reality, there are few real data available regarding the accuracy or otherwise of the techniques

Table 8.13 Comparison of the human reliability quantification techniques reviewed

Method	Qualitative insights	Evidence base	Versatility	Risk reduction guidance	Resource requirements
THERP	Low	Medium	Low	Medium	Medium
HEART	Low	High	Low	High	Low
SLIM	Medium	Medium	High	High	Medium-high
IDEAS	High	Medium	High	High	Medium-high

that have been reviewed. Although the predictions made by some of the techniques discussed in this Section have been evaluated against a small database of 'objective' human error probabilities (See Kirwan [3]), there were a number of methodological problems with this study that made it difficult to draw firm conclusions regarding the absolute accuracy of the techniques considered.

8.8.1 *Evaluation criteria*

The criteria considered in the table are as follows:

Qualitative insights

This is the degree to which the technique is able to assist the analyst in developing a comprehensive knowledge of the factors that will drive the human error probabilities in the situation being assessed. As has been emphasised in earlier sections of this Chapter, the qualitative insights regarding what needs to be done to maximise human reliability in the situation being assessed, should, in many ways, be regarded as the primary goal of human reliability assessment. However, the engineering requirements to develop absolute error probabilities for use in traditional probabilistic safety analyses or quantitative risk assessments may often obscure this objective. Nevertheless, if a comprehensive model of the factors influencing human error in a particular domain has already been developed as part of a safety assessment, this can be used as an input to the quantification process, at the very least to make the analysts aware of the contextual factors that may need to be considered during quantification.

Evidence base

This criterion refers to the extent to which a technique either contains or is able to utilise data regarding the factors influencing human error. Techniques which rate highly on this dimension will normally be preferred to those which rely solely on individual expert judgements of the analyst.

Versatility

This refers to the extent to which the technique is applicable or configurable such that it can be applied to all types of situations and error types (including slips, mistakes and violations).

Risk reduction guidance

This is the extent to which the technique provides guidance in suggesting which interventions are likely to be the most cost-effective in reducing the risk of human error.

Resource Requirements

This final criterion is one which most frequently determines the choice of methods. Unfortunately, human reliability assessment is often performed to satisfy a requirement (for example in a safety case) by analysts with an engineering background who are not necessarily sympathetic to the need to comprehensively address human factors issues. Despite the fact that in most safety critical systems the majority of the risks will arise from human errors rather than hardware failures, the resources that are allocated to human reliability modelling and assessment are typically less than five per cent of those devoted to minimising engineering risks. For this reason, techniques that will satisfy the requirement to have conducted a human reliability analysis but which require least resources may be chosen, even if they are not necessarily appropriate to the situation being evaluated. If the results are to stand up to expert scrutiny, it is important that this temptation is avoided by Project Managers.

8.8.2 *Evaluation of the reviewed methods against the criteria*

THERP

This technique is rated low on qualitative insights because it only provides a limited set of factors in its database that are assumed to be the primary drivers of human error in all situations. In reality, there may be many domain specific factors that will need to be taken into account when applying THERP in situations other than the nuclear domain where it was first developed. The quality of the evidence base in the form of the database that is provided with THERP is difficult to assess, since the THERP Handbook [4] does not provide any information regarding its sources. However, as discussed earlier, it is believed that the original THERP data was derived from experimental sources such as [9], modified by the factors in the Handbook. The versatility of THERP is judged to be low, as its databases and applications guidance is primarily oriented to performing assessment in nuclear safety applications. Once the THERP event trees have been constructed, it is possible to insert a range of error probabilities into the critical nodes to see which changes have the greatest effect on the output probability for the overall scenario. However, since a relatively small number of PIFs are considered by THERP, this limits its capability to provide a wide range of error reduction strategies. The resources required to perform a THERP analysis are likely to depend strongly on the complexity of the task being assessed. However, due to the requirement to quantify at the level of individual task steps, the resource requirements may often be quite considerable.

HEART

The qualitative insights arising from HEART are likely to be low, as the technique provides its own list of PIFs, and hence there is little incentive for the analyst to extend his or her consideration of the drivers of human error outside this list. If the analyst does perform a comprehensive analysis to identify some context specific factors, a problem arises with regard to how to incorporate them in the quantification process. Since HEART contains its own database of PIFs, based on experimental data, its evidence base is judged to be high, as long as the error probability under consideration is driven only by these factors. The versatility of HEART is judged to be low in that it is limited to situations for which its PIF database is applicable. It provides good facilities for 'what if' analyses, where the effect of changing the factors influencing human error is investigated, as shown by the case study in Section 8.7.4.1. The resources required to conduct HEART analyses are generally regarded as low. Because HEART provides a range of factors that may influence the probability of error, the analyst only has to match these factors to the scenario, rather than having to develop a specific model.

SLIM

Because SLIM requires a context specific set of PIFs to be applied to the situation being assessed, the analyst is required to identify the relevant PIFs. This ensures that the analyst gains some understanding of the context and hence SLIM is assessed as medium on the dimension of providing qualitative insights. The SLIM technique provides no built-in information in the form of a data base (as in THERP) or a set of suggested PIFs (as in HEART). However, SLIM is able to utilise external data sources such as the experimental literature, as well as the knowledge and experience of the analyst as a source of potential PIFs. It is therefore rated as medium on this criterion. Because of its ability to be applied to all types of errors and situations, it is rated high on the criterion of versatility. The direct relationship between the calculated error probability and the factors assumed to drive this probability allows sensitivity and cost effectiveness evaluations to be easily performed, giving a rating of high on the risk reduction dimension. The versatility of SLIM comes with the cost that more effort is required from the analyst to develop a context specific model relating PIFs to error probabilities. However, once the model has been developed, it is relatively easy to perform a large number of assessments of tasks to which the model is applicable. SLIM is therefore rated as being medium-high on the criterion of resource requirements, depending on whether an existing model can be utilised, or a new one has to be developed.

IDEAS

IDEAS can be regarded as a more sophisticated version of SLIM, to address the indirect as well as the well as the direct PIFs that drive performance. For this reason, its evaluations on most of the criteria are identical to those for SLIM. However, because of its capability to consider the underlying factors which drive error probabilities the qualitative insights, criterion is assessed as higher for IDEAS compared with SLIM.

8.9 Summary

This document has provided an overview of a framework for the assessment of human error in risk assessments. The main emphasis has been on the importance of a systematic approach to the qualitative modelling of human error. This leads to the identification and possible reduction of the human sources of risk. This process is of considerable value in its own right, and does not necessarily have to be accompanied by the quantification of error probabilities.

Because of the engineering requirement to provide numerical estimates of human error probabilities in applications such as safety cases, examples of major quantification techniques have been provided, together with case studies illustrating their application. It must be recognised that quantification remains a difficult area, mainly because of the limitations of data. However, the availability of a systematic framework within which to perform the human reliability assessment means that despite data limitations, a comprehensive treatment of human reliability can still yield considerable benefits in identifying, assessing and ultimately minimising human sources of risk.

8.10 References

1 WILLIAMS, J. C.: 'HEART – a proposed method for assessing and reducing human error', in 9th Advances in Reliability Technology Symposium, University of Bradford, 1986
2 WRIGHT, K., EMBREY, D., and ANDERSON, M.: 'Getting at the underlying systematic causes of SPADS: a new approach', Rail Professional, August 2000
3 KIRWAN, B., and AINSWORTH, L. K. (Eds): 'Guide to task analysis' (Taylor & Francis, London, 1992)
4 SWAIN, A. D., and GUTTMANN, H. E.: 'Handbook of human reliability analysis with emphasis on nuclear power plant applications', NUREG/CR-1278 (Nuclear Regulatory Commission, Washington, DC, 1983)
5 KIRWAN, B.: 'A guide to practical human reliability assessment' (Taylor & Francis, London, 1994)
6 KIRWAN, B.: 'Human error analysis', in WILSON, J. R., and CORLETT, E. N. (Eds): 'Evaluation of human work, a practical ergonomics methodology' (Taylor & Francis, London, 1990)
7 MEISTER, D.: 'Human reliability', in MUCKLER, F. A. (Ed.): 'Human factors review: 1984' (Human Factors Society Inc., Santa Monica, CA, 1984)
8 SWAIN, A. D.: 'Accident sequence evaluation program human reliability analysis procedure', NUREG/CR-4772 (US Nuclear Regulatory Commission, Washington, DC, 1987)

9 MUNGER *et al.*: 'An index of electronic equipment operability: data store'. Report AIR-C43-1/62-RP(1) (American Institute for Research, Pittsburgh, PA, 1962)

10 EMBREY, D. E.: 'Incorporating management and organisational factors into probabilistic safety analysis', *Reliability Engineering*, 1992, **38**, pp. 199–208

11 PONTECORVO, A. B.: 'A method of predicting human reliability', *Annals of Reliability and Maintenance*, 1965, **4**, 337–342. 4th Annual Reliability and Maintainability Conference

12 PHILLIPS, L. D., EMBREY, D. E., HUMPHREYS, P., and SELBY, D. L.: 'A socio-technical approach to assessing human reliability', in OLIVER, R. M., and SMITH, J. A. (Eds): 'Influence diagrams, belief nets and decision making: their influence on safety and reliability' (Wiley, New York, 1990)

13 WOODS, K.: 'The prevention of intrathecal medication errors', a report to the Chief Medical Officer, Department of Health (www.doh.gov.uk/imeprevent/index.htm)

14 IDEX (Influence Diagram Expert) SOFTWARE: Available from Human Reliability Associates Ltd (www.humanreliability.com)

Chapter 9
Control room design

John Wood

9.1 The human operator in the control loop

Most of the population will be unaware of the operators seated in front of rows of screens controlling the services we take for granted – metro systems, sewage works, electricity distribution and security, for example. These operators are at the core of an electronic network of which the control room is the hub. This network will be pulsing electronic information to and from their workstations, linking them with an infrastructure that is unseen from their work positions.

The amount of information potentially accessible to each operator is tending to increase as economies of scale drive organisations to centralise their control functions and cover greater geographic areas. This increasing coverage, coupled with a drive to bring more of the network back to an individual operator, is raising new challenges as far as workloads – or more precisely – overload and underload are concerned. The response of systems designers and specifiers has been to use automation as a means of minimising operator workload with the underlying assumption about workload being 'less is better'. This issue of underloading operators and maintaining them in the 'control loop' is now emerging as the latest challenge for human factors engineers in control systems design and control room design.

The programme to upgrade the UK rail network illustrates how conflicting demands must be addressed by the human factors engineer. The rail network runs with a high level of safety but does include a mix of control systems, some of which are labour intensive. The need to provide extra capacity and improve speeds on the network requires that old-fashioned manual signal boxes with levers are replaced by electronic based systems (cable actuation replaced by electronic links and electronic control). Each new generation of control brings with it a new level of automation, potentially disengaging the operator from the control loop.

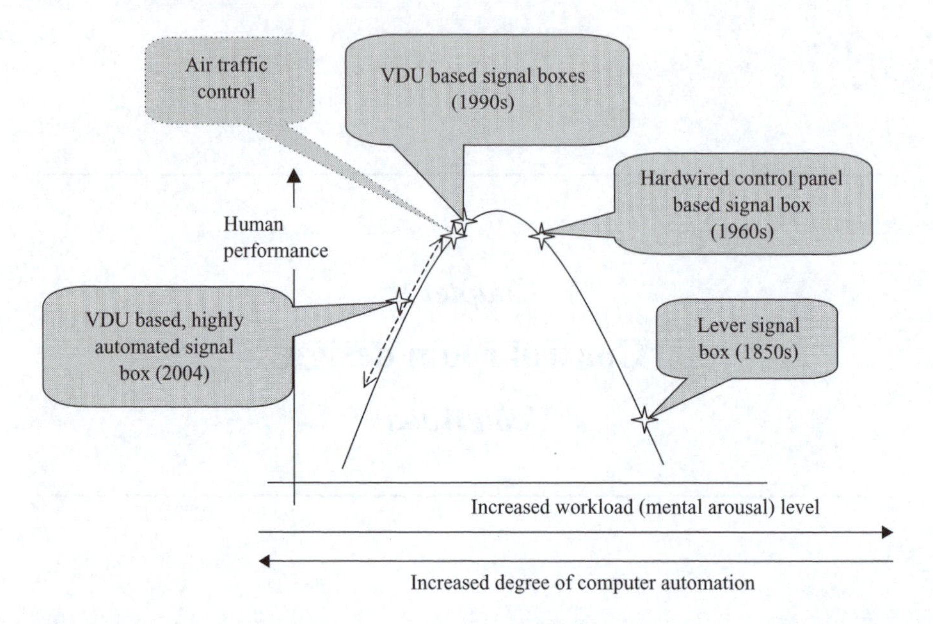

Figure 9.1 Human performance related to arousal levels and degrees of automation

The complex relationship between operator performance, levels of automation and levels of workload are illustrated in Figure 9.1. A hypothetical curve is drawn linking human performance against arousal. In this instance the range of existing rail signal box types has been plotted.

For calibration purposes the likely position of current air traffic control systems in this spectrum of rail control systems has been included.

Whatever the complexity of a semi-automated controlled process, the control loop can be reduced to a very simple model. The block diagram, Figure 9.2, illustrates the importance of maintaining a 'closed loop' with the operator providing a vital link in the chain.

Where the operator works alongside semi-automated processes the ergonomist can apply some practical measures to reduce the likelihood of human failure. The fact that most modern systems demand very little input for 95 per cent of the time does not help – operators get bored and their attention wanders. Also, these very systems, when things do go wrong, can escalate demands on the operators dramatically and without notice. The ergonomist needs to find the right mix of challenging and interesting tasks that will neither overload nor underload the operator at any time.

It is possible to minimise the likelihood of the operator 'dropping out' of the control loop by applying ergonomic thinking to equipment design, the working environment and working practices. Examples of these are presented in Table 9.1.

Failure to successfully mitigate against these potential breakdowns can be costly in process or safety terms – responses to alarms are delayed or incorrect action taken. (A detailed review of human error and human reliability can be found in Chapters 7 and 8.)

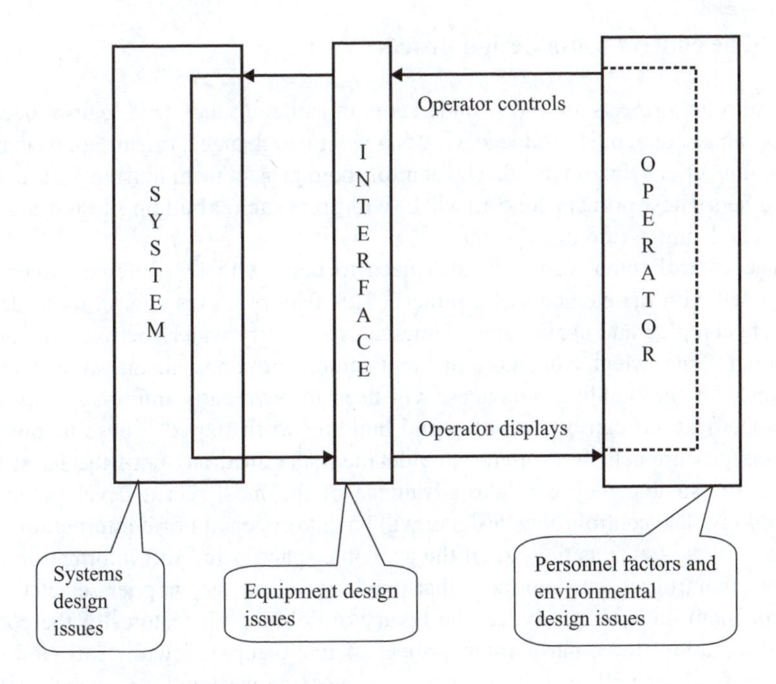

Figure 9.2 Operator in the closed loop

Table 9.1 Ergonomic countermeasures against operator-related open loop failure

Feature	Sample ergonomic countermeasures
Personnel factors	Minimise sleep deprivation due to inappropriate shift patterns Avoid designing boring jobs where skills are under-utilised Aim to avoid selection of over-skilled or over-qualified personnel
Systems design	Introduce secondary tasks to raise levels of activity Avoid highly automated systems Introduce regular operation of system in 'failure' modes
Equipment design	Avoidance of hypnotic effects – picture autocyling or repetitive auditory signals
Environmental design	Introduction of windows to create 'change' of visual environment Avoidance of environments that are too quiet Avoidance of environments that are too warm Introduce plants Steer clear of interior schemes that are too 'calm' and 'neutral'
Equipment layout	Creation of layouts that require movement Layouts that encourage direct verbal contact and team working

9.2 The control room design process

Control room projects are rarely undertaken in isolation; they tend to arise because new systems are being introduced, working practices changed, organisations merged or equipment is being replaced. The control room project team must expect to work with others whose primary concern will be with areas such as building design, systems design and human resources.

The control room team will also need to cope with the different timescales associated with the elements of a project. This feature makes control room design projects complex and challenging. Timescales will vary widely between the building programme, which can take years, and equipment procurement, which can take months. For the building, architects will need to have early information in order that specifications can be drawn up and building work started. Those involved in systems procurement will often leave detailed specifications until the latest possible stage so that they can take advantage of the most recent developments in technology. The control room designer will have to cope with firm information from some sources – such as the size of the available space – and soft information from others – control system interfaces that may depend on the supplier selected. The control room designer rarely has the luxury of defining all features of the control room. To allow the control room project to move forward with 'soft' information the designer will have to make use of 'working assumptions', which will be discussed later.

The flowchart in Figure 9.3 is based on that in the standard on control room design [1] and summarises the typical stages through which a control room design project might pass.

If life were as simple as that portrayed in the flowchart, control room projects would be a lot easier to handle. However, time, available resources, technical information and a host of other factors restrict the ergonomist from sailing through these stages in a single sweep. It is anticipating, managing, and overcoming these obstacles that gives such satisfaction in a successful control room project. The main ergonomic tasks to be undertaken at each of the main stages are now summarised.

9.2.1 Phase 1: establishing the brief

This stage should spell out the context within which the ergonomics project is being undertaken including resources and constraints. Ergonomics methods should include:

- Documentation review.
- Interviews.
- Audits of comparable control rooms.
- Technology reviews.
- Requirements preparation and review.
- Ergonomic cost benefit reviews.

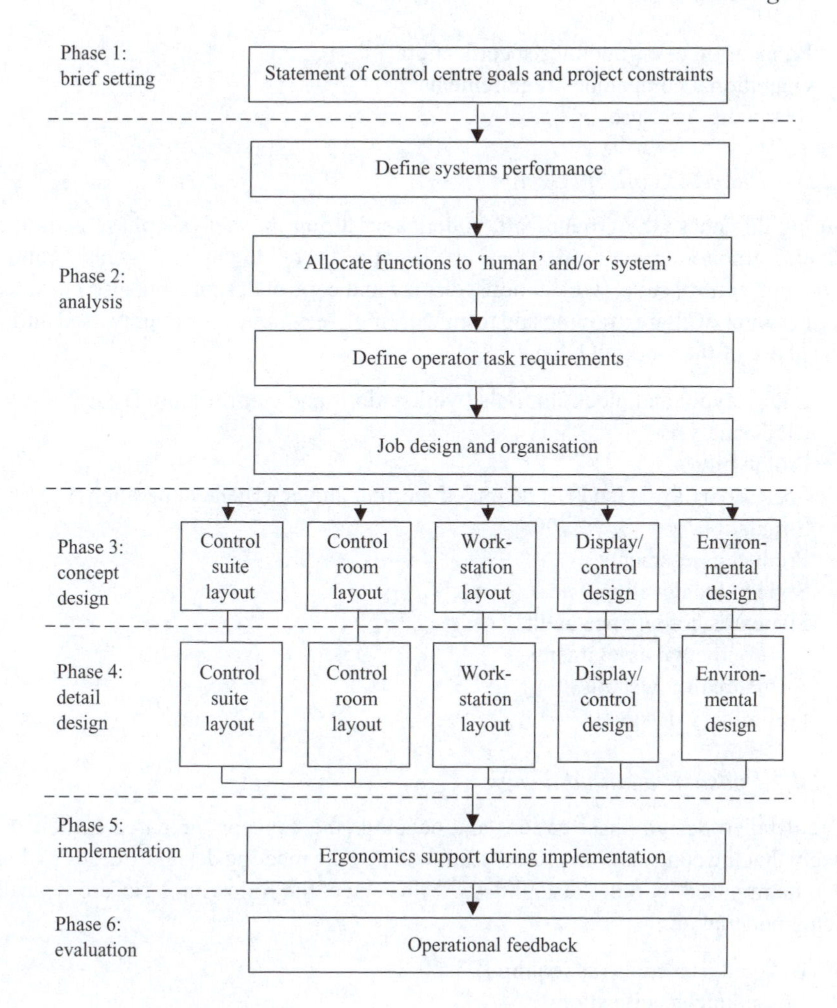

Figure 9.3 Control room design process

9.2.2 Phase 2: analysis

The ergonomics programme will address the strategic issues including performance requirements, allocation of functions and job design. Ergonomics methods are likely to include:

- Task analysis.
- Functional analysis (normal, degraded and maintenance).
- Operational simulations.
- Hazard and risk analysis.
- Systems interface analysis.
- Workload analysis.
- Human error analysis.

- Preparation of ergonomics specification.
- Operational competency requirements.
- Setting up 'user groups'.

9.2.3 Phase 3: concept design

During this phase the groundwork undertaken during the analysis phase is used to develop ergonomic solutions. The scope is not restricted to physical issues – control room and control suite, workstations, display and control design – but also considers organisation, welfare, training and team working. Techniques commonly used during this phase of the work include:

- Initial layout and mock-up trials (workstations and control room layout).
- LINK analysis.
- Prototyping.
- Scenario analysis (such as degraded, normal and emergency operations).
- Simulation.
- Product assessments.
- Style guide development.
- Standards compliance audit.
- Environmental assessments.
- Virtual reality visualisation.
- Preliminary designs.

9.2.4 Phase 4: detailed design

The detailed design phase refines and develops the concepts prepared earlier. It is likely that the control room project team will now be meeting design 'freezes', which will be imposed by other disciplines. At this stage, the following tasks are typically being undertaken:

- Detailed user trials (as required).
- Ergonomic specifications.
- Procurement ergonomics.

9.2.5 Phase 5: implementation

All too often control room design is considered 'finished' after detailed design – any late changes are not passed in front of the ergonomist for review and approval. It is essential that an ergonomic input is maintained throughout this implementation phase so that ergonomic solutions are not diluted or compromised during late modification. Ergonomics support typically includes the following:

- Monitoring for deviation from agreed ergonomic solutions.
- Ergonomics advice and support during on-site modifications and adjustments.
- Procurement ergonomics for equipment, environmental and furniture specifications.
- Test and acceptance.

9.2.6 Phase 6: evaluation

Ergonomics should be part of the verification and validation of design solutions. It should also continue into post control room commissioning assessment and provide feedback for future control room projects. In order for this to happen resources must have been allocated at early budgeting stages and incorporated within the overall project schedule. Any ergonomics compromises made during the design process should be fully documented. Ergonomics methods include:

- Operational safety assessment.
- Human error analysis.
- Job satisfaction and user opinion assessment.
- Operational audits and workshops.

A cornerstone in the process is the development of an 'Ergonomics Specification' during the analysis phase. Of all documents produced by the control room design team, this is the blueprint for all the later work. It is within this document that key features are summarised such as outline job descriptions, staffing numbers and rostering, equipment lists, likely job aids and potential linkages between different operating functions. A wise control room designer will seek 'sign-off' for the 'Ergonomics Specification' from the rest of the team for the items spelt out, since it will contain both 'hard' and 'soft' information. An example of the latter is where control room design needs to race ahead of equipment procurement. Assumptions will need to be made to allow for the production of workstation arrangements and development of room layouts. The Ergonomics Specification allows these assumptions on soft information to be made, documented, and agreed by all and the risks to be shared.

The control room design process should be logical to the client and provide an audit trail with deliverables. Often the process will have a starting point part way through – existing equipment is to be reused or existing job descriptions are to be retained – and the control room design team will need to work with these constraints. It is still wise to spell out the steps to be taken, however modified from the ideal presented in Figure 9.3, and the outputs that will be produced at each stage. For an external consultant these 'deliverables' may well be the milestones for payment but will also be part of an incremental process whereby the whole team contributes to the control room design process.

Client organisations need to understand that an ergonomic process for control room design imposes certain requirements on themselves as well. Management time will need to be made available for advising on new systems to be introduced or changes in working practices or policy – certainly on their aspirations for their new centre and the image they wish to present. Time will also be required for 'users' to participate in ergonomic tests during the control room design programme and in workshops held at strategic points during the development process. These need to be planned and resourced at the outset of the programme.

The size of a control room project team varies from a single individual to a multi-disciplinary team depending on the range of issues to be covered and the scope of the project. The organogram presented in Figure 9.4 illustrates the organisation of a large

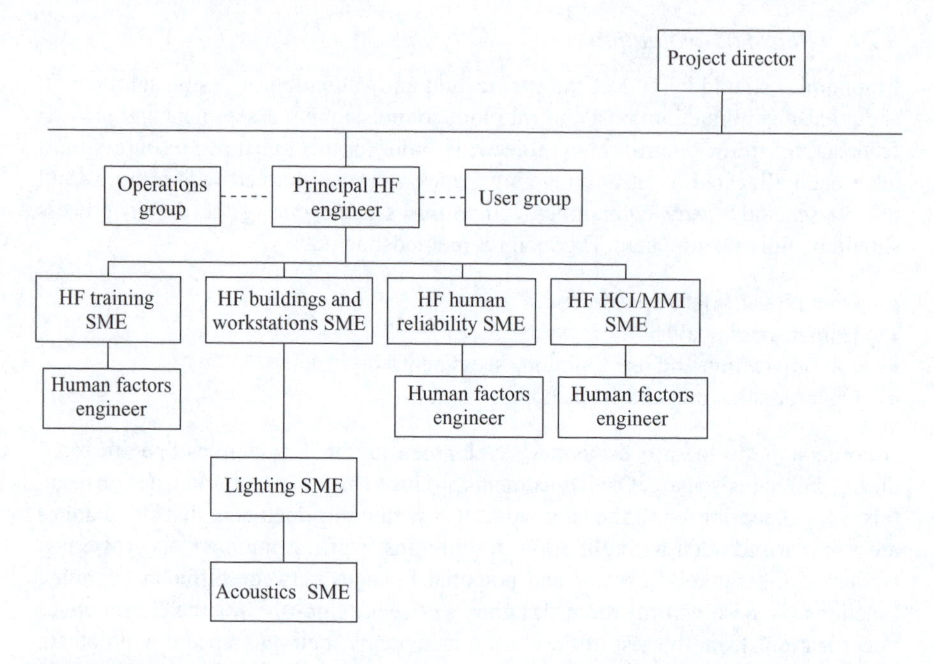

Figure 9.4 Control room design team for major infrastructure project

control room team and is based on an actual major infrastructure project running over a number of years. With a team involving a single individual, that person may well have to address the whole range of topics at various stages of the programme.

A requirement of the ISO control room design standard is that the design process should be properly documented. It has long been recognised that control room project teams break up once a new centre is up and running. Any built-in design assumptions are easily lost though they may be absolutely essential for future modifications – for example, which were the areas on the control room floor allocated for future expansion? It is for this reason that the standard requires that the process by which the control room was designed is summarised, together with all working assumptions, and any key underlying ergonomic principles adopted; for example, that a certain workstation has wider circulation space around it to accommodate wheelchair access.

9.3 The control suite and its layout

When first examining a control centre project it is natural to focus on the control room because this may seem to be the place where 'all the action is'. There are, however, a number of closely related rooms that provide support for staff and equipment which must also be carefully designed – for example, rest areas, equipment rooms, offices and kitchens. It is this entire group of areas that is defined as a control suite and within it lies the control room itself.

In some instances the location of the control suite on a site may be critical to success. A process control may need to have views over the plant as well as being easily accessible for visitors and security checks. For example, an airport control tower must have views over the airfield in order to function, whilst for others location may be unimportant. 'En-route' air traffic controllers will never be looking directly at the aircraft they are controlling.

Even if no external view is required, windows should be incorporated into control rooms for psychological reasons. Windows can create their own ergonomic problems and these should be minimised by environmental control measures such as introducing blinds and filters. The orientation of the control building, in respect of the rising and setting sun, can do much to avoid potential heat gain problems and disabling glare from windows.

The operational links between different elements of the control suite should be teased out by a LINK analysis. For some control suites the links may be dictated by staffing arrangements. A 24 hour water treatment centre may well have only one individual on duty in the control room during the night and kitchens and rest areas need to be within easy reach of the main workstations for this arrangement to succeed.

The elements of a typical control suite are illustrated in Figure 9.5.

In the examples of an airfield control room and a process plant the location of the control suite is dictated by external views. In other instances there is a much looser link between the control suite and the system – for example, there will usually be more flexibility with the siting of a control room for a metro system or a national electricity distribution centre.

For other installations there may be a definite decision to locate a control room, and its supporting functions, remote from the site it controls. Examples occur in plants manufacturing highly explosive materials or where, for physical security reasons, a remote centre is preferred.

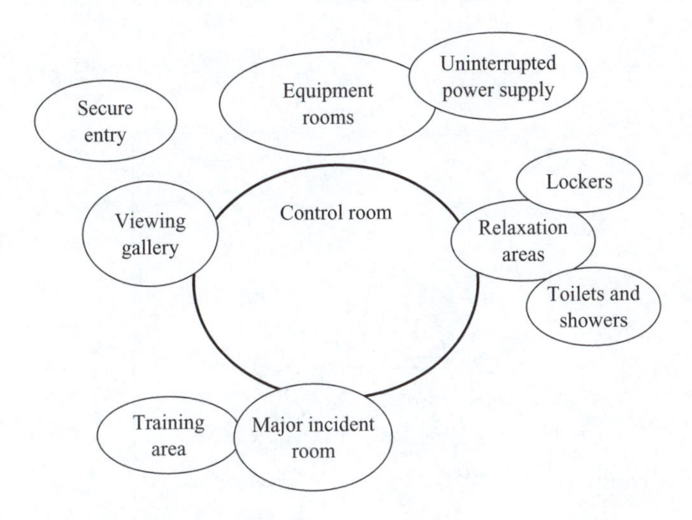

Figure 9.5 Elements of a control suite

9.4 The layout of control rooms

The actual shape of a space can have a significant impact on its suitability for control room use. As a general rule, the squarer the shape the more flexibility it offers for alternative layouts of workstations. In Figure 9.6 the same area has been drawn up but using different shapes – the flexibility of the space increases from a room with an 'L'-shape to those with squarer proportions.

Apart from the shape, judgements based purely on the floor area allocated for a control room can also be potentially misleading. There may be a substantial difference between the *allocated space* and the *usable space*. Workstations cannot be located over ventilation grilles, vertical ducts will require access by maintenance engineers, awkward corners will be unusable, structural columns create dead areas around them, and doors take up floor space for opening clearances. When all these constraints are fully taken into account, what might on paper to have seemed perfectly adequate can in reality be too small (Figure 9.7).

By distilling the experience of a large number of control room layout programmes it is possible to identify some core principles. Whilst they may not be applicable in all cases they provide the designer with a sound basis from which to start this work.

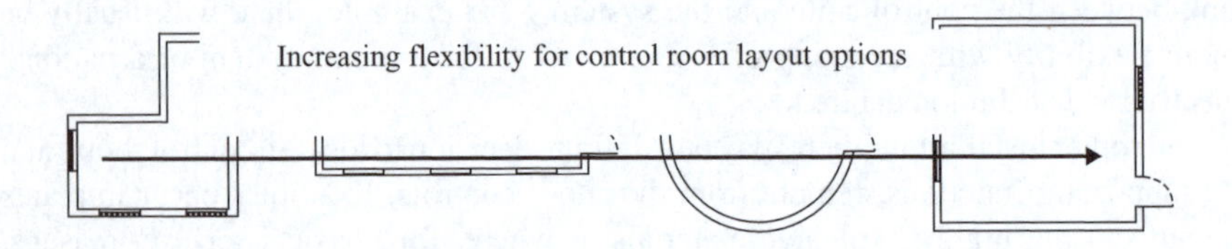

Figure 9.6 Control room shape and room layouts

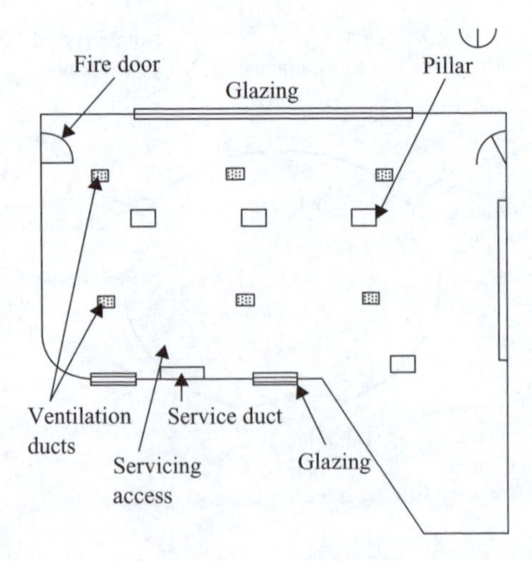

Figure 9.7 Usable and non-usable space

9.4.1 General principles for control room layout

- Room layouts should be based on a task analysis and an understanding of the worker population – including workers with disabilities.
- Where control rooms house a number of operators the layout should facilitate team working opportunities and social interaction.
- Control room layouts should reflect the allocation of responsibilities and the requirements for supervision (e.g. appropriate grouping of team members).
- Operational links, including sightlines or direct speech communication, should be optimised when developing control room layouts.
- Control rooms that exhibit either overcrowding of workpositions, or widely dispersed workstations, are not recommended. Layouts should allow, wherever practical, direct verbal communications between the control room operators but also avoid excessively short separation between adjacent operators.
- Control rooms with similar functions, and in the same plant or facility, should adopt the same ergonomic principles of room layout to facilitate decision-making and transfer of training.
- Where physically disadvantaged control room operators or visitors are expected to use the control room adequate facilities should be provided – this may require ramps for access.
- There are ergonomic benefits in varying posture during periods of work and it is recommended that control workstation layouts and work regimes allow control room operators to change their posture at the control workstation and to move from their workstation from time to time. This ergonomic requirement should not interfere with the primary control duties or need to be undertaken when time-critical tasks need to be undertaken.
- Room dimensions and control workstation layout dimensions should take account of the anthropometric dimensions of the user population (e.g. view of off-workstation, shared overview displays).
- Control room operators using visual displays should not be facing windows unless these windows are a primary information source.
- The layout of the control room should ensure that all off-workstation visual displays, necessary for the control room operators' task, are visible from all relevant control workstations.
- Circulation of control room staff, maintenance staff and all visitors should be achieved with minimum disruption to the work of the control room operators.
- Main entrances and exits should not form part of the working visual fields of the control room operators unless they have specific responsibilities for checking the flow of staff into and out of the control room – e.g. supervisors.
- Control rooms should provide for future expansion – this should take account of such factors as the planned lifespan of the control room and predictions on increases in workloads; a figure allowing for a 25 per cent increase in the number of working positions has been used as a 'rule of thumb'.
- For existing rooms (refurbishments) operators should not be located close to ventilation system inlets and, for new buildings, the design team should ensure that air vents are not located close to operator positions.

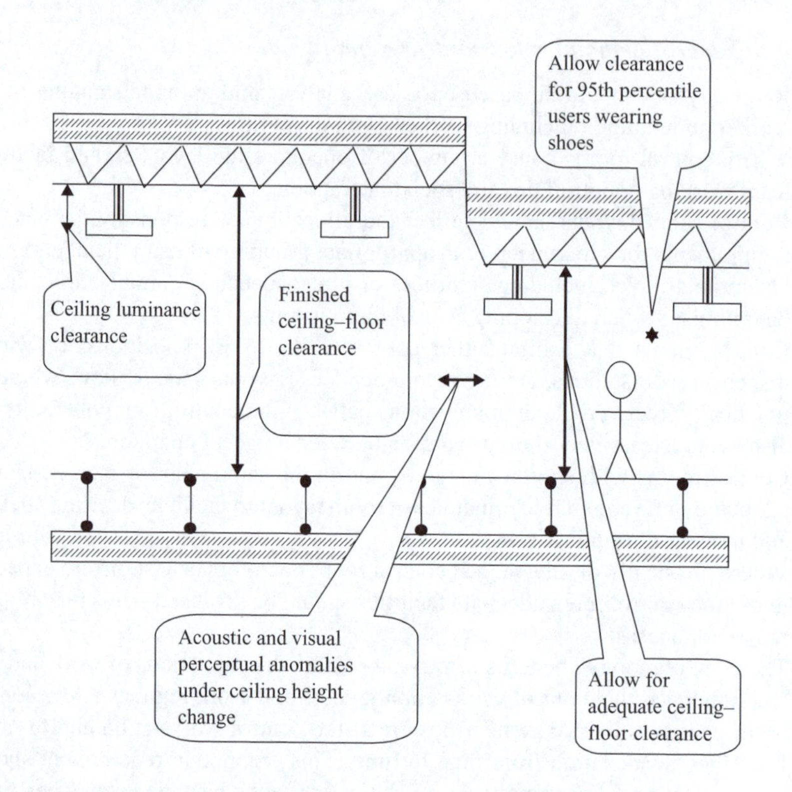

Figure 9.8 Vertical dimensions and control room space

- Room layouts should take into consideration the transmission, and potential build-up, of noise.

In selecting a suitable space the vertical dimension needs also to be considered. Major acoustic problems have been encountered with drum-shaped control rooms combined with a high domed ceiling. When fitting control rooms into converted office space the limited floor to ceiling heights can be a constraint on the use of shared overview displays or raised control room areas. Figure 9.8 illustrates some of the difficulties and uncertainties that can be generated.

Many traditional control room layouts used different floor levels to achieve sightlines to supervisors and wall mounted displays, being raised on a plinth even being synonymous with a supervisory function. Plinths, however, can give rise to their own problems. The additional space required around a plinth, for circulation purposes, needs to be considered, as well as the potential difficulties created for wheelchair access. Where vertical heights are constrained, typically with conversions, this approach is unlikely to provide a solution. The careful use of sightlines and strategically locating workstations in a layout can achieve most of what is sought with variations in floor levels, without the built-in constraints on future change.

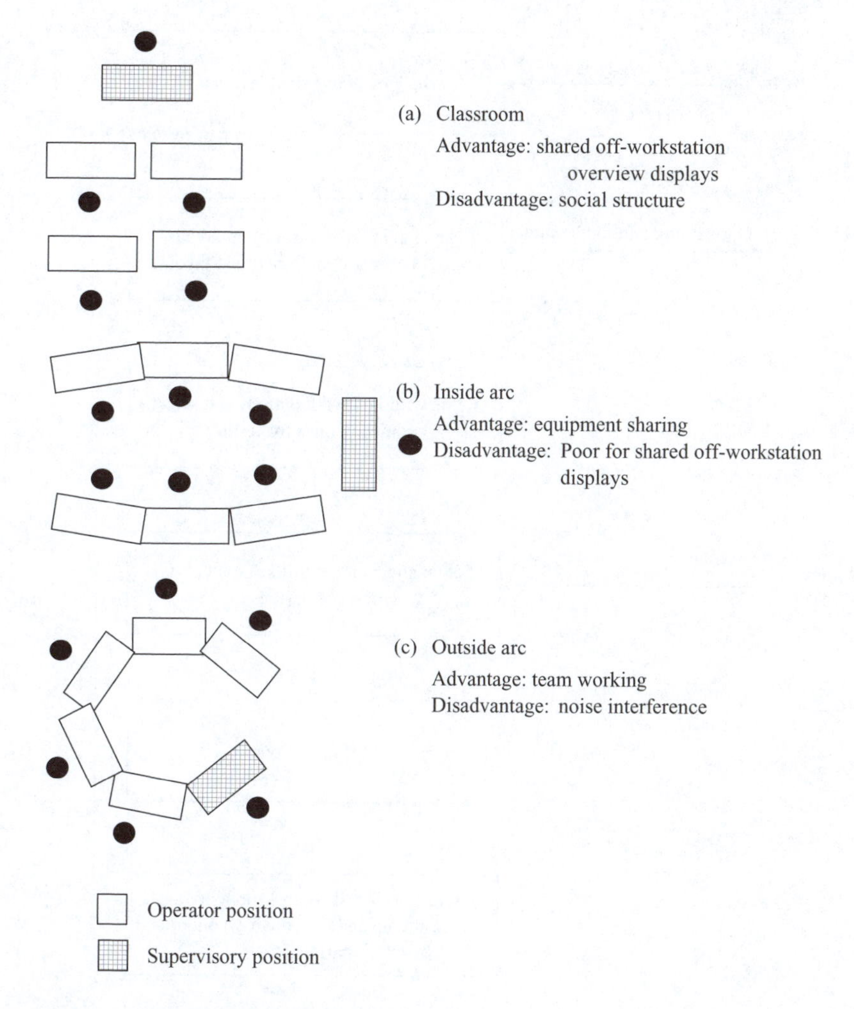

Figure 9.9 Alternative groupings of workstations

The most appropriate way in which individual workstations can be grouped within a control room should be primarily informed by operational requirements expressed through the task analysis. Whilst workstation layout options may appear to be very different they often fall into one of three stereotypes. First, there is the classic 'classroom' arrangement in which the workstations are all facing forward and arranged in a series of rows. Here the supervisory position may either be at the front, facing the control team, or at the rear looking into the backs of the operators but with sight of their screens (Figure 9.9(a)). Second, there are 'social' arrangements where there is no common focus and workstations are grouped to maximise the opportunity for team working or social interaction (Figure 9.9(b)). Finally, there are arrangements where a shared, off-workstation overview display dominates the layout and

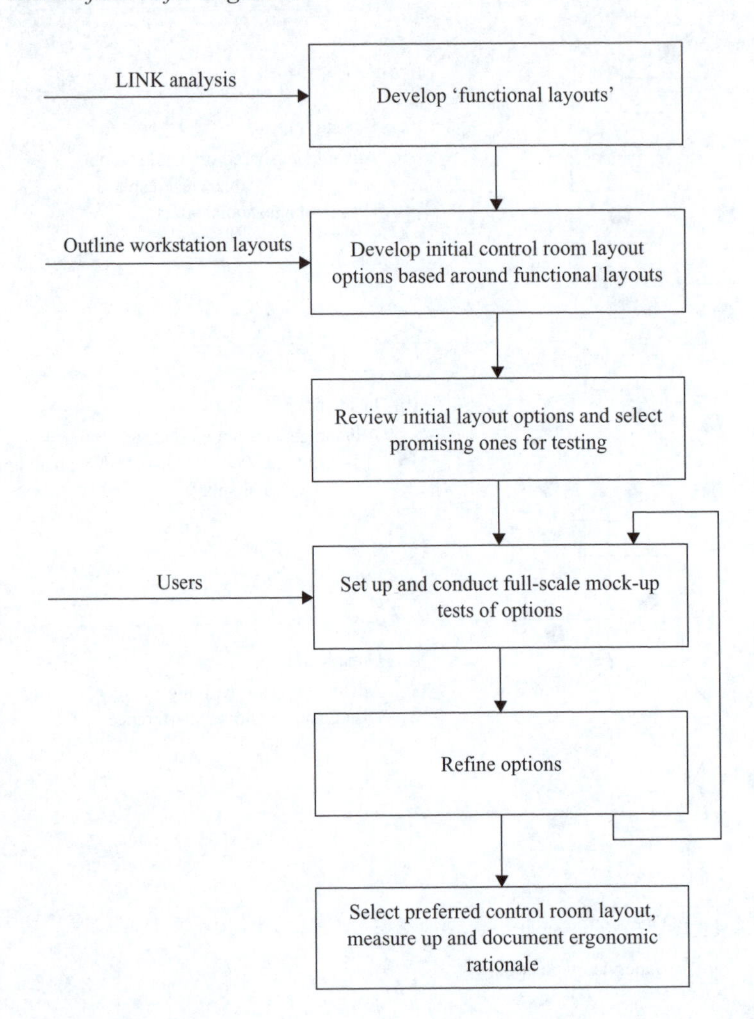

Figure 9.10 Control room layout process

all operators are angled to view this item including the supervisor (Figure 9.9(c)). Sometimes layouts will be a blend of each of these options because there are no 'correct' solutions. A layout should emerge that matches the operational requirements as closely as possible.

The actual process of control room layout is iterative and it is unlikely that a completely satisfactory arrangement will be achieved in the first instance. The flowchart presented in Figure 9.10, based on that in the International Standard, provides a process for control room layout. Using it will provide an audit trail and should minimise the risks of generating unworkable solutions.

The first step in the process is the generation of functional layouts within the available control room space. Functional layouts are akin to architects' 'bubble diagrams'

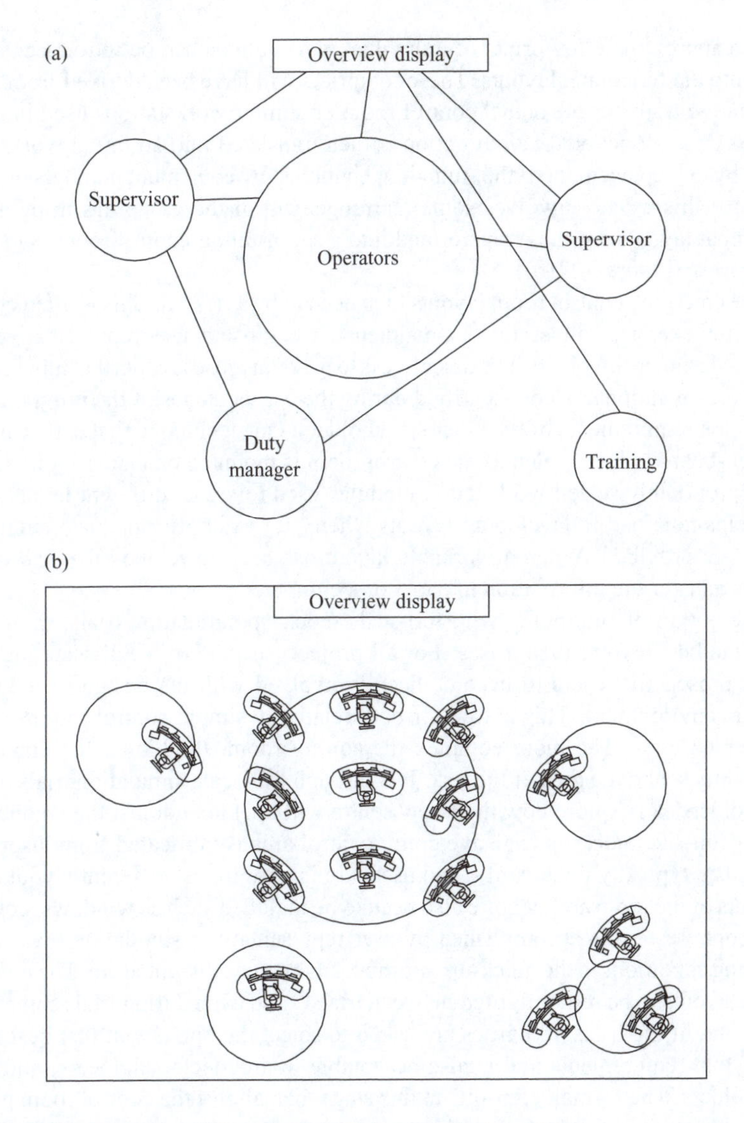

Figure 9.11 Functional layouts to initial control room layouts

where discrete functions are represented by circles roughly corresponding to the numbers of workstations associated with that function (see Figure 9.11(a)). The relationship of the functions to each other should be based on the task and LINK analysis work. For example, groups of operators engaged in similar activities may wish to be close to each other to share information whereas the supervisory function may wish to have links with all functions in the control room. There is unlikely to be a single solution that drops out of this exercise as being overwhelmingly superior and it is more likely that three or four functional layout options will emerge.

The approximate footprints of individual workstations can be introduced at this stage into the functional layouts. These footprints will have been derived from initial workstation trials for the actual control room or similar workstations used in earlier projects. Each functional layout option is then translated into an initial workstation layout by the replacement of the 'functional bubbles' by equivalent numbers of workstations – this usually involves some rearrangement of the groupings in the space. Functional layouts are thus transformed into a set of initial control room layouts by this process (Figure 9.11(b)).

The emerging control room layouts then need to be assessed. This is often done as a table-top exercise with architects, maintenance engineers, user representatives and systems designers involved. The objective is to pick out those layouts that most closely match the operational needs identified during the earlier stages of the programme as well as the requirements of the other stakeholders. One technique that can be used is a 'block-board' where a plan of the control room is mounted on a board, gridded up, and appropriately scaled workstation modules used to create different layouts. The grid helps to assist in developing layouts where, for example, adequate circulation spaces are provided. When a workable layout has been developed the grid can be used to transfer the information to paper or a computer.

The option of continuing with full-scale mock-up simulation trials, in order to select the best layout, then arises. For all projects the use of a full-scale mock-up allows prospective users to become deeply involved with the design of their new working environment. This is true for both relatively simple control rooms as well as complex ones. The more complex the control room, the higher the number of uncertainties, and the greater the risk. It is through full-scale simulation trials that all stakeholders can explore how their new centre will operate, not just the ergonomist.

For full-size mock-up trials the entire control area is fabricated from disposable materials – typically thick card – and users are invited to test different layouts. Key elements of the room architecture also need to be included such as windows, columns and doors. Testing is accomplished by user representatives simulating a variety of operating conditions in the mock-up – normal, emergency, disrupted etc. The mock-up allows layouts to be re-configured between trials – and even during trials sometimes. At the end of the trials all parties are asked to select the one layout that best meets operational requirements and is also acceptable to architects, engineers, and other stakeholders. The formal 'sign-off' at this stage then allows the control room project team to move forward confidently with detail design work. As a detail, where there is an option for specifying the location of external windows, this is the stage of the process at which it should be done. Within the general arrangement of the control room fixed sites can be selected for windows that will give operators a view out without them being a source of glare or unwanted reflections.

Visitor handling is a particular function often neglected in this design stage. Common complaints by control room operators often centre around visitors and their intrusiveness. Notwithstanding specially designed visitor galleries, chief executives tend to take it upon themselves to dispense with such facilities and walk straight onto the operating floor. The control room project team would be wise to establish the policy on visitors – typical numbers, what they should not see, what impression they are to gain from their visit, and so on – and then factor these into the early control room layouts.

There is a soft issue that needs to be looked at which is vital to the ultimate success of the control room. Whilst it is relatively easy to tease out the operational links that need to occur in a control area, defining the 'social' structure that should be created is far more difficult. Effort spent on determining how teams of operators wish to work together – particularly during periods of low demand – and who would like to talk to whom will pay dividends. After all, even if the ergonomics is all spot-on, an environment needs to be created in which individuals choose to work because they genuinely like working there.

The final stage in this layout and mock-up process is the documentation of all ergonomic assumptions made in achieving the layouts, thus allowing for future change to be conducted in an informed way.

9.5 Workstation layout principles and workstation design

Control room tasks often involve time-critical elements and an operators' entire job may be focused on a single workposition housing a variety of equipment. The combination of time-criticality and equipment densities contributes to the challenges faced by the ergonomist developing suitable workstation layouts.

Whilst most workstations in control rooms are predominantly for seated operation there are occasions where both seated and standing operators need to be accommodated and some workstations may even be operated only by standing operators – permit-to-work workpositions can fall into this category.

Simple models of seated posture at a workstation tend to assume that the operator is sitting more or less upright, with their arms stretched out in front of them on a keyboard, the soles of their feet flat on the ground and their thighs parallel with the floor. For anyone who has seen operators reclined back, heels on the workstation leading edge and a newspaper in their lap, this model is clearly inadequate and misleading. It is now recognised that there are a multiplicity of postures adopted by individuals and, indeed, ergonomics advocates that postures are changed at regular intervals – though possibly not to include the 'heel's on workstation' posture!

Obviously the range of postures has an effect on the location of the operators' eyes both in relation to the height off the floor and also the distance from the screens as in Figure 9.12.

Published guidance on general viewing distances to desk-mounted display screens has been largely based on the use of a single VDU, placed more or less directly in front of the user, and usually involves a figure of 500 mm from eye to screen [2]. However, in many control room applications the operator is presented with an array of screens, and viewing distances of between 750 and 1000 mm are more typical. At these distances operators can easily glance over the entire array and, if necessary, move forward to scrutinise more closely an individual display. The greater viewing distances associated with some control room work will obviously have an impact on the minimum recommended sizes of characters and icons on screens. These may

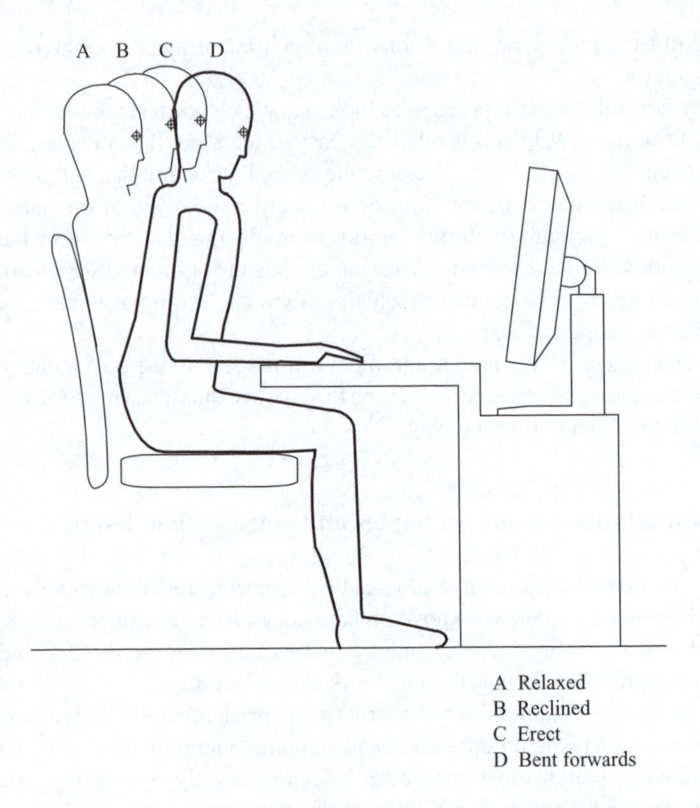

Figure 9.12 Variation in eye position with posture

require that minimum sizes are increased to compensate for the greater viewing distances.

9.5.1 Workstation design process

The flowchart, Figure 9.13, summarises the process whereby workstation layouts can be generated from a task analysis or task synthesis. As with the control room layout process it is likely to involve a number of iterations, though a structured approach should minimise abortive work.

9.5.1.1 Plan and section considerations

Plan arrangements of workstations should allow operators to comfortably scan their array of displays and easily reach controls. In practice there is usually not enough room to place all key displays in a prime position and work-surface constraints limit the location of controls. Where these conflicting demands occur a method of prioritisation needs to be used – frequency of use or mission importance, for example. Log books and reference material can often be placed at arm's length from the operator and do not need to occupy prime space on the work-surface. The assumptions associated

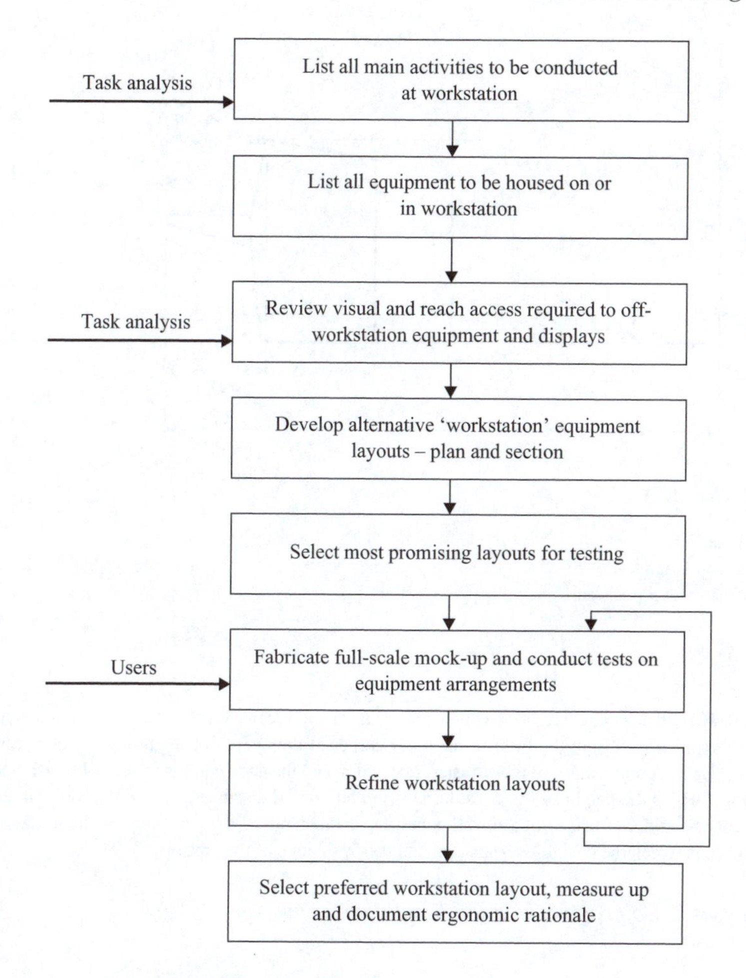

Figure 9.13 Process of workstation layout

with the use of the equipment should be documented before workstation layouts are started – they may well change during the project as new facts emerge but they at least provide a starting point for this activity. Allocation of equipment to primary and secondary visual and reach zones should be based on the priorities established and conflicts resolved by discussion or through workstation mock-up trials. Where no resolution is possible equipment specifications may need to be revisited, though this is not always practical.

Workstation sections tend to fall into two main categories – those where the operator has a view over the equipment in front of them and those where there is no requirement to see over. In either case adequate clearance needs to be allowed under the worktop for the largest operator to be comfortably seated. Obstructions under the work-surface, whether equipment or workstation structure, are to be avoided at all costs.

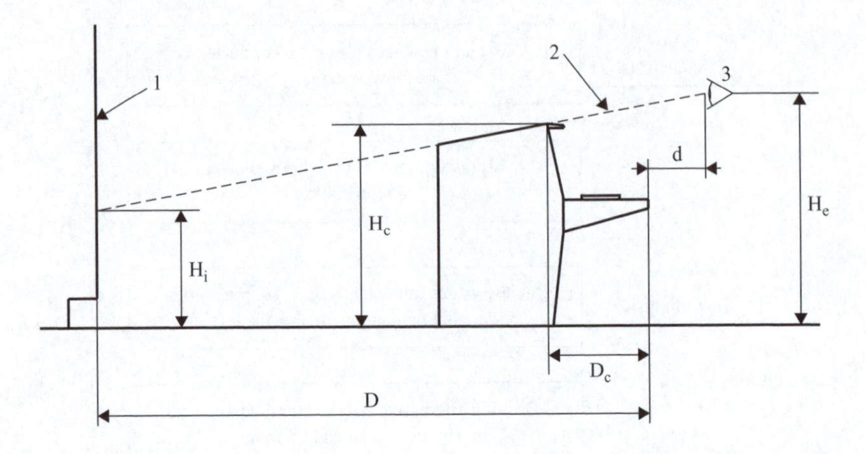

Key

1 Panel
2 Line of sight
3 Design-eye-position

$$H_i = H_e - (D + d)\frac{H_e - H_c}{D_c + d}$$

where H_i is the lowest height at which the visual display can be seen; H_e is the design-eye-position, measured from the floor to the outer corner of the eye; 5th percentile shall be applied (Note: H_a is a combination of the adjusted seat height and the anthropometric data of 'eye height, sitting'); H_c is the height of the console; D is the horizontal distance between the front edge of the console and the surface of the wall panel; D_c is the depth of the console; d is the horizontal distance between the design-eye-position and the front edge of the console.

Figure 9.14 Control workstation height and view over

The dimensions for maximum heights of view-over workstations should be dictated by the seated eye-height of the shortest operator. It is also wise to factor in a degree of 'slump' when considering the maximum overall heights of equipment, as shown in Figure 9.14.

View over equipment arrays may not be just for operators to be able to see overview displays or to see each other, it may also play a vital role in supervision. Supervisors often supplement the information that they get from the system by looking around a control room to see how individuals are coping. They look for hunched shoulders and animated expressions as indications of potential problems. They like to be able to see operators' mouths to see whether they are talking before pressing the intercom button to offer to help. In workstation profile terms this means more than just seeing the top of the head down to the eyes – to benefit from these visual cues supervisors will need to see the entire head and the shoulders. Where the number of display screens demands that they are stacked in front of an operator, a double bank may be a solution, for a view-over arrangement, with the lower array sunk below the work-surface. However, this may not provide an ideal solution in all cases since it imposes constraints on operator posture and viewing distances are further increased with an impact on HCI design.

The workstation maintenance engineer should be treated as a user in the same way as the operator seated at the workstation. Checking for adequate space for removal of equipment, appropriate access for in-situ checks and that equipment can be safely and comfortably lifted is most important. In some cases it may not be practical to achieve safe lifting due to the sheer bulk of the equipment and for these cases alternatives may need to be sought such as the use of individual trolleys for large items of equipment.

9.5.1.2 Mock-up trials

Full-scale mock-up trials allow for alternative workstation configurations, developed on paper, to be tested in a meaningful way with operators. Typically this will involve the fabrication of the major items of equipment in card that can be supported on card bases of a suitable height. These space models can be enhanced with photographs presenting a typical interface arrangement. Subjects are asked to assess workstation options by simulating the main tasks they are likely to undertake at the workstation – both under normal and abnormal conditions. Crucially other items can be introduced into these trials including paper-based job aids – such as log books or permit-to-work clipboards – and the layout checked for its ability to accommodate these. In a similar vein, if refreshments are allowed at the workstation safe places must be found for drinks so that the likelihood of spillage over keyboards or other equipment is minimised. Should viewing distances, or equipment juxtaposition, appear to be close to critical ergonomic thresholds it may be advisable to introduce some real equipment. Substituting real equipment for card mock-ups allows for more precise judgements to be exercised by subjects.

Mock-up trials provide a medium through which layouts can be refined such that they marry up as closely as practical to operational demands. The most important factor in any set of trials is the experience that the operators bring to the trials. The quality of the trials' output is largely dependent on the operators' understanding of the demands that will be placed on the workstation user and anticipating how they will cope. Subjects should be selected for their knowledge about operations and also reflect the age and gender of the predicted user population, and this should include those with physical disabilities where such individuals are expected to be part of the operational team. Mock-up trials should also involve maintenance engineers, systems suppliers and building services specialists. At the end of the trials, an agreed layout should have been achieved which, in a similar way to the full-scale room trials, can be signed off by all parties.

9.5.2 General design considerations

Careful attention to the detailing of workstations can also make its own contribution to a satisfactory ergonomic solution. Visible finishes should relate to the visual tasks – for example, work-surfaces are best in mid-tones if paperwork needs to be read, and paint finishes should be selected such that they minimise local task lighting being reflected onto screens.

The rear of workstations can be made more interesting by the use of more highly saturated colours and patterns, though one needs to remember that the back of one workstation may be part of the peripheral visual field to another operator. Areas around displays benefit from mid-tones although the inside of service areas are best in a lighter tone so as to improve the dispersion of light and hence increase visibility for the engineer. Generous radii on worktops and other protruding surfaces will minimise the likelihood of injury.

9.6 Control room interface design (HCI)

For safety-critical control room environments a poorly designed interface can have serious consequences – aircraft heights misread, valves in process plants left open, a garbled radio message that is unreadable. Since the general background aspects of HCI design are discussed in detail in Chapter 10 this section dwells only on those elements specific to control rooms.

9.6.1 Principles for control room HCI design

What follows is a set of high-level principles considered to be essential in the satisfactory design of the control room HCI.

- The human operator must at all times be the highest authority in the human–machine system.
- The operator has to be provided with all the information s/he needs to accomplish the task.
- The user must be able to maintain a comprehensive and robust mental model of the system and its associated sub-systems.
- The system should not ask the operator to input data that is already available within it.
- Information should be presented according to known ergonomic principles so as to ensure that it is conveyed quickly and accurately to users.
- Objects that are to be recognised at various levels of display hierarchy should be designed so they are clearly legible and recognisable at all levels of magnification or zoom.
- Only valid information in terms of time, origin and appropriate resolution should be displayed and where this is not practical it should be indicated (e.g. time of last measurement).
- The information presented by the system should be offered in such a way that it can be easily and unambiguously understood without any additional help.
- The level of attention getting – applied to a particular item of information – should be matched to the importance of that information to the operator and the safety of the system.

- Operators should not be expected to 'fly blind'. Controlled objects should be displayed whenever possible. (Note that this may not be possible with certain systems such as dispatch of resources by radio.)
- Interactions with the system should be kept simple and complex exchanges kept to a minimum.
- The system should aid the operator in inputting information efficiently, correctly, and with a minimum risk of errors. For frequent control actions, short cuts should be made available to the operator.
- Appropriate feedback must be provided to the operators at all times.

Compliance with these fundamental principles should always be checked during the design process.

9.6.2 Overview displays

For many traditional industries an 'overview' display used to be some form of mimic of the area or system being controlled. Whether it was electricity distribution, rail network signalling, or emergency telephones on the motorway network, information was presented on large wall-mounted panels with the most important variable elements being self-illuminated lamps. These large displays offered an entire overview of the network at a glance but were expensive to fabricate and needed labour intensive modification whenever the infrastructure changed.

With the replacement of these systems with ones that do not require such mimics there are strong financial arguments to provide the same information from desk monitors only, thus doing away with the wall-mounted mimic and its associated costs and space requirements. Where they are retained the ergonomic challenge is to determine the most effective distribution of information between shared overviews and desk-based displays. The basis for any design should be the preparation of a functional specification in which the information presentation requirements are made explicit, content priorities established and coding protocols agreed.

The financial savings from scrapping overview mimics need to be balanced against known ergonomic benefits. Overview displays can provide immediate status information which can be shared by all without the need to take control actions or searching. Supervisors often see an overview as an advantage, particularly when away from their own workstations. In addition they do have a public relations function in relation to visitors and the media.

9.6.2.1 Ergonomics of CCTV displays

Declining costs of CCTV equipment, coupled with improvements in picture transmission, have resulted in ever larger schemes. Whereas 20 years ago the use of CCTV was relatively limited it is now to be found offshore, in nuclear processing plants, controlling motorways, as an essential part of security and surveillance systems and in city centre monitoring schemes.

As with any other system the most appropriate CCTV interface can only be specified if the actual CCTV task is understood. Table 9.2 presents a framework that has been developed for the classification of CCTV-based tasks [3].

Table 9.2 Task classification for CCTV pictures

Task type	Example	Task pacing	Key HF issues
Respond to complex deviations in an essentially static background picture	X-ray machine	Usually self-paced	Absolute size of targets. Search pattern.
Respond to change of state within complex and changing background picture	Crowd control at football matches. Railway platform crowding. Prison yard.	Paced	Degree of change. Percentage of image affected. Time displayed.
Respond to simple changes of state in an essentially static background picture	Traffic density on motorways	Paced	Percentage of image affected. Time displayed.

The levels of performance associated with the CCTV observation tasks need to be clearly understood before an appropriate ergonomic solution can be determined. So, for example, whilst 100 per cent detection might be the performance required from an X-ray baggage inspector, a lower threshold may well be acceptable for detecting the build-up of traffic on a motorway. These performance requirements will have a direct impact on the most appropriate ergonomic solution including working practices – for example, 100 per cent detection requirements in X-raying baggage results in work cycles limited to 20 minutes.

Some of the ergonomic factors to be considered by the designer in presenting CCTV information to operators are summarised in Table 9.3.

9.7 Environmental design of control rooms

Environmental design of control rooms is about making sure that the acoustics, lighting, control room finishes and thermal conditions support the jobs that operators are carrying out. Decisions about these factors should be task related, though aesthetic considerations have a role to play in an ergonomic solution. This section provides some guidance, though there has been some debate in the international ergonomics community about how prescriptive such guidance should be. It is argued that, for example, control room temperature in hot countries should be higher than in temperate zones – others argue that with air-conditioning such subtle differences are irrelevant. The figures presented are ones generally considered suitable for control rooms though they may need to be refined from case to case. Further information can be found in Reference 4.

Table 9.3 CCTV picture presentation and ergonomic considerations

CCTV system characteristic	Ergonomic issues
Number of monitors	Detection rates fall once operators need to view more than one monitor. Assume that scores of less than 60% detection will be the norm if more than six monitors are observed.
Size of targets	General rule of thumb is that targets need to be a certain proportion of overall screen size for the following tasks to be undertaken: Monitor – 5% Detect – 10% Recognize – 50% Identify – 120%
Picture presentation	For movement detection avoid banks of 'autocycling' pictures. Splitting a large monitor into tiny 'mini' pictures is normally useless for monitoring and detection tasks.

9.7.1 General principles

The following high-level principles offer an overall framework within which the design of the control room environment should be executed.

- The needs of the control centre operator should be paramount. (It is easy for the heating and ventilation engineers to focus on the needs of the equipment rather than the operators.)
- Levels of illumination and temperature should be adjustable so that operators can optimise their performance and comfort.
- Where conflicting demands exist a balance shall be sought that favours operational needs.
- Environmental factors invariably work in combination and should therefore be considered in a holistic way (e.g. interaction between air conditioning systems and the acoustic environment).
- Environmental design should be used to counterbalance the detrimental effects of shift work (e.g. use of temperature increases to compensate for diurnal rhythms).
- The design of environmental systems should take account of future change (e.g. equipment, workstation layouts, and work organisation).

9.7.1.1 Lighting

Visual tasks are the cornerstone of most control room operations, whether scanning arrays of desk-mounted electronic displays or wall displays or looking at paperwork. Some of these will be time, or safety, critical and the control room designer needs to ensure that the quantity and quality of lighting provided supports these activities.

Table 9.4 Lighting parameters and control room design

Key parameters	Suggested values	Comments
Work-surface illumination for paper-based tasks	200–750 lux	Dimming to minimum of 200 lux
Work-surface illumination where VDUs used	Upper limit of 500 lux	Dimming to minimum of 200 lux
Glare index	UGR of 19 or less	
Contrast ratio between self-illuminated equipment and immediate surround	10 : 1 or less	

Some of the key parameters concerning the design of control room lighting are presented in Table 9.4 with some suggested values. More detailed information can be found in BS 8206-2 [5].

Lighting arrangements, whether natural, artificial or a combination, should be appropriate to the visual demands of the task and take account of normal and emergency working – the latter may involve a switch from a VDU-based task to a paper-based replacement. Local lighting of an operating position may be necessary where, for example, an operator's duty involves the use of self-illuminated equipment. The age of the operators also needs to be taken into account – the 50 year old supervisor will require substantially more light for the same paper-based task as the 20 year old controller he is supervising.

Glare can reduce visual abilities as well as being a source of discomfort; the disabling impact of glare can be seen when looking at a window during daytime and attempting to see any detail of the window frame. Inappropriately placed lights over a large wall mimic, or locating a mimic next to a window, can have a similar effect on the ability to distinguish details on the mimic surface. Lighting schemes for control rooms should avoid any form of glare and should minimise reflected glare off screens. Where the requirements point to the provision of task lighting particular care should be taken that the light provided for the local user is not a source of glare to others in the control room.

In the past, artificial lighting for areas where VDUs are used has been based around downlights, where light shining either directly into an operators' eyes or on to display screens has been controlled by louvres. This form of illumination could achieve its original design objectives, provided desks were not subsequently moved around underneath the carefully fixed light fittings. It has a tendency to produce high-contrast pools of light and dark areas. Alternative schemes using indirect lighting (where the light source is shone upwards onto a white ceiling) have been found to minimise reflections on screens, and allow for generally higher levels of ambient lighting and an improved visual environment. Most recent control room schemes have been based on a combination of direct and indirect lighting at a ratio of 20 : 80 per cent, which allows for some modelling of the visual environment. Counteracting the rather bland environment created by an indirect lighting scheme is discussed further in the section on interior design.

Windows in control rooms can be present for two reasons – operational and psychological. Where they are used for operational needs, such as in rail shunting yards or control towers on airfields, the information they provide should be analysed in the same way as their electronic counterparts. For most situations windows are not required for operational reasons but should be included for psychological reasons. The natural reference provided by knowing whether it is day or night, raining or sunny, overcomes the disorientation felt by many who work in windowless environments.

Windows, and their immediate surroundings, need to be treated to reduce glare by the use of filters or blinds. Where they are sources of information the impact of these controlling measures will need to be considered, e.g. loss of scene brightness caused by the use of neutral filters.

9.7.1.2 Thermal environment

What is considered to be a comfortable thermal environment by an individual is influenced by such things as the amount of clothing worn, the time of the day and the degree of physical activity. A glance around a typical office highlights the problems when trying to find an acceptable norm – some individuals will have achieved comfort with rolled-up shirt sleeves, others will have pullovers over shirts and a few may find comfort is only achievable wearing a jacket. The 24 hour operation of most control rooms raises a further complication of diurnal rhythms where body temperature is known to dip at certain stages of the cycle making the individual feel cold and shivery.

Unlike lighting, which can be controlled in zones, the thermal environment is a 'whole room' factor with only limited control by groups or individual operators. In some installations small vents have been incorporated into the workstation to allow individual operators to control air movement in their immediate surroundings.

Wherever possible control rooms should be air-conditioned with the staff having some element of control over the temperature levels. The air conditioning should also be programmed to raise the temperature by a couple of degrees in the early morning to compensate for the lowering of the body temperature of the occupants.

Air movement, air quality, mean radiant temperatures and relative humidity are other factors that have an impact on operator comfort. Some of the key parameters that need to be addressed in the design of a control room are presented in Table 9.5 together with some suggested values that have been used in UK control rooms.

9.7.1.3 Auditory environment

The importance of voice communications, and the ability to distinguish auditory alarms, imposes special requirements on the design of the auditory environment of many control rooms. A careful balance needs to be struck between damping everything down – to the point of creating an echoing chamber – to allowing a level of acoustic 'liveliness' that results in a 'cocktail party' effect occurring. Supervisors,

Table 9.5 Thermal environment parameters and control room design

Key parameters	Suggested values	Comments
Winter room temperatures	20–22°C	These are suggested for a temperate country
Summer room temperatures	21–23°C	These are suggested for a temperate country
Vertical air temp difference between head and ankle	Less than 3°C	
Mean air velocity	Less that 0.15 m/s	
Relative humidity	30 to 70%	
Air quality	10 litres per second per person	From an outdoor air supply For non-smoking environment
Carbon dioxide concentration	Not to exceed 1.8 g/m^3 when room at full occupancy	

for example, may find it helpful to 'overhear' conversations – not necessarily every word – but not to be drowned in a cacophony of sound because of a lack of auditory control. Also, operators within a functional group may want to hear each other but not an adjacent groups' discussions. Striking the appropriate balance should always be dictated by the underlying task demands of all the operators in the control room.

Acoustic considerations may also play a part in selecting the most appropriate position of the control area within a control suite or even on a site. External noise sources such as vehicles, noisy process units, and car parks are all potentially intrusive and are best avoided rather than being controlled acoustically by costly special glazing or wall treatments.

Some of the key parameters associated with the design of the acoustic environment are presented in Table 9.6 together with some suggested values. It is worth stressing the interdependence of the different parameters because this makes it highly inadvisable to simply select from the ranges independently.

9.7.2 Interior design considerations

9.7.2.1 General interior design approach

The interior design of a control room should support its operational requirements. Natural and artificial lighting, colours, materials and finishes, for example, need to be carefully specified within ergonomics parameters to allow visual, auditory and cognitive tasks to be performed efficiently and effectively.

The line between a bland environment that meets the performance requirements and a visually interesting interior that does not detract from communications and control is a fine one. There is room for choice of materials, colours and textures and

Table 9.6 Acoustic environment and control room design

Key parameters	Suggested values	Comments
Preferred background noise level range	30 dB to 35 dB L_{Aeq}	Maintains a degree of aural privacy
Maximum ambient noise level	45 dB L_{Aeq}	
Minimum ambient noise level	30 dB L_{Aeq}	
Auditory alarms	10 dB above background levels	Should be less that 15 dB above background to avoid startling and interference with speech communication
Reverberation	Preferred level 0.4 sec. Not more than 0.75 sec.	Dependent on the size of the room For mid frequency reverberation times

the general approach should be to provide a pleasant working environment which is a calming backdrop for operational activities.

9.7.2.2 User participation in interior design

Although the various elements of an interior design scheme, such as wall finishes, carpets, chair fabrics, luminaires and window blinds should be carefully selected by professional designers, there is scope for control room users to be involved in the choice of colours and finishes. Sample boards of design scheme options should be prepared and presented to both managers and users for comment. These should, however, only present options that are ergonomically acceptable.

9.7.2.3 Aesthetics and image

The choice of finishes for a control room should support the image that senior management wishes to convey, both to their own staff and to visitors. The aesthetics of the control room should to some degree reflect the function of the industry or organisation of which it is part – the design for a secure military environment will be approached somewhat differently to that for a highly public centre for, say, national travel information.

9.7.2.4 Ergonomics parameters and colour schemes

The ergonomics process will typically specify reflectance values for walls, floors and ceilings. Within these tonal limits there is scope for choice. An overall 'warm' colour scheme of beiges and yellows is likely to meet the ergonomics specification as is a 'cool' scheme of greens and blues.

9.7.2.5 Space provision and aesthetics

Minimum space requirements have been reviewed earlier. There will be, however, some choice over and above these requirements which relates to the 'feel' of the overall space. These include room height in relation to plan size, spacing between individuals and distances from windows that will contribute to a feeling of comfort. It is important that the design scheme does not neglect to take account of the quality of space as well as basic space provision.

9.7.2.6 Workstations, furniture and architectural details

Furniture and room details, such as door frames and skirtings, can be finished in timber to soften and humanise the control room environment. A figured grain will add character and appeal, and the use of timber for furniture will be warmer to the touch than steel or plastic. Timber species should be selected for reasons of density and durability and finishes should be matte.

9.7.2.7 Seating fabrics

Whilst the selection of control room seating should be based on performance criteria the choice of chair fabric provides an opportunity to use colours of stronger chroma than those used for other objects in the primary visual field of operators.

The choice of colour and fabric type should also take account of 24 hour usage and associated wear and soiling. Darker chair fabrics with a 'fleck', rather than a plain colour, are more successful in the longer term.

9.7.2.8 Wall finishes

Wall finishes should have a surface reflectance value of between 0.50 and 0.60 and have colour of a weak chroma. Within these limits designers are able to select from a large range of products, especially vinyl coverings, with a great diversity of colour, pattern and texture. Textured finishes help to reduce reflective glare, but also add a quality and warmth to an interior scheme that is not possible through the use of flat paint finishes alone.

9.7.2.9 Skirtings

These provide protection to the base of walls, but by using natural hardwoods some quality and warmth can be added to the environment.

9.7.2.10 Carpets

Carpets will generally contribute to the control rooms' reverberation times, and should be of a heavy contract grade. The recent trend for replacing hard control room floors, such as vinyls and linoleum, with carpet tiles has greatly improved the quality of the visual environment as well as improved acoustic control.

Carpets should be chosen that will remain good in appearance in the long term and a small random pattern with subtle colour variation should meet both practical and visual interest requirements. Larger patterns, and geometric designs, are not appropriate for control rooms and should be avoided.

9.7.2.11 Interior refurbishments in older buildings

Many control rooms are established in older buildings that may contain aesthetically pleasing architectural details.

The designers should consider retaining original elements such as windows, ceiling covings, doors and joinery details within the overall interior scheme, providing there is no detrimental effect on operations.

9.7.2.12 Daylight

As previously discussed, people like to have a view of the outside world, even if that view is not a particularly pleasant one! The external view allows the eyes to relax, particularly when tasks involve scrutinising detail at close distances. Daylight also tends to make control rooms feel brighter and more pleasant places to work.

9.7.2.13 Plants

As well as providing visual rest areas for control room users, the irregular form and variation in the texture of plants, and their colour, provide a pleasant visual contrast to the geometry of workstations, cupboards, whiteboards, ceiling grids and lighting schemes. For completely windowless control rooms the appropriate use of plants will play an even more important role in overcoming a monotonous and 'lifeless' environment.

9.8 References

1 BS EN ISO 11064: 2001. 'Ergonomic design of control centres (parts 1–3)'
2 BS EN ISO 9241 (1992–2000) 'Ergonomic requirements for office work with visual display terminals (VDTs) (parts 1–14)'
3 WOOD, J.: 'The ergonomics of closed circuit television systems'. Proceedings of the 13th Triennial Congress of the International Ergonomics Association, Tampere, Finland, 1997
4 WALLACE, E., DIFFLEY, D., BAINES, E., and ALDRIDGE, J.: 'Ergonomic design considerations for public area CCTV safety and security applications'. Proceedings of the 13th Triennial Congress of the International Ergonomics Association, Tampere, Finland, 1997
5 BS 8206-2: 1992. 'Lighting for buildings. Code of practice for daylighting'

Chapter 10

Human–computer interfaces: a principled approach to design

Robert D. Macredie and Jane Coughlan

10.1 Introduction

The purpose of this chapter is to delineate the many facets of developing interfaces in order to provide fruitful design directions for practitioners to take in the area. To this end, the chapter is divided into four distinct sections. The first, Section 10.2, provides a brief history of interface design along with defining its scope and why it is important, especially in light of the continual problems in usefulness and ease of use of many interfaces that are designed today. Based on this understanding, Section 10.3 categorises the area according to the broad treatment of three key topic areas in interface design, covering human factors issues (see Chapter 1 for formal definitions), interface types and design principles that are important considerations in the development of interfaces. Practical applications of the knowledge gained from the discussions thus far will be demonstrated in Section 10.4, through two interface design cases. These cases are analysed with consideration to the issues drawn out from previous sections, thus spotlighting the various pitfalls and benefits to design. Section 10.5 concludes the chapter with a final reflection on the nature and inherent difficulties in producing effective interfaces.

10.2 Interfaces: a story about design

In this section, the topic of interface design is introduced in an explanation of its background, definition, user requirements and models before an outline of the later sections of the chapter.

10.2.1 Background to interface design

The concept of the human–computer interface emerged from research in the field of Human–Computer Interaction (HCI). Despite numerous definitions, the field can be said to be formally concerned with the design of computer systems that support people in the safe conduct of their activities (see [1]), where productivity is increased if the system is useful and usable to all [2]. Recently, Faulkner described HCI as 'the study of the relationships that exist between human users and the computer systems that they use in the performance of their various tasks' ([3], p. 1). This places more of an emphasis upon the dynamic interactions that take place between the user and computer, which is mediated by the human–computer interface, and therefore portrayed throughout this chapter as a tool in communication (see [4] for a discussion on these points related to safety).

In light of this, many different disciplines have been brought to bear within HCI, namely computer science, psychology (cognitive, social and organisational) ergonomics, sociology and anthropology. On one hand this serves to provide a valuable analysis of the field in a multitude of ways, but on the other can also point to the difficulties of multi-disciplinary working in the co-ordination of these different perspectives (see [5]). Indeed, creating effective interfaces is a highly challenging and complex task [6, 7]. Many reasons exist to account for this inherent difficulty, not least a wish to understand the tasks and the users in the environment in which the system will be placed, so that users can interact with it in such a way that it will be deemed 'friendly' enough to be accepted and ultimately successful in terms of its usability. In fact, in more recent times, there has been a call to change the focus of HCI from human–computer to human–system interface (HSI – see [8]), or user–system integration (USI – [9]). The commonality between these two approaches is the focus on the system, which in essence is concerned with the wider issues regarding system design, rather than just the straightforward one-to-one interaction between a single human and a computer. Given this call for an extension in the commonly held view of the interface, the scope of the interface as addressed in this chapter is described next.

10.2.2 The interface defined

Physically, the interface is the visible part of the computer system with which users have first contact. The interface includes such things as the keyboard, mouse or joystick, the display screen – menus and windows – that is seen, heard (e.g., 'beeps'), touched or pointed to. In addition, accompanying documentation (e.g., manuals) can also be seen as part of the interface [10]. However, the scope of the interface is seen to stretch beyond physical boundaries with inclusion of the people responsible for shaping interactions with the interface such as management [11]. Altogether, the interface provides the necessary information channels for the user and computer to communicate [12]. But an important question is *how* this communication actually occurs, because what have been described so far are the basic elements of an interface and it is clear that there are other aspects equally as important that require co-ordination within design. These aspects are related to the context into which the system will be placed, such as the organisation, which is highly social in nature. For instance,

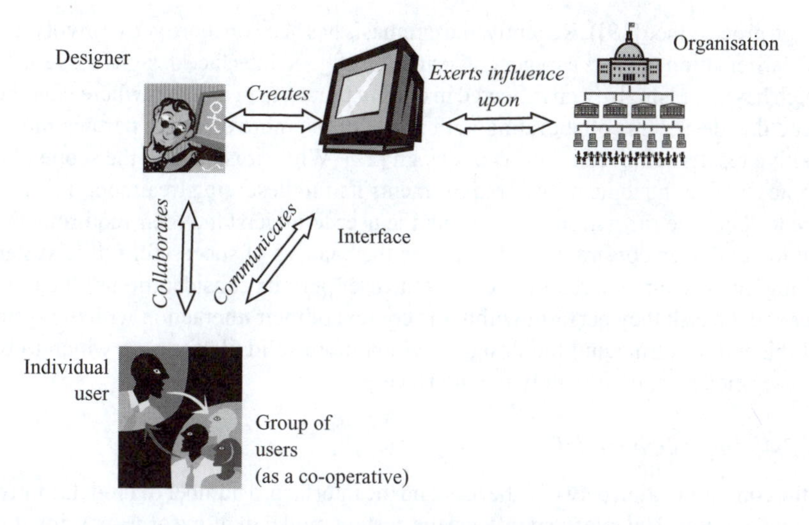

Figure 10.1 Relationship between the user, interface and organisation

work within organisations tends to be co-operative with groups of people interacting together and with the technology. This interaction can sometimes be determined by social constructs, such as the culture of the organisation, which has been shown to have a bearing on design, so much so that Marcus [13] has proposed that it would be useful for user interface developers to be skilled at cultural analysis, and therefore attentive to some crucial determinants of design.

Essentially then the interface stands as the face or even 'personality' of the system. It is also a symbol of organisational policy, which, as has been asserted, constitutes the culture and politics and also the co-operative nature of work groups. As Ackerman [14] indicates, computer systems often fail to support this aspect (see Section 10.4 for design cases that make explicit reference to the background influences within which the system is to be embedded). Design input is charged with the responsibility of translating user needs and organisational requirements, which culminates in the interface. This relationship is shown in Figure 10.1.

10.2.3 User requirements

Interface design practices are generally based on user models, as Berrais [15] has pointed out (see Section 10.2.4). However, it is also the case that the user requirements for an interface are equally influential though unclear from the beginning of the development process. Moreover, there is a gap between the analysis phase in which the requirements are gathered and the design of the end product [16]. This means that the interface style (see Section 10.3.2), for example, is determined by the designer rather than the user, a factor that will be demonstrated at the arguably late stage of usability evaluation [17]. Wood [18] has paid particular attention to this gap between front and back-end activities and highlights a number of methods in which this process can be improved, all of which involve a focus on the user as part of a user-centred design process (see [19]). Recently, the emphasis has been on more active involvement and participation of users as a general principle of good interface design (see [20, 21]), which has increasingly been reflected in current practices in industry where it has been found that designers collaborating with users form a more effective partnership with positive results achieved in terms of design [22]. While it is beyond the scope of this chapter to give the topic of user requirements its full deserving treatment, it is raised here to reinforce the notion that helping the user to understand their requirements in relation to design constraints will increase the chances of successful future systems. Taking into account the requirements of the user permits consideration of the type of user and the task they perform within the context of their interaction with the system. All this will in turn equip the designer with a more solid platform on which to base the user models and ultimately the interface.

10.2.4 Interface models

In the communication between the user and the interface, a number of models motivate that interaction. The interface reflects the system model in terms of its possible functionality and the user, in turn, forms what is often termed a mental model (e.g., [23]) as to how the application works, through for example experience, knowledge, needs, etc., and this will form the basis of future interactions with the system. Immediately, then, it is clear how important it is for the design of the interface to be able to help users (of all levels of expertise) to gain an accurate reflection of the computer system, through an appropriate mental model (see [24]). Therefore, the onus is on the interface designer to understand the user and the task being performed so that all parties can understand the system model. In many ways the interface designer acts as an intermediary between the user and the programmer with his own conceptual model that will create a system that can be adapted to user needs, as Mandel [10] has pointed out (although, in certain projects, the programmer and designer may be one and the same person). The designer's model has been traditionally described as a tripartite iceberg structure (see [25]) shown in Figure 10.2.

As can be seen from Figure 10.2, the most important part to the iceberg structure is the third (and largely hidden) layer of the user model, which represents the efforts of the user's matching mental models to the tasks they are trying to achieve, most commonly by the use of metaphors (e.g., desktop) that aim to induce a high level of understanding in the user and aid use of the system. However, as explained in Section 10.2.2, there is more to a system than just its functional appearance in terms of look and feel. The environment in which it will be placed has a strong impact on the design, and its dynamic and mutable nature means that the system (as based on models) will have to manage a number of contingencies, that is individual differences in user requirements and interactions with the system, in order to perform to its specification in real working life.

10.2.5 Scope of the chapter

To explore the significant issues of interface design further the chapter is additionally divided into three clear sections. The first is presented in Section 10.3, and explains

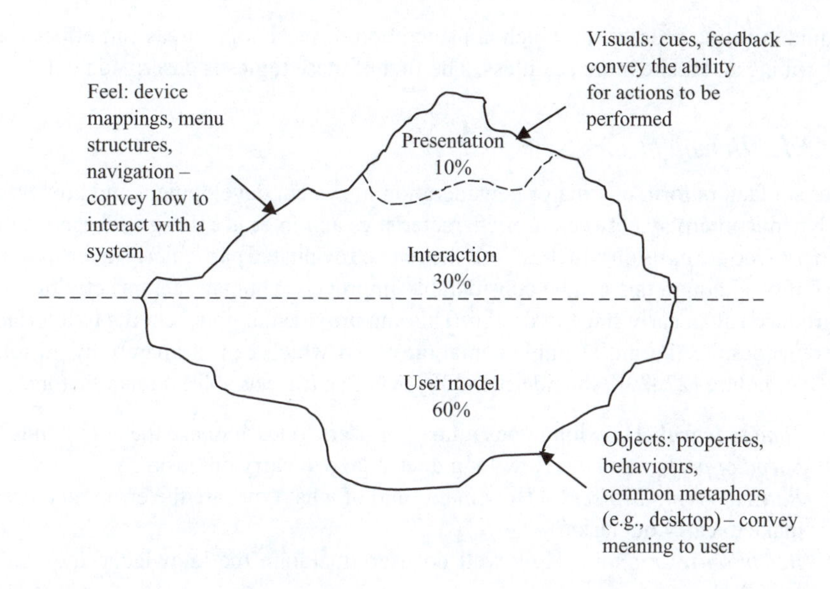

Figure 10.2 Aspects of the user interface (adapted from [25])

in some depth the significance of key topic areas to the design of interfaces. The second, Section 10.4, 'actions' the issues that have been outlined by presenting two design cases that provide an explicit illustration of different interface examples and how problems in design can be overcome with attention to the key areas that have been previously outlined. The third and final part (Section 10.5) consolidates the knowledge gained with reflections on the discussion points that have been raised in relation to successful interface design.

10.3 Key topic areas

In order to expand upon the definition of the interface presented in Section 10.2.2, different facets of the design of interfaces lead to inclusion of the following (not mutually exclusive) topic areas: (1) human factors; (2) interface types; and (3) principles of interface design (usability). Consequently, in all topics, the features specifically relevant to the design of interfaces will be highlighted with emphasis resting on the principles for design, which will be taken up again in a practical application of two design cases (see Section 10.4). Nevertheless, categorisation of the interface design arena in this way, with the topic of interface styles sandwiched between human factors and design principles, reinforces the prevailing (socially and communication oriented) view of this chapter. The interface should be regarded as representing more than just a boundary point between the user and the system (see also Figure 10.1). As Laurel [26] has suggested, the interface can be seen as a *shared context for action*, whereby the user engages with the system as part of a highly interactive and heavily communicative process, to which consideration of these topic areas can effectively contribute and make more seamless. The first of these topics is described next.

10.3.1 Human factors

Human factors form one major consideration in system development and constitute a dynamic interplay between user characteristics and interface design. These factors can provide the basis on which an interface can be evaluated particularly in terms of its usability, although this can be considerably improved if human factors activities are introduced at an early stage in design. This can provide tangible benefits to interface design (despite the initial implementation costs) which can ultimately avoid total system failure [27–29]. Shneiderman [30] lists five (measurable) human factors:

1. *Time to learn* – How long does it take for users to learn to use the commands?
2. *Speed of performance* – How long does it take to carry out tasks?
3. *Rates of errors by users* – How many, and of what type, are the errors that users make to carry out tasks?
4. *Retention over time* – How well do users maintain the knowledge they have learnt?
5. *Subjective satisfaction* – How much do users like using the system and its look?

As Shneiderman [30] acknowledges, the importance of these human factors varies between different types of system. For example, in safety-critical systems (e.g., air traffic control), the avoidance of hazards is of great concern and will need to be strictly maintained to ensure safety at all times. In industrial/commercial systems (e.g., banking) the speed of performance requires extra attention in order to deal effectively with the sheer volume of customer transactions that require processing. The way that these human factors are successfully incorporated into design is reflected in part by the type of interface and the features of use.

10.3.2 Interface types

In terms of type, the interface has evolved considerably over time. The interface type has a bearing on the interaction style with the user, although it is more common nowadays to combine a number of styles for users to meet their goals (see [31]). In brief, the different interface types range from the early linguistic command line interface to iconic ones involving direct manipulation. As the name implies command line interfaces are where command lines are input to the system as a way of exchanging responses between the machine and user, though this is generally suited to the expert user with knowledge of such commands. Direct manipulation interfaces are ones where graphical icons in the user interface can be moved around by the user, typically known as a graphical user interface (GUI). In between these broad interface types are styles that involve the selection of menus (e.g., often arranged in a hierarchy in a series of drop-down menus) and form filling (e.g., inputting information). In some cases, natural language and gesture interfaces have been attempted, although many problems in voice recognition and gesture analysis have been experienced (see [32]).

Design developments are even moving towards interfaces that can trade in emotions (e.g., [33]) – when considering that user frustration with today's technology is very common, such an interface style might be an interesting development. There are, of course, many advantages and disadvantages to designing different types of interface. The main problems arise in the way that different interface styles support the mental models of key players in the design process, namely the user, programmer and the designer. For example, there is some evidence that a direct manipulation interface induces more positive perceptions among users relating to ease of use than a command-based interface [34]. Usability problems with interface types such as menus can be enhanced with sound to alert the user to errors [35], or totally innovated by the use of 'dynamic lists', which allow the user the flexibility of customising list items in the menu [36]. The way the interface style is manipulated will determine the quality of the interaction that occurs between the user and the system. Certain interface styles could, for example, afford an easier level of learning for the user so that the system will be mastered much sooner (see [37]). This is just one of many considerations that will need to be balanced against other, often competing factors, which may be significant for each individual design.

10.3.3 Interface design principles

Numerous interface design principles have been advanced for the development of user interfaces (see Smith and Mosier [38] for a vast and comprehensive compilation). The value of these principles lies in the way that they bring different issues into a collective whole, but need not be necessarily applied rigidly. Indeed, Grudin [39] has argued against adherence to the principle of consistency for certain designs, which is commonly heralded as a standard for interface design. Rather, the principles should be evaluated as part of a general design philosophy to assist designers to understand and question, but most importantly consider their design in relation to the users and the context of use. The design principles presented below are taken from Norman [23] who offers a springboard on which a more user/human factors oriented design approach can be based to improve the usability of interfaces. Therefore seven key principles for screen design layout can be framed as follows.

10.3.3.1 Constraints

Constraints refer to making use of object properties to tell the users what they can and cannot do. Physical constraints are often used in the design of artefacts within the real world to limit the way in which a user can interact with the artefact. Such constraints have been similarly employed in interface design. For example, a radio button, often used in the design of computer interfaces, can be 'pushed' in and out, taking two states (usually used to denote 'on' and 'off'). There are, however, other types of constraints. Semantic constraints are those where the meaning of the situation in which the object of interest is used scopes the set of possible actions that might be taken. Such semantic constraints rely on the knowledge that we have of the world and the situation of the object's use. For example, it is only meaningful to sit facing one way on a motorbike. Cultural constraints are ones that rely on conventions, which form an accepted part of the culture in which the object resides and in which the user is operating. In Europe, for example, we accept that the colour red means 'stop' in travel situations. Logical constraints are those that dictate what should be done through logic. For example, if a dialogue box appears that contains a message like 'input/output violation error has occurred' and an 'OK' button, clicking anywhere outside of the box produces no action. It may be logically concluded that we are constrained to press the 'OK' button if we want to proceed, despite the fact that we may not think that it is 'OK' that the error has occurred. There is clearly a great interplay between the different types of constraints and the categories are not as clearly demarcated as the preceding discussion implies.

10.3.3.2 Mappings

Mappings entail making links clear and obvious and are closely related to constraints, and are often argued to reduce the reliance on other information held in memory. Mappings are concerned with the relationship between different things. For example, when using word processors the scroll wheel on a mouse can be used to scroll up and down a document, as can the scroll bar down the right-hand side of the document window. In both cases moving the interface object one way produces a corresponding movement type in the document. So, pushing the wheel up (i.e., away from the user) moves you up in the document (i.e., towards the beginning/top). This represents a natural mapping between the action of the user and the outcome in the application.

10.3.3.3 Visibility

Visibility refers to making clear what is happening to the user. It is about providing the user with obvious, usually visual, information about the system, its use and the mappings between action and outcome. For example, it should be clear to the user what actions are open to him/her at any time. Feedback is an associated, often overlooked issue – when a user does something, there should be some form of feedback to indicate that the action has had an effect, and so to avoid multiple button presses before an effect is registered.

10.3.3.4 Consistency

Consistency refers to making things work in the same way at different times. It is important that when a user acts in a particular way in different contexts, the object or system responds consistently. For example, in all circumstances the scroll wheel should act in the same way – it should never scroll down in the document when it is moved upwards. There are many examples of inconsistency in interface design (and in the 'real world') and they tend to cause great frustration for users.

10.3.3.5 Experience

Experience refers to the importance of making use of what users already know. This is obvious but can cause difficulties in that it tends to lead designers to make assumptions about users, which may be problematic. For example, the exploitation of experience may be problematic where systems are used across cultural boundaries. An example of where experience is drawn on heavily in interface design is in the use of standard interface objects (e.g., dialogue boxes) across multiple applications and platforms. The argument is that once you have the experience of them, you will be able to deal with them in varying contexts. There is clearly an interaction here with the issue of consistency – if you are drawing on the experience of users by, for example, using standard interface objects, you must ensure that the objects act in the same way in different contexts, otherwise you will have dissatisfied users.

10.3.3.6 Affordance

Affordance refers to making use of the properties of items to suggest use. In many ways it is a common sense concept, but one that is actually very complex when looked at in detail. For example, the shape of a chair affords sitting on; its nature does not suggest that you should, for example, use it to slice bread. There are great debates in psychology about affordance and its relation to 'knowledge in the world' versus 'knowledge in the head'. For our purposes, it is enough to think of affordance in a simple way: buttons afford pressing; sliders afford sliding, etc. There are links to visibility, constraints and mapping when considering affordance.

10.3.3.7 Simplicity

Simplicity refers to the most fundamental issue in interface design – making tasks as simple as possible. Simple things are easier to use than complex things. However, we often want to build in vast functionality to computer systems and this sometimes leads to complex interfaces and interaction design. Keeping simplicity at the heart of the approach to design is a really good starting point but is often easier said than done. Simplicity is an overarching concept, and we can think of the other issues covered above as contributing to it. If we can design in a way that is sensitive to the issues above it is more likely that the system will be simple.

These basic design principles provide a grounding for any approach to design. The next section applies the knowledge that has been expounded up to this point with the aid of two practical examples of interface design.

10.4 Design cases

In order to illustrate the facets of interface design that have been presented thus far, this section will focus on two design cases. Both examples focus on real-life systems exhibiting as many design features as interface problems. The first is a piece of technology that has proliferated in public places in recent times, such as tube and rail stations – a ticket machine – with which a user (or passenger) interfaces personally in the task of purchasing a ticket for travel. The second piece of technology is a customer-facing system that has been used by a bank as a tool for selling mortgages – a point-of-sale (POS) system – which mediates the interaction between a user (or customer) and a mortgage interviewer who guides the customer through the system. The analysis of the attendant design problems in each case is limited to a critique framed in terms of the pertinent user interface issues presented in preceding sections of this chapter (see Section 10.3.3).

10.4.1 The ticket machine interface

The focus on public technology is topical, given the fact that it is becoming increasingly commonplace in many different settings in today's hi-tech society (see [40]). The business idea behind the instigation of such public technology for transport is in the short term to act as 'queue-busters' by speeding up efficiency with faster processing times and in the long term to prove to be a sound financial investment over the employment of human ticket clerks. However, the ticket machine interface presented here (see Figures 10.3 and 10.4) proved inappropriate to the fulfilment of our fictional character's task, which was travelling (see also Reference 41 for a fuller account of a study of a ticket machine interface). Presentation of the design case is firstly headed by a standard (and seemingly innocuous) request by a typical traveller (called Julie). The case has been divided into two scenarios (A and B) and is organised by the intended task to illustrate the nature of the user interaction with a system, along with a proposed context for the execution of the task. The task interface design problems as encountered by the user are then related and finally the usability issues that are inevitably violated by such a design are highlighted.

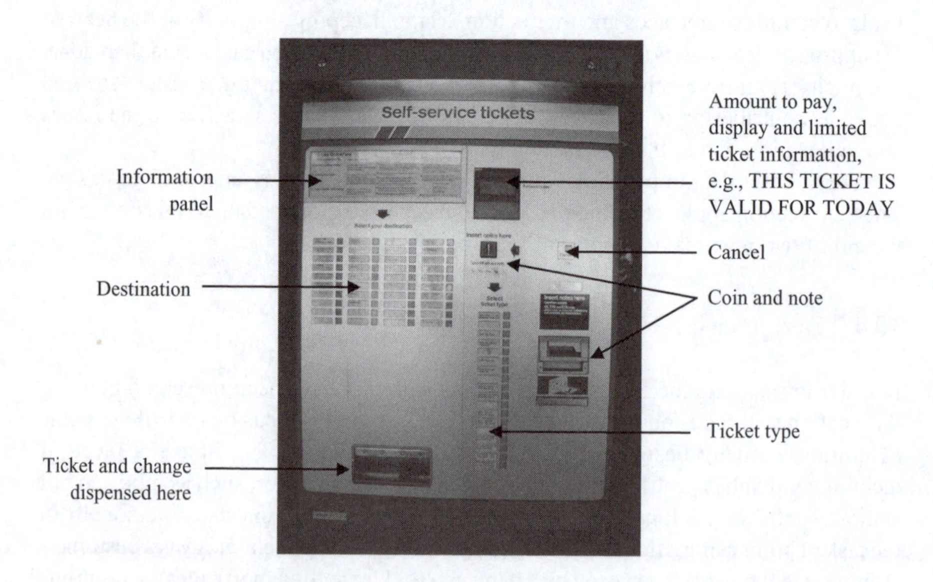

Figure 10.3 The ticket machine interface

Figure 10.4 *The ticket machine 'in use'. People queuing past the machines, which are in clear working order, to make their purchase at the ticket office*

10.4.2 'Good morning, could I have a return ticket to Leytonstone, please?'

Scenario A

Task: Buying a single traveller return ticket.
Context: Julie wants to buy a return train ticket from Hayes and Harlington (Thames Train station) to Leytonstone (London Underground tube station). She has a Disabled Persons Railcard, which entitles her to certain discounts.

Task interface problems

There are, then, three important elements for Julie to successfully buy the ticket she requires to commence her journey (and in no particular order): (1) destination selection; (2) return ticket type; and (3) discount application. However, this ticket buying process is hindered by the fact that the interface does not easily allow for a combination of these elements.

1. Julie is forced to select her destination first, but Leytonstone is not featured on the destination list, which features local stations on the Thames Trains network and tube stations in the form of a variety of travelcard types and zones. However, the interface provides no link between these tube travelcards and the destination that she is aiming towards, as it presupposes a knowledge of the London Underground

zone system. In the end, she is forced to ask someone as to the best travelcard type to buy for this journey.

2. Locating the Disabled Persons Railcard option is difficult, as there appears to be no designated button for such a selection. All the other available discount cards appear under ticket type.

Associated task interface problems

Buying a return ticket for a group (e.g., Julie, her husband and two children) is not straightforward as every individual in the group is forced to buy their ticket separately.

Scenario B

Task: Buying a ticket at the expensive rate, before 9.30 am.
Context: Julie's friend, Stuart who is unused to travelling into London, arrives at the station at 9.20 am.

Task interface problems

Stuart's unfamiliarity with commuting into London (Leytonstone) at peak time, before 9.30 am, is not helped by the interface. It does not inform him fully of the wider range of (much cheaper) options available to him, options that a human ticket clerk would normally provide to make the journey more economical.

1. Following the steps to ticket buying (see Scenario A), the interface does not at any point indicate to Stuart that waiting another 10 minutes or more would allow him to travel at a considerably cheaper rate; it simply tells him the price of the ticket at the time of selection.
2. Alternatively, travelling part of the journey at the more expensive rate, with a cheaper ticket to cover the rest of the journey made after 9.30 am is also not possible.

Associated task interface problems

Julie and Stuart decide to spend the weekend travelling around London. The purchase of a weekend travelcard (available from the human ticket clerk) would be the most economical for their trip as it saves on buying two individual travel cards, but is not an option offered by the machine.

User interface principles revisited

The usability of the ticket machine interface is analysed according to the seven principles of design discussed in Section 10.3.3. The results of this analysis are displayed in Table 10.1.

In summary then, the ticket machine is relatively successful in terms of exhibiting an interface that would be usable by a fair proportion of the commuting populace, as they are used to the opportunities for travel from this particular station and therefore the tickets available. It is perhaps this prior knowledge that is possessed by

Table 10.1 *Correspondence of the ticket machine interface features to design principles*

Design principles	Interface features
Constraints	The ticket machine had a display panel that lit up alerting the user to 'USE COINS ONLY NOTES NOT ACCEPTED'. However, observation of the machine in use revealed that the users appeared not to notice this message.
Mappings	There was an association between the user action and outcome in the application as the user was guided around the system by a series of flashing numbers (1–3) so that the user had to press a button in order to move around the system.
Visibility	General navigation around the system is fairly straightforward (see Mappings) but there are a number of redundant buttons on the screen. Out of 58 buttons on the screen shared between destination list and ticket type, 12 (20 per cent) are blank and appear not to be in use. However, while in some way this can be seen as a grouping technique, where ticket types for example are separated by a blank button (Figure 10.3), it could also mean to a user that an option that should be available is now missing.
Consistency	The interface does appear to act consistently within the limits of the ticket types that it offers.
Experience	The interface does appear to fit in with the general user's level of experience. However, where the machine fails is in not offering ticket types that the users have been able to obtain from other machines or even from human ticket clerks.
Affordance	The interface is based on a button press style of communication for selecting options, although inoperable buttons can confuse the user (see Visibility).
Simplicity	The ticket buying process has been greatly simplified by the machine breaking down the sub-components of the task of buying a ticket, but the order in which buying a ticket is enforced by the machine is perhaps not natural to every user. A short period of casually observing the machine in use revealed that people did not seem to navigate very quickly around the system despite a limited number of options available to them.

users of the ticket machine that allow for certain design flaws to be overlooked and are thus encountered by those users who are less familiar with London travel. In design terms, this interface then raises some interesting questions as to how best to manage the wealth of information (in terms of ticket types, available for different people at particular times) that requires representation in such a public system for the sale of tickets for travel. Ultimately the system stands or fails by whether people can or are prepared to use it. The fact that customers shun the machine in favour of the human ticket clerk (as illustrated by Figure 10.4) demonstrates that it appears to fail on some front. Some of the reasons for this may lie in the flouting of certain design principles as shown in Table 10.1, or a blatant clash of designer and user models in terms of how a user thinks about the process of making a journey, but all of these reasons create a sense of urgency to creating an interface *for* a user in the achievement of his task, in this case travelling.

10.4.3 The point-of-sale machine interface

In a concerted effort to realise the vision of the paperless office, financial institutions such as banks are trying to adopt a more computerised approach to certain client application processes, such as with the design of on-line banking systems (e.g., [42]). Mortgage applications in particular tended to generate a large amount of paper as forms are often made in triplicate for administration purposes and to meet legal requirements. The development of a point-of-sale (POS) system reflected the bank's desire to speed up the mortgage application process and thus increase revenue from this source, store the information in electronic form without fear of losing valuable hard copies and provide a customer-facing system, making the mortgage interview a more social and interactive event for both parties, rather than the traditional and rather mechanical pen-and-paper approach. The system was intended to encapsulate a clear structure to the process with the incorporation of the lending guidelines that presented information, which it was mandatory for the interviewer to pass to the customer.

So, the legal technicalities would be catered for in the system in the form of a series of presentations: the interviewer could focus on their sales techniques and the customer could see every step of the mortgage application process through informative screen shots. However, in practice the mortgage interview as encapsulated by POS proved to be a highly idealised version, which faltered in the face of real customer enquiries (Figure 10.5).

Presentation of this second design case is again headed by a standard opening request by a prospective house buyer. The case has been organised in the form of a typical interview that occurs with the support of the system. The task interface design problems as encountered by the customer and interviewer are then compared to the usability design principles for a final evaluation.

10.4.4 'Hi, I would like some information on the mortgage deals you have available, please'

The process that a customer goes through in a typical (hour long) mortgage interview with the aid of POS is relayed below in a series of ten steps. These comprise a number of presentations and questions that are delivered by the interviewer to explain the procedures and general information for the customer seeking to borrow money in such a way.

1. When a customer is seated in the interview room, the first step is to carry out a client search on the CLIENT SEARCH screen, which includes a search on the client address by postcode. Here the monitor is turned away from the customer

Figure 10.5 *Set-up of the mortgage interview room. The point-of-sale system sits between the mortgage interviewer (on the right) and the customer (on the left). The customer orientates herself towards the interface, which the mortgage interviewer uses as a tool to mediate and explain the mortgage application process at every step*

for data protection reasons in case details of a client with the same name are brought up.

2. The next screen is the NEW INTEREST screen where the monitor is turned to face the customer so as to involve them as soon as possible, although a screen takes a while to load. The whole purpose of the interview is that the customer is seeking a loan from the bank but the system does not underwrite loans so first a check has to be made on whether the money can be offered to them. Here part of the manual system resurrects itself with the filling in of a form, details of which are input to the system at a later date.
3. Next begins a series of presentations:
   Presentation 1: In Today's Meeting We'll . . . introduces the structure of the interview.
   Presentation 2: Why We Want to Share . . . outlines the background and reasons for this financial institution's existence.
   Presentation 3: Taking Care of Our Members . . . explains what this bank is going to do for the customer.
   Presentation 4: Why Choose a Mortgage With Us . . . reiterates previous information.
   Presentation 5: The Code of Mortgage Lending Practice . . . includes the three level of services on offer, although the interviewer is encouraged to take the method of service that involves an overall discussion of all the products on offer and hence going through every step of the system.

Presentation 6: Association of British Insurers Code of Practice . . . explains the insurance for the mortgage.

4. Next in the interview are a whole series of questions:
   Question 1: Are you a member of Unison?
   Question 2: Do you hold any savings accounts? The customer would expect the system to have this information, if they also save with this bank, having input their details at stage 1.
   Question 3: What is the purpose of the loan and how will the deposit for the house be raised? Asking the purpose is nonsensical in a mortgage interview, but one that the system does not allow to be avoided.
   Question 4: What do you do for a living?
   Question 5: Have you ever been made bankrupt? A highly sensitive question.
   Question 6: Have you any annual commitments?
   Question 7: Will you be staying in the same house for the term of the mortgage?
5. A presentation follows next:
   Presentation 7: Which Type of Mortgage Should I Choose . . . introduces all the mortgage products (although curiously not capped mortgages which have to be explained on a bit of paper).
6. The customer's preference is input to the system.
7. A graph is shown to the customer to illustrate the difference between a repayment and interest only mortgage.
8. More questions begin:
   Question 8: What is the amount to be borrowed?
   Question 9: What type of house evaluation do you require? The interviewer will explain the different types of evaluation at this point.
   Question 10: What type of insurance do you require? The interviewer presents on the bank's insurance product, which is a big commission earner.
   Question 11: What is the type of insurance product cover? The bank only offers one.
   Question 12: How much would it cost to rebuild your house? A difficult question for the customer to answer.
   Question 13: What type of security does your property have? Namely, is it in a neighbourhood watch area, though new residents will not often know.
   Question 14: What is the name of the insurance product? Again, there is only one product on offer.
9. Finally the details are saved, which takes some time and creates a gap in the conversation.
10. The end of the interview is marked by the quote, which is saved and then sent to print. Here again there is a delay (20 minutes is not unusual).

User interface principles revisited

The usability of POS is analysed according to the seven principles of design discussed in Section 10.3.3. The results of this analysis are displayed in Table 10.2.

Table 10.2 Correspondence of the POS interface features to design principles

Design principles	Interface features
Constraints	POS is very structured and flexibility is limited from the outset. This is reflected in the way that information can be input. For example, question 4 asks what the customer does for a living, but the interviewer is not permitted by the system to input a person's occupation 'freehand'; they have to search by the first three letters, which often does not bring up the customer's occupation straightaway.
Mappings	Progress through the interview proceeds through a series of mouse clicks on the on-screen buttons to bring up the presentations for the customer and move the interview along through the various stages. There is an obvious relationship between the button clicking and the screens that pop up although they may not be what the interviewer would have naturally jumped to had the interview been manually based (see Visibility).
Visibility	Navigation is straightforward but some screens occasionally pop up that disrupt the flow of the interview or ask for information of a sensitive nature (e.g., on bankruptcy) or repeat information (e.g., on insurance products) that the interviewer cannot hide from the customer or override despite the fact that the information is not necessary at certain junctures.
Consistency	The system does act consistently, within the limits of the application stages that it goes through, but inconsistently in comparison to the way a traditional interview would proceed in terms of rules of dyadic conversation (see Mappings and Visibility).
Experience	Both customer and interviewer have to get to grips with a computerised system but generally it is faithful to the manual system, allowing the interviewer to make use of their experience to make the interview process as smooth as possible.
Affordance	The interface is one that faces the customer and certain presentations serve to engage the customer. For example in step 7 a graph appears that affords a customer an at-a-glance explanation of different mortgage repayment options. Such a graphic acts as a mechanism, which allows the interviewer to purposefully invite the customer to view the system and become more of a participant instead of a passive recipient of the features that the interface has to offer.
Simplicity	The interface is simple for the interviewer to use and renders the mortgage interview process more transparent, but the difficulty comes in being able to manage awkward screens that disrupt the flow, and which could portray a poor image of the bank in terms of an incompetent interviewer.

In summary, this design case presents a very different example of an interface from that of the ticket machine, in terms of having two types of users of the system, one that actively uses the system and one that is affected by it. In terms of interface design the remit is to design a system that will fulfil the requirements of the bank for completing application forms and the user requirements in explaining the mortgage products on offer so that the user leaves the bank satisfied with the service that has been offered. Therefore the system has to be both sophisticated enough for the skilled user (i.e., the interviewer) to be able to input all the necessary information but simple enough for the customer to be able to follow the process through the exchanges that take place through the interface. In general, the interface was well received by both interviewer and customer and seemed to fit the joint user task of processing an application and applying for a loan. An interesting aside to the interface issues would be questioning if a better version of POS could improve business revenue given that customers are so highly engaged in a system and therefore are placed in a prime position to evaluate the bank's services and the possibility of future dealings. Ultimately, the most important criterion that the design of this interface has to meet is that it is an effective communication tool and that it can adapt to the vagaries of human interaction. This interface did exhibit obvious problems in this direction in the interruption of the flow conversation, although the design of such customer-facing systems represents an important future challenge for designers in managing the user mental model of how 'money' is conceptualised in all its various forms of asking for a loan or saving, etc.

10.5 Conclusions

The development of interfaces today has made huge leaps and bounds in design terms, although the problems that have been encountered or documented over time have still not abated. The chapter has illustrated this with two contrasting design cases, which broke the design rules to the same extent. This would imply successful interface design is either a lofty ideal or that the design principles are simply hard to follow. The reality is that interface design needs to be more of a holistic enterprise with attention to the user (for example regular/irregular traveller, first time buyer/remortgaging), the task that they wish to achieve (for example buying a travel ticket, applying for a mortgage) and the various contexts in which these tasks are performed (general transport, banking settings). The system does not only have to be understood and designed in terms of performance specifics but also in terms of the pattern of social relationships that it impacts upon, namely outward communication with the individual/group of users and inwardly to the organisation 'behind' the interface.

Problems

1. In consideration of the problems presented by the ticket machine (see Section 10.4.1), what are the design implications in terms of the user requirements and mental models?

2. Looking at Figure 10.3 of the ticket machine interface, how would you approach the design of such an interface, in terms of interface style?
3. With regards to the problems presented by POS (see Section 10.4.3), what are the most important features that an interface requires so that it (a) meets the requirements of the interviewer for capturing client information, and (b) is understandable to the customer, and why?
4. Looking at Figure 10.5 of POS, how would you approach the design of such an interface, which would appeal to both types of user (interviewer and customer) in terms of interface style?

10.6 References

1 PREECE, J., ROGERS, Y., SHARP, H., BENYON, D., HOLLAND, S., and CAREY, T.: 'Human-computer interaction' (Addison-Wesley, Reading, MA, 1995)

2 LANDAUER, T. K.: 'The trouble with computers: usefulness, usability and productivity' (MIT Press, Boston, MA, 1995)

3 FAULKNER, C.: 'The essence of human computer interaction' (Prentice Hall, London, 1998)

4 SANDOM, C., and MACREDIE, R. D.: 'Analysing situated interaction hazards: as activity-based awareness approach', *Cognition, Technology and Work*, 2003, **5**, pp. 218–228

5 GRUDIN, J., and POLTROCK, S. E.: 'User interface design in large corporations: coordination and communication across disciplines'. Proceedings of CHI'89 Conference on *Human Factors in Computing Systems*, 1989, pp. 197–203

6 MYERS, B. A.: 'Why are human-computer interfaces difficult to design and implement?' Technical Report CMU-CS-93-183, Computer Science Department, Carnegie Mellon University, Pittsburgh, PA, 1993

7 ALM, I.: 'Designing interactive interfaces: theoretical consideration of the complexity of standards and guidelines, and the difference between evolving and formalised systems', *Interacting with Computers*, 2003, **15** (1), pp. 109–119

8 GUIDA, G., and LAMPERTI, G.: 'AMMETH: a methodology for requirements analysis of advanced human-system interfaces', *IEEE Transactions on Systems, Man, and Cybernetics – Part A: Systems and Humans*, 2000, **30** (3), pp. 298–321

9 SHACKEL, B.: 'People and computers – some recent highlights', *Applied Ergonomics*, 2000, **31** (6), pp. 595–608

10 MANDEL, T.: 'The elements of user interface design' (John Wiley & Sons, New York, 1997)

11 GRUDIN, J.: 'INTERFACE – an evolving concept', *Communications of the ACM*, 1993, **36** (4), pp. 112–119

12 LEWIS, C., and RIEMAN, J.: 'Task-centered user interface design: a practical introduction' 1993 [Online]. Available: http://www.acm.org/~perlman/uidesign.html

13 MARCUS, A.: 'Culture class vs. culture clash', *Interactions*, 2002, **9** (3), pp. 25–28
14 ACKERMAN, M. S.: 'The intellectual challenge of CSCW: the gap between the social requirements and technical feasibility', *Human-Computer Interaction*, 2000, **15** (2/3), pp. 79–203
15 BERRAIS, A.: 'Knowledge-based expert systems: user interface implications', *Advances in Engineering Software*, 1997, **28** (1), pp. 31–41
16 STARY, C.: 'Shifting knowledge from analysis to design: requirements for contextual user interface development', *Behaviour and Information Technology*, 2002, **21** (6), pp. 425–440
17 HIX, D., HARTSON, H. R., SIOCHI, A. C., and RUPPERT, D.: 'Customer responsibility for ensuring usability – requirements on the user-interface development process', *Journal of Systems and Software*, 1994, **25** (3), pp. 241–255
18 WOOD, L. E. (Ed.): 'User interface design: bridging the gap from user requirements to design' (CRC Press, Boca Raton, FL, 1998)
19 NORMAN, D., and DRAPER, S. (Eds): 'User centered system design' (Lawrence Erlbaum, Hillsdale, NJ, 1986)
20 GOULD, J., and LEWIS, C.: 'Designing for usability – key principles and what designers think', *Communications of the ACM*, 1985, **28** (3), pp. 300–311
21 MULLIGAN, R. M., ALTOM, M. W., and SIMKIN, D. K.: 'User interface design in the trenches: some tips on shooting from the hip'. Proceedings of CHI'91 Conference on *Human Factors in Computing Systems*, 1991, pp. 232–236
22 KUJALA, S.: 'User involvement: a review of the benefits and challenges', *Behaviour and Information Technology*, 2003, **22** (1), pp. 1–16
23 NORMAN, D.: 'The psychology of everyday things' (Basic Books, New York, 1988)
24 MAYHEW, D.: 'Principles and guidelines in software user interface design' (Prentice-Hall, Englewood Cliffs, NJ, 1992)
25 BERRY, D.: 'The user experience: the iceberg analogy of usability', IBM, October 2000 [Online]. Available: http://www-106.ibm.com/developerworks/library/w-berry/
26 LAUREL, B.: 'Computers as theatre' (Addison-Wesley, Reading, MA, 1993)
27 ROSSON, M. B.: 'Real world design', *SIGCHI Bulletin*, 1987, **19** (2), pp. 61–62
28 MANTEI, M. M., and TEOREY, T. J.: 'Cost/benefit analysis for incorporating human factors in the software lifecycle', *Communications of the ACM*, 1988, **31** (4), pp. 428–439
29 GREATOREX, G. L., and BUCK, B. C.: 'Human factors and system design', *GEC Review*, 1995, **10** (3), pp. 176–185
30 SHNEIDERMAN, B.: 'Designing the user interface: strategies for effective human-computer interaction' (Addison-Wesley, Reading, MA, 1998, 3rd edn.)
31 PREECE, J. (Ed.): 'A guide to usability: human factors in computing' (Addison-Wesley, Reading, MA, 1993)

32 CARBONELL, N., VALOT, C., MIGNOT, C., and DAUCHY, P.: 'An empirical study of the use of speech and gestures in human-computer communication', *Travail Humain*, 1997, **60** (2), pp. 155–184
33 PICARD, R. W.: 'Toward computers that recognize and respond to user emotion', *IBM Systems Journal*, 2000, **39** (3/4), pp. 705–719
34 WEIDENBECK, S., and DAVIS, S.: 'The influence of interaction style and experience on user perceptions of software packages', *International Journal of Human-Computer Studies*, 1997, **46** (5), pp. 563–588
35 BREWSTER, S. A., and CREASE, M. G.: 'Correcting menu usability problems with sound', *Behaviour and Information Technology*, 1999, **18** (3), pp. 165–177
36 HALL, L. E., and BESCOS, X.: 'Menu – what menu?' *Interacting with Computers*, 1995, **7** (4), pp. 383–394
37 DAVIS, S., and WEIDENBECK, S.: 'The effect of interaction style and training method on end user learning of software packages', *Interacting with Computers*, 1998, **11** (2), pp. 147–172
38 SMITH, S. L., and MOSIER, J. N.: 'Guidelines for designing user interface software', Technical Report ESD-TR-86-278, The MITRE Corporation, Bedford, MA, 1986
39 GRUDIN, J.: 'The case against user interface consistency', *Communications of the ACM*, 1989, **32** (10), pp. 1164–1173
40 MAGUIRE, M. C.: 'A review of user-interface design guidelines for public information kiosk systems', *International Journal of Human-Computer Studies*, 1999, **50** (3), pp. 263–286
41 THIMBLEBY, H., BLANDFORD, A., CAIRNS, P., CURZON, P., and JONES, M.: 'User interface design as systems design'. Proceedings of the HCI '02 Conference on *People and Computers XVI*, 2002, pp. 281–301
42 VANDERMEULEN J., and VANSTAPPEN, K.: 'Designing and building a user interface for an on-line banking service: a case study', *Ergonomics*, 1990, **33** (4), pp. 487–492

Further Reading

There is a wealth of material available on the design of interfaces. An excellent resource for (extended) further reading is available on-line at: http://world.std.com/~uieweb/biblio.htm. It provides an annotated A–Z bibliography of books, which should prove invaluable for providing designers with a thorough and comprehensive grounding in all aspects of interface design. However, given the rather bewildering list of sources available, some additional references have been selected, which in the authors' opinion are extremely useful for the follow up of specific issues raised in this chapter and offer good practical advice for conducting future interface design activities in a more assured way.

JACKO, J. A., and SEARS, A. (Eds): 'The human-computer interaction handbook: fundamentals, evolving technologies and emerging applications' (Lawrence Erlbaum, London, 2003)

Contains a varied collection of readings on the subject human–computer interaction.

LIM, K. Y., and LONG, J.: 'The MUSE method for usability engineering' (Cambridge University Press, Cambridge, 1994)

Details an example of a human factors methodology, which covers the design process from user requirements to user interface design.

MACAULAY, L.: 'Human-computer interaction for software designers' (Thomson, London, 1995)

Provides a clear step-by-step approach for the novice designer.

NORMAN, D. A.: 'The invisible computer: why good products can fail, the personal computer is so complex, and information appliances are the solution' (MIT Press, Cambridge, MA, 1998)

Anything published by Donald Norman would prove useful for the design of interfaces and this latest offering is no exception.

RUDISILL, M., LEWIS, C., POLSON, P. B., and MACKAY, T. D.: 'Human computer interface design: success stories, emerging methods, and real-world context' (Morgan Kaufmann, San Francisco, CA, 1996)

Replete with case studies, illustrating interface design in action.

www.baddesigns.com

Michael Darnell's site hosting a collection of bad human factors designs for a light-hearted look at why some things are hard to use.

Chapter 11
Usability
Martin Maguire

11.1 Introduction

Most computer users will have had experience of poorly designed interactive systems that are difficult to understand or frustrating to use. These may range from a consumer product with many functions and fiddly controls, to a call centre system that is hard for the telephone operator to navigate to obtain the information that a customer requires. System success is therefore very dependent upon the user being able to operate it successfully. If they have difficulties in using the system then they are unlikely to achieve their task goals.

Different terms have been used to describe the quality of a system that enables the user to operate it successfully, such as 'user friendly', 'easy to use', 'intuitive', 'natural' and 'transparent'. A more general term that has been adopted is 'usability'.

11.1.1 Definition of usability

Usability has been defined in different ways. One approach, from the software engineering community, is to define it in terms of system features. The ISO/IEC 9126-1 Software Product Quality Model [1] defines usability as a quality with the attributes of understandability, learnability, operability and attractiveness.

Another approach from the human factors community is to define it in terms of the outcome for the user. Shackel [2], for instance, proposed a definition as 'the capability in human functional terms to be used easily and effectively by the specified range of users, given specified training and user support, to fulfil the specified range of tasks, within the specified range of environmental scenarios'.

This idea that usability should be specified in terms of the users, tasks and environment is important because it emphasises that the system should be designed with a target group of users, performing particular tasks, in a certain environment. This is the basis for the definition of usability contained within the ISO 9241 standard [3], which is now widely accepted:

> Usability is the extent to which a product or system can be used by specified users to achieve specified goals with effectiveness, efficiency and satisfaction in a specified context of use.

Thus usability is defined in terms of what the user can achieve with the system and the level of satisfaction they gain in using it.[1]

11.1.2 Usability benefits

Based on previous United States experience, the business case for allocating resources to the development of a usable system is as follows:

Increased revenue based on:

- Increased user productivity.
- Improved customer satisfaction.
- Improved brand image and competitive advantage.

Reduces costs due to:

- Reduced system development time.
- Reduced user error.
- Reduced user training time and costs.
- Support call savings (several dollars per call).

It is not a simple process to identify the cost-benefits of usability work in strict monetary terms although some authors such as Bias and Mayhew [4] have been able to show cases where the cost of usability work can be recovered many times. In one such study, usability activities resulted in the time required for a frequently performed clerical procedure to be reduced slightly. The accumulated time saved within the organisation produced cost savings of millions of dollars.

11.2 Usability perspectives and role in system development

Usability can be considered from two perspectives each of which has a useful role to play in system development. These are:

- **Usability features**
  If a system is designed to have good usability features, such as large keys, clear text, and easy navigation, it is more likely to be successful. The specification of usability principles can thus help to guide the development of the system to make it usable. Such principles can also form a useful checklist when evaluating the system informally. The idea of enhancing a system for the user while it is under development is known as *formative usability*.

[1] In ISO/IEC 9126-1 [1], the above definition is termed 'quality in use' to distinguish it from the use of the term usability to cover specific software attributes.

- **User experience and performance**
  The user's experience with the system reflects what they can achieve with it, how quickly they can perform tasks, and how much they enjoy using it or how acceptable they find it. The specification of usability goals for achievement and satisfaction can help to set usability targets that can be tested for their achievement. This is called *summative usability*.

Figure 11.1 shows how each of these two perspectives can form complementary activities within the system development process.

The following sections of this chapter describe the major usability activities that may take place within system development. These include sections on formative and summative usability, which are complementary to each other:

11.3: Scoping, planning and context of use.
11.4: Formative usability.
11.5: Summative usability.

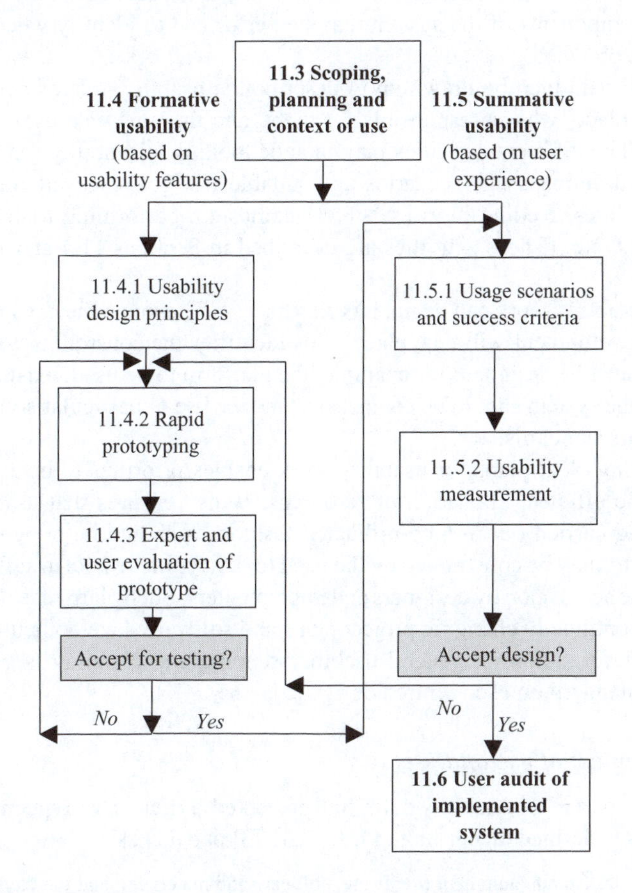

Figure 11.1 Two forms of usability work in system development

11.6: User audit of implemented system.
11.7: Examples of usability projects.
11.8: Conclusions.
11.9: References.

11.3 Scoping, planning and context of use

11.3.1 Usability scoping and planning

The first step in implementing a usability programme within the system development process is to scope the work involved and to plan for it to take place. A usability scoping and planning meeting should be held with the key project personnel who have an interest in the usability of the system. The meeting will review the project aims and specify which aspects of the system require an investment in usability work. This is achieved by listing the areas of the current system where the users appear to have most problems. An analysis, in business terms, of the cost-benefits of improving individual components of the system may be performed to identify where the most value can be obtained.

A plan should then be drawn up to describe the usability-related activities that should take place, who is responsible for them, and the timescale over which they should run. The different activities may include looking at usability issues for each user group, defining usage scenarios and suitable test goals, developing usability design guidelines, feeding in to user interface design, performing usability testing of prototypes, etc. (These activities are described in Sections 11.4 and 11.5 of this chapter.)

Before usability work can begin, it is also necessary to know the *design context* in which the development will take place. This identifies the constraints within which the system must be designed, for example, the platform to be used, existing systems with which the system should be compatible, and the use of particular style guides or human factors standards.

Developing clear plans for usability work enables priorities to be assessed and facilitates the efficient allocation of resources. It also ensures that usability work is visible and carried out in a co-ordinated fashion. It should be remembered that usability plans may be constrained by the need to deliver the results in sufficient time for them to be acted upon by designers and implementers. Such plans may not be stable because of continually changing project plans and software development schedules.

Further information on general usability scoping and planning is given in the ISO 13407 standard on user-centred design [5].

11.3.2 Context of use analysis

An important part of the usability definition presented earlier, is 'in a specified context of use'. This is defined (according to ISO 9241 [3] part 11) as:

> The users, tasks and equipment (hardware, software and materials), and the physical and social environments in which the system or product is used.

Consideration of all of the users, tasks and environments is important in understanding the usability requirements for a system. If the user group has little experience with Information and Communications Technology (ICT), then the system must be designed to be simple and self-explanatory. If the user is likely to be given different tasks with a lot of variability, then the system must be designed to work in a flexible way. If the user is working in a specific environment then the system must be designed accordingly. For in an unstable physical environment such as on board a ship, it may be better to provide a roller ball to use with a computer (which can be fixed down) rather than a free moving mouse. Bank machines are often affected by sunlight on the screen so need to be provided with non-reflective glass or a protective hood. If a system is likely to be used at home by the public, who will have a wide range of computer screen sizes, the pages must be designed for comfortable viewing on each one. If the system is being used in a busy working environment where interruptions are common, then it may be necessary to provide the user with a facility to suspend the current task and concentrate on a new task.

The different elements of context that may affect a system design are shown in Table 11.1. However, for a particular system, only a subset of these will be important.

Further information on context of use in usability work is available in Maguire [6] and Thomas and Bevan [7].

Having identified the context of use factors for the system, these may lead to user requirements for design features to meet those contextual conditions. Examples of the kinds of usability features that are needed to match contextual factors are shown in Figure 11.2, which describes four contrasting application areas.

Table 11.1 Context of use factors

User group	Tasks	Technical environment	Physical environment	Organisational environment
• System skills and experience • Task knowledge • Training • Qualifications • Language skills • Age and gender • Physical and cognitive capabilities • Attitudes and motivations	• Task list • Goal • Output • Steps • Frequency • Importance • Duration • Dependencies	• Hardware • Software • Network • Reference materials • Other equipment	• Auditory environment • Thermal environment • Visual environment • Vibration • Space and furniture • User posture • Health hazards • Protective clothing and equipment	• Work practices • Assistance • Interruptions • Management/ communications structure • Computer use policy • Organisational aims • Industrial relations • Job characteristics

Classroom software

- Single or group working so allowance for several users at once.
- Classroom noise may obscure sounds from package.
- Quiet classroom should not be disturbed by sounds from software package.
- Teacher will need to learn software package quickly and relate it to the curriculum so teacher's notes and introduction would be needed.

Call centre system

- System needs to be flexible to meet wide range of user enquiries.
- System needs to provide rapid response to avoid customers waiting on the phone.
- To help share workload, system should allocate calls equally rather than just to next free operator.
- User should have means of communicating with supervisor to deal with complex situations.

Aircraft flying support systems
US Department of Defense (DOD)

- Controls and displays need to be readable in unstable conditions.
- Controls and displays need to be visible at night.
- Space and posture require the pilot to be able to see all important displays directly in front.
- Controls must be operable when pilot wears protective clothing, e.g. gloves, helmet.

ATM or bank machine

- Wheelchair users require recess under bank till.
- Speech output and large buttons may be considered for visually impaired users.
- Customisation features may be used to support rapid access for users in a hurry.
- To reduce fraud, biometric techniques such iris scan or finger print ID may be used.
- Visors should be provided for sunny conditions and lit buttons for night time use.

Figure 11.2 Usability features to meet contextual needs for different applications

11.4 Formative usability – enhancing usability within design

Formative usability is the process of enhancing a system while it is being designed (i.e. is in formation). Specifying usability design principles can help to guide the development of a system to make it usable. These might be general principles or more specific guidelines tailored to the application domain. An organisation might also produce a style guide that lays down design rules for screen layouts, navigation, text styles, fields and labels, buttons, menus, error and system messages, and general corporate style. While the style guide needs to be prescriptive and offer clear design rules, it also must cater for the wide range of subsystems that the organisation may develop.

The following section presents a set of general design principles that can form the basis for the design of a usable system. They are based upon the work of writers such as Shneiderman [8] and Nielsen [9] and this author's own experience.

11.4.1 Usability design principles

Important principles to guide the design of usable systems are as follows:

- accessible
- suitable for the task
- simple and easy to learn
- efficient and flexible
- consistent and meeting user expectations
- status and feedback information
- helpful
- avoiding and tolerating errors
- customisable.

The following sections present a set of design principles that, if followed, will help produce a use system.

11.4.1.1 Accessible

Clearly the user can only use a system successfully if they can access it. If the user's workstation has technical limitations such as a small screen or slow network link, they may have difficulty in using the system. Similarly, in a noisy environment the user will have trouble in hearing any auditory signals that present information to him or her. Another aspect of accessibility is whether the system is flexible enough to support the needs of users with the range of abilities of the user community. If the user has a visual impairment (a condition that affects about two million people in the UK) then it should be possible for them to enlarge the characters on the screen. (The RNIB advise that a font size of 16 point can be read by 80 per cent of the UK population.) If the user has some loss in colour vision, then the system should provide redundant coding of information so that they can still receive it. There is now clear recognition of the need for Web sites to be accessible (see www.rnib.org/digital) for example in supporting screen readers. Similarly a set of accessibility tools has been defined by Microsoft (www.microsoft.com/access) to enable PC applications to support users with physical disabilities who, for example, can only use a keyboard (possibly customised to their needs) and not a mouse. Principles and guidelines on accessibility are also available from the Trace centre (http://trace.wisc.edu/) and the North Caroline State University (www.design.ncsu.edu/cud/index.html).

11.4.1.2 Suitable for the task

A usable system must provide relevant functions to support the task. Many systems have failed in this respect. One example was an early mobile phone system that could only receive calls, while outgoing calls had to be made while in the vicinity of a telephone base-station. The take-up of videophones at the desktop has also had its development hampered since early devices provided a poor-quality picture and office workers did not perceive the need to be seen as well as heard when on the phone. Systems may also be used to allocate jobs each day to field engineers. If the system automatically makes appointments and does not allow flexibility to easily change them, or to reallocate them to other engineers, this creates pressure on the engineers and poor service to customers.

11.4.1.3 Simple and easy to learn

Many computer systems suffer from over-complexity and the provision of too many functions. This can occur when the development team try to incorporate facilities to meet all possible user needs. The '80-20' law often seems to apply, i.e. that 20 per cent of the functions will satisfy 80 per cent of the needs. The remaining functions can either be integrated into the systems or excluded. Video-editing software, for example, can have many facilities that are rarely used which over-complicate the user interface and make it difficult to learn. If the system structures the user's task into simple stages, this will make it easy to learn and use. For example, a video editing package might split the task into guided stages of capturing clips from raw video, assembling a storyboard, adding effects, and saving completed film in an appropriate format. Simple uncluttered screen displays also help to make the system easier to use. Simplification can also be achieved by customisation, as described later.

11.4.1.4 Efficient and flexible

As users learn to use a system, they will want to minimise effort and use it in different ways. Typically users of a mouse-based graphical system will prefer to use keyboard short cuts such as the Windows key functions 'Alt-f' for File, 'Ctrl-c' for copy, 'Ctrl-s' for save, etc. Web user interfaces should provide similar features although it will take some time before this is common practice. Provision of palettes or toolbars can make function selection quicker than the use of pull-down menus. Flexibility in searching for information is often critical. If a customer cannot remember their account number when phoning a call centre, it is important that the account information can be called up by other means. In a process where the user is entering data in a step-by-step manner, it is important to be able to review previous steps without needing to cancel the whole transaction. Providing suitable defaults for commands or data entry, e.g. the current year in a date field, will help make the system more efficient to use. Web portals can offer facilities to improve efficiency when working with several applications such as single sign-on, rapid access applications and integration of application databases to avoid data re-entry and support intelligent linking of data.

11.4.1.5 Consistent and meets user expectations

Users will have expectations about how a particular system will work. If for instance they are using an e-commerce site to buy a mouse, they will be surprised if they are listed under 'rodents'. When the user puts an electronic document into the 'wastebin', they will be frustrated if it is immediately deleted and is not recoverable. It is important that as the user proceeds through the user interface, a feature does not confront them that conflicts with what they would naturally expect. Equally the system should be internally consistent. If one data entry field requires a length in centimetres, it will be confusing if another field requires the length in metres. If a name is entered as 'Forename', 'Surname' in one part of the system, another screen that requires the entry order as 'Surname', 'Forename' will be confusing. However there may be good reasons for some differences, e.g. people's heights may be specified in feet and inches while other measurements are in metric units. If so, the screen should give clear prompts (e.g. field labels) to reflect the differences.

11.4.1.6 Status and feedback information

Status information such as screen title, page number in a sequence or levels of message priority are important to inform the user of the current state of the system. A customer transaction system might display the customer name and reference number at the top of each screen. On Web pages, a status or 'breadcrumb' line may show progress through a purchase or the position in a menu hierarchy. Similarly, good feedback on user actions is important to give the user confidence that the system is active and they are proceeding in the correct way. Display of a progress bar to show the percentage of a task complete has become a standard technique for providing feedback on system operation.

11.4.1.7 Helpful

Providing online help is a convenient way for the user to obtain assistance at the point of need. Simple guidance or messages at certain points in the user interface can act as useful prompts of how inputs should be made to the system. Such help is often called contextual help. More comprehensive help showing the user how to perform certain tasks or recover from certain problems is also important. Provision of much help text online can be daunting for the user and should be supplemented with screen shots or images showing the interactions required. Simple interactive tutorials, which offer structured learning on the use of a system, are a useful supplement to formal training and helping the user learn to use the system more fully.

11.4.1.8 Avoiding and tolerating errors

Users will always make errors so careful design to help avoid and overcome them is essential. Providing pick lists for data entry, rather than requiring the user to type the information, can help reduce errors. However, such lists can be clumsy and slow if, for example, the user has to specify a time and date by picking from several lists to specify hours, minutes, date, month and year. If, instead, the user is required to enter the time and date in character form, they should be given appropriate format prompts, e.g. mm:hh, dd/mm/yyyy, to help reduce the likelihood of error. Validation of the data entered will also help ensure that any errors are corrected. Providing an 'undo' key to reverse errors is also important and is expected by experienced users.

11.4.1.9 Customisable

A customisable system can provide important benefits. If the user can select the required toolbars for their task, excluding those which have no relevance, this will reduce the clutter at the user interface. Selection of screen colours can help to overcome colour vision impairment. Customisation can also help to overcome user disabilities. Being able to choose larger characters can also assist those with poor vision. Also it is useful if the user can change minor behavioural elements of the system, such as whether a list is ordered by newest item first or by oldest, whether sound prompts are turned on or off, or whether printing can take place in the background.

Designing the system following the above principles will help ensure that users can operate it successfully. The amount of effort that is put into creating a usable system design will also widen the range of potential users for the system and the market for it.

11.4.2 Rapid prototyping

The next stage in the formative development process is to produce a rapid prototype of the system to demonstrate the system concept and as a basis for obtaining user feedback. The idea is to develop a representation quickly and with few resources to present a vision of the future system design. Users and experts can then view the prototype and provide comments on the usefulness of its functions and level of usability (see Section 11.4.3). New versions of the design can then be produced quickly in response to the feedback. When the prototype is regarded as satisfactory from the users' point of view, it can then be implemented and a formal usability test performed on it (summative usability).

Two common forms of prototyping are paper prototyping and software prototyping.

11.4.2.1 Paper prototype development

With paper prototyping, designers create a paper-based simulation of interface elements (menus, dialogues, icons etc.) using pencil, paper, card, acetate and sticky post-its. Alternatively, the prototype might consist of a sequence of paper screens clipped together into a booklet, which can be accessed in sequence or randomly to reflect different interaction choices. When the paper prototype has been prepared a member of the design team sits before a user and 'plays the computer' by moving interface elements around in response to the user's actions. The difficulties encountered by the user and their comments are recorded by an observer and/or audio/video tape recorder.

The main benefits of paper prototyping are that communication and collaboration between designers and users is encouraged. Paper prototypes are quick to build and refine and only minimal resources and materials are required to convey product feel. Paper prototypes do, however, have their limitations. Since they are simple and approximate representations of the system, they do not support the exploration of fine detail. Due to the use of paper and a human operator, this form of prototype cannot be reliably used to simulate system response times. Also the individual playing the role of the computer must be fully aware of the functionality of the intended system in order to simulate the computer. Further information can be obtained from References 10 and 11.

General guidelines for developing and using paper prototypes are as follows:

- Only simple materials are required to create the elements of the prototype. These include paper, acetate, pens and adhesives. Typical a sheet of paper or card is used for each screen. 'Post-its' or smaller pieces of paper are used to represent interface elements such as system messages, input forms and dialogue boxes and any element that moves or changes appearance.
- Two evaluators should take part in the user session, one to manipulate the paper interface elements (acting as the system), and the other to facilitate the session. Ideally a third evaluator should observe, and make notes of user comments and the problem areas identified.
- Prepare realistic task scenarios for the user to perform during the evaluation.
- Have spare paper, post-its, etc. available to simulate new paths of interactions that the user would expect to make.
- Practise the possible interactions with a pilot user to make the interactive process as slick as possible.
- Select a small number of appropriate users to test the prototype. Try to cover the range of users within the target population.
- Conduct each session by manipulating the paper prototype in response to the interactions that the user specifies as they perform each task.
- Work through the paper-based interactions as fully as possible and try to cover the different paths that users may wish to follow.
- Conduct post-session interviews with the user, drawing upon pre-set questions and issues raised during the prototype evaluation.

Again, further information can be obtained from References 10 and 11.

11.4.2.2 Software prototype development

With software prototyping computer-based simulations are developed to provide a more realistic mock-up of the system under development. As such the representations often have greater fidelity to the finished system than is possible with simple paper mock-ups. The prototype may be shown to end-users for comment or users may be asked to accomplish set tasks and any problems that arise are noted.

The main benefits of software prototyping are that they give users a tangible demonstration of what the system is about. If appropriate tools are chosen (e.g. Macromedia Director, Microsoft Powerpoint or a Web page editor), they permit the swift development of interactive software prototypes. Software prototypes have a high fidelity with the final product and thus support metric-based evaluations. Limitations of software prototypes are that they require software development skills. Although rapid, the method is generally more time consuming than paper prototyping.

Guidelines for developing software prototypes are as follows:

- Avoid spending too long on the development of the first prototype as users may require substantial changes to it. Similarly, try not to put too much effort into particular features (e.g. animations) that may not be required.
- Avoid making the prototype too polished as this may force users to accept it as finished.
- Avoid putting features into the prototype that will raise the users' expectations but which are unlikely to be achieved with the real system (e.g. too fast response times, too sophisticated graphics).

Further information is available from References 12, 13 and 14.

11.4.3 Expert and user evaluation of prototype

Having developed a prototype for the system, it should be reviewed with usability experts, task experts and typical end-users to provide informal feedback and to propose suggestions for improvement. There are many techniques for conducting reviews but two common approaches are heuristic evaluation and informal user evaluation.

11.4.3.1 Heuristic evaluation

Heuristic evaluation, also known as expert evaluation, is a technique used to identify potential problems that operators can be expected to meet when using a system or product. Analysts evaluate the system with reference to established guidelines or principles, noting down their observations and often ranking them in order of severity. The analysts are usually experts in human factors or HCI, while task or domain experts will also provide useful feedback.

The main benefits of heuristic evaluation are that the method provides quick and relatively cheap feedback to designers and the results can generate good ideas for improving the user interface. The development team will also receive a good estimate of how much the user interface can be improved. There is a general acceptance that the design feedback information provided by the method is valid and useful. It can also be obtained early on in the design process, whilst checking conformity to established guidelines helps to promote compatibility with similar systems. If experts are available, carrying out a heuristic evaluation is beneficial before actual users are brought in to help with further testing.

The main limitations are that the kinds of problems found by heuristic evaluation are normally restricted to aspects of the interface that are reasonably easy to demonstrate: use of colours, layout and information structuring, consistency of the terminology, consistency of the interaction mechanisms. It is generally agreed that problems found by inspection methods and by performance measures (generated by summative testing) overlap to some degree, although both approaches will find problems not found by the other. The method can seem overly critical as designers may only get feedback on the problems within the interface, as the method is normally not used for the identification of the 'good' aspects. It can be very time-consuming to check conformance to voluminous written guidelines so it is often necessary to rely on the expert's knowledge of those guidelines and his/her ability to identify non-conformances 'on-the-fly'. As such the quality of results depends on the knowledge and capability of the experts who conduct the evaluation.

General guidelines in running a heuristic evaluation are as follows:

- Two to four human factors experts or task experts are required. It has been shown that three is the optimum number of experts required to capture most of the usability problems in an interface [15]. Using larger numbers of experts can be expensive and may fail to reveal significantly more problems, thus leading to diminishing returns. Under some circumstances a developer may be employed to demonstrate the system but the experts should have hands-on access. Ideally the experts should assess the system independently of one another, although working as a group can also be used. The experts will need approximately half a day each for their assessment plus one to two more days to document and report their findings depending upon the complexity of the system being investigated.
- As well as access to the prototype to evaluate, all participants should have available the relevant documents, including the standards and heuristics against which conformance is being checked. They should also be provided with a set of typical task scenarios and an agreed set of evaluative principles (such as the principles listed in Section 11.4.1).
- The experts should be aware of any relevant contextual information relating to the intended user group, tasks and usage of the product. A heuristics briefing can be held to ensure agreement on a relevant set of criteria for the evaluation although this might be omitted if the experts are familiar with the method and operate by a known set of criteria. The experts then work with the system preferably using mock tasks and record their observations as a list of problems. If two or more experts are assessing the system, it is best if they do not communicate with one another until the assessment is complete. After the evaluation, the analysts can collate the problem lists and the individual items can be rated for severity and/or safety criticality. Theoretical statistics regarding the number of identified versus unidentified problems can also be calculated if required.
- A report detailing the identified problems is written and fed back to the development team. The report should clearly define the ranking scheme used if the problem lists have been prioritised.

Further information on heuristic evaluation can be obtained from References 4 (pp. 251–254), 15 and 16.

11.4.3.2 Informal user evaluation

Informal user evaluation is an alternative means of evaluating a prototype with users rather than experts. Here users employ a prototype to work through task scenarios. They explain what they are doing by talking or 'thinking aloud' and this is recorded on tape and/or captured by an observer. The observer also prompts users when they are quiet and actively questions the users with respect to their intentions and expectations. The main benefits of co-operative testing are that usability problems can be detected early in the design process. Information on the user's thought processes as well as their actions can be obtained. User comments provide compelling feedback to the development team. A limitation is that large quantities of user utterances may be recorded and these can be very time consuming to analyse.

General guidelines in informal user evaluations are as follows:

- Two evaluators are required, one to act as the facilitator and another to record any observations. If resources are constrained a facilitator can be used alone as long as adequate recording facilities are in place.
- The prototype must be suitably functional to evaluate a range of tasks. A location for the evaluation is also needed and some means of recording the user's comments and actions will also be needed (e.g. audio or video recording, system logs, notebooks).
- Identify and recruit a sample of users (identified from the context of use analysis). A sample size of at least four or five is recommended to provide sufficient coverage to identify most of the usability defects.
- Select realistic and appropriate tasks for the users to perform that test those features of the system people will actually use in their work, and those features implemented in the prototype. Write task instructions for the users that are clearly expressed.
- Conduct a pilot session and work through the tasks and instructions. This will also indicate how much time is required from each user. Prepare any pre-set questions to be asked and clear instructions for those running the evaluation.
- Each session is conducted by observing the users as they work through the tasks and recording what they say, and by exploring their impressions and intentions through relevant questions. Make notes of unexpected events and user comments but keep up a dialogue so that the user is encouraged to explain their actions and expectations of the system. Carry out post-session interviews with the users, drawing upon pre-set questions and issues raised during the evaluation.
- Analyse information obtained, summarise unexpected behaviour and user comments. Consider the themes and severity of the problems identified. Summarise design implications and recommendations for improvements and feed back to the system/process design team. Tape recordings can be used to demonstrate particular themes.

Further information can be obtained from References 12, 17, 18, 19 and 20.

11.5 Summative usability – usability in terms of user experience

The second approach for ensuring that a system is usable (summative usability) is to specify user performance and satisfaction criteria and to test the system prototype with users to determine whether the criteria have been achieved. This section describes the activities underlying summative usability.

11.5.1 Usage scenarios and success criteria

When a system is being specified or evaluated, it is helpful to define some summative usability requirements. These are operational statements that define the level of usability that the system must reach if it is to be regarded as successful. The process of specifying usability criteria should be to consider the user goals for the system and from these to develop usability criteria.

Examples of usability test criteria are listed in Table 11.2.

Specifying usability goals in this way, determines how the system should be tested and gives a clear design target that must be met if the system is to be regarded as sufficiently usable.

11.5.1.1 Choosing usability objectives and criteria

Focusing usability objectives on the most important user tasks is likely to be the most practical approach, although it may mean ignoring many functions. Usability objectives may be set at a broad task level (e.g. perform a claims assessment) or a narrow task level (e.g. complete customer details) or a feature level (e.g. use the search facility). Setting usability objectives at the broad task level is the most realistic test of usability, but setting objectives at a narrow level may permit evaluation earlier in the development process.

The choice of criterion values of usability measures depends on the requirements for the overall system and the needs of the organisation setting the criteria. They are often set by comparison with other similar products or systems. When setting criterion values for several users, the criteria may be set as an average across the

Table 11.2 Examples of usability test criteria

Type of measure	Example criteria
User performance	Sample data entry staff will be able to complete the task with at least 95% accuracy in under 10 minutes.
	90% of all executives should be able to use all the core facilities with no more than 5% of inputs in error after 30 minutes training.
User satisfaction	The mean score for all the questions in the satisfaction questionnaire will be greater than 50 (midpoint of scale).

users (e.g. the average time for completion of a task is no more than 10 minutes), as a target for individuals (e.g. all users can complete the task within 10 minutes), or for a percentage of users (e.g. 90 per cent of users are able to complete the task in 10 minutes). It may be necessary to specify criteria both for the target level of usability and for the minimum acceptable level of the quality of use.

If a task requires several users to collaborate to achieve a goal (group working), usability objectives can be set for both individual and the group.

11.5.2 Usability measurement

When usability is measured, it is important that the conditions for a test of usability are representative of important aspects of the overall context of use. Unless evaluation of usability can take place in conditions of actual use, it will be necessary to decide which important attributes of the actual or intended context of use are to be represented in the context used for evaluation. For example, in testing a DTV service, it may be appropriate to hold the evaluation sessions in a simulated lounge with a sofa, decorations and TV and to recruit users with an interest in DTV services. Particular attention should be given to those attributes that are judged to have a significant impact on the usability of the overall system. A key difference between usability measurement and informal testing (Section 11.4.3) is that for measurement, the evaluator should not intervene and assist the user. If the user gets completely stuck, they should move on to the next task.

11.5.2.1 Choosing tasks, users, environments

The choice of an appropriate sample of users, tasks for the users to perform and the environment for the test depends on the objectives of the evaluation, and how the system or product is expected to be used. Care should be taken in generalising the results of usability measurement from one context to another that may have significantly different types of users, tasks or environments. For a general-purpose product it may therefore be necessary to specify or measure usability in two or more different contexts. Also, if usability is measured over a short period of time it may not take account of infrequent occurrences which could have an impact on usability, such as interruptions or intermittent system errors. These aspects need to be considered separately.

11.5.2.2 Choosing usability measures

A specification of usability should consist of appropriate measures of user performance (effectiveness and efficiency), and of user satisfaction based on ISO 9241 [3]. It is normally necessary to provide at least one measure for each of the components of quality of use, and it may be necessary to repeat measures in several different contexts. There is no general rule for how measures can be combined.

Thus usability measures, based on the terms used in the definition, can be defined as in Table 11.3.

Table 11.3 Types of usability measure

Definitions of usability measurement types	
Effectiveness	The accuracy and completeness with which users achieve specified goals. An effectiveness measure may be defined as the percentage of the task(s) achieved.
Efficiency	The resources expended in relation to the accuracy and completeness with which users achieve goals. An efficiency measure may be the effectiveness achieved per unit of time e.g. 5% of the task achieved per minute.
Satisfaction	The comfort and acceptability of use. This may be measured by using rating scales in a validated questionnaire.

These measures may be defined either when specifying usability requirements for the system or usability evaluation criteria, e.g. it must be possible for a specified sample of users to be able to complete a test task with a mean of 80 per cent effectiveness.

11.5.2.3 Effectiveness

Effectiveness measures relate to the goals or sub-goals of using the system to the accuracy and completeness with which these goals can be achieved. For example if the desired goal is to perform a search on a specified topic, then 'accuracy' could be specified by the proportion of relevant hits in the first 20 listed. 'Completeness' could be specified as the proportion of key references that are shown in the list. Accuracy and completeness may be measured separately or multiplied together to give a combined measure.

11.5.2.4 Efficiency

Efficiency measures relate the level of effectiveness achieved to the resources used. The resources may be time, mental or physical effort or financial cost to the organisation. From a user's perspective, the time he or she spends carrying out the task, or the effort required to complete the task, are the resources he or she consumes. Thus efficiency can be defined in two ways:

$$\text{Efficiency} = \frac{\text{Effectiveness}}{\text{Task time or effort or cost}}$$

From the point of view of the organisation employing the user, the resource consumed is the cost to the organisation of the user carrying out the task, for instance:

- The labour costs of the user's time.
- The cost of the resources and the equipment used.
- The cost of any training required by the user.

The nature of the system and tasks being specified or evaluated will determine the range of the measures to be taken. For instance, the usability of a new animation creation feature will be influenced by a comparatively limited and well-defined context, while a new staff time-recording system will be influenced by a wide context that may include other users and organisational issues. For the animation feature efficiency might be based on the disk space or time required for a user to create an animation, while for the staff time-recording system, efficiency might be based on overall cost savings for the whole staff in the organisation.

11.5.2.5 Satisfaction

Measures of satisfaction describe the perceived usability of the overall system by its users and the acceptability of the system to the people who use it and to other people affected by its use. Measures of satisfaction may relate to specific aspects of the system or may be measures of satisfaction with the overall system. Satisfaction can be specified and measured by attitude rating scales such as SUMI (Software Usability Measurement Inventory, [21]), but for existing systems, attitude can also be assessed indirectly, for instance by measures such as the ratio of positive to negative comments during use, rate of absenteeism, or health problem reports. Measures of satisfaction can provide a useful indication of the user's perception of usability, even if it is not possible to obtain measures of effectiveness and efficiency.

11.5.2.6 Relative importance of measures

The choice of appropriate measures and level of detail is dependent on which context of use characteristics may influence usability and the objectives of the parties involved in the measurement. The importance each measure has relative to the goals of the overall system should be considered. Effectiveness and efficiency are usually a prime concern, but satisfaction may be even more important, for instance where usage is discretionary. For more information, see Quesenbury [22] who considers how weightings can be applied to what she defines as the '5Es of usability: effective, efficient, engaging, error tolerant, and easy to learn'.

11.5.2.7 Derived measures

Some usability objectives will relate to the effectiveness, efficiency and satisfaction of interaction in a particular context. But there are often broader objectives such as learnability or flexibility. These can be assessed by measuring effectiveness, efficiency and satisfaction across a range of contexts.

The learnability of a product can be measured by comparing the quality of use for users over time, or comparing the usability of a product for experienced and inexperienced users. (One of the prerequisites for learnability is that the product implements the dialogue principle 'suitability for learning', which refers to the attributes of a product that facilitate learning.)

Flexibility of use by different users for different tasks can be assessed by measuring usability in a number of different contexts. (One contribution to flexibility in use is that the product implements the dialogue principle 'suitability for individualisation', which refers to attributes of the product that facilitate adaptation to the user's needs for a given task.) Usability can also be assessed separately for subsidiary tasks such as maintenance. The maintainability of a product can be assessed by the quality of use of the maintenance procedure: in this case the task is software maintenance, and the user is the software maintainer.

11.5.2.8 Usability testing process

This section describes the process for measuring usability in a controlled and realistic manner. The benefits of the approach are that user trials will: indicate how users will react to the real system when built; provide experimental evidence to show the problems that users might envisage with the future system; enable the design team to compare existing products as a way of considering future options. Limitations are that: if the trials are too controlled, they may not give a realistic assessment of how users will perform with the system; they can be an expensive method, with many days needed to set up the trials, test with users and analyse results. Inputs to the design process may be too slow for the intended timescales.

Usability testing depends on recruiting a set of suitable subjects as users of the system. It also requires equipment to run the system and record evaluation data. This may include a video camera, video recorder, microphones, etc. Also it will be necessary to set up the tests in a suitable environment such as a laboratory or quiet area.

To prepare for controlled testing, the following items are required:

1. The description of the test tasks and scenarios.
2. A simple test procedure with written instructions.
3. A description of usability goals to be considered and criteria for assessing their achievement.
4. A predefined format to identify problems.
5. A debriefing interview guide.
6. A procedure to rank problems.
7. Develop or select data recording tools to apply during the test session (e.g. observation sheets) and afterwards (e.g. questionnaire and interview schedules).
8. Distribution of testing roles within the design team (e.g. overseeing the session, running the tests with the user, observation, controlling recording equipment etc.).
9. Estimate of the number of subjects required (possibly firmed up after the pilot test session).

Once trials are run, data is analysed and problem severity is prioritised in an implications report.

Practical guidelines for running user tests are as follows:

- Conduct the tests in a friendly atmosphere.
- Explain to the user that you are testing how usable the system is and are therefore not going to assist them. If they get stuck, they should move on to the next task. It may be necessary to intervene to set up the system for each task.
- Let the user decide when they want to move on to the next task.

- Allow enough time between user sessions for overruns.
- Make arrangements for telephone calls to be taken by someone else rather than interrupting the session.
- Make it clear that it is the system being tested.
- Make it clear beforehand how much subjects will be paid for the whole session. Avoid flexible payment based on time spent in the session.

For further information see References 9, 12, 17, 19 and 20.

11.6 User audit of implemented system

Collecting real usage data from users when the product is in service can be the most valuable test of whether a product is usable. This data can be useful to both user organisation and supplier. If such data can be gathered it can be reported back to marketing and engineering for use in planning improvements, and for the next generation of products.

The most common audit method involves collecting logs of usable data and following these with surveys and interviews to establish the reasons for usage and non-usage. Usage logs must be declared to avoid the charge of undisclosed monitoring. The data often reveals that some core facilities are in widespread use but that many others are rarely used. The survey work should identify whether non-use is because of no perceived benefit or because of usability problems. These distinctions may be significant in planning the next product. The following example illustrates the type of questions that an audit of usage might ask users:

- Have you used similar systems before? If so, which and for how long?
- How experienced a user of the system are you?
- How experienced are you at your job?
- How often do you use the system?
- Which functions do you use most often?
- What training have you received?
- Do you need further training?
- What form did your training take?
- Which features of the system do you like most? How do these help you?
- Which features of the system do you dislike? Why? What problems do they cause?
- Have you experienced any serious problems with the system? What did you do? Did you know how to get help?

11.7 Examples of usability projects

This section describes a series of case studies of usability work in practice to show how the activities described in this chapter were carried out. The studies mainly cover work that the author has been involved in at the HUSAT Research Institute (now part of ESRI) at Loughborough University. Each study is labelled as 'formative' or 'summative'.

11.7.1 Expert evaluation of training opportunities software (formative)

An evaluation was carried out on a website [23], which provided information about business-related courses to SMEs (Small and Medium Enterprises). This was part of a programme of work to develop user requirements for a web-based e-learning service or 'virtual campus'. An evaluation was performed by two human factors experts who spent time reviewing each of the main parts of the system from their own experience, knowledge of typical tasks and usability principles. When providing comments on the system the aim was not to improve the current system but to identify features and implications for the new system. Inputs, from a usability perspective, were made to the user specification of the new virtual campus system. These included elements such as: the inclusion of functions to cover course providers as well as users (as this stakeholder had not previously been considered); suggestion for a mechanism to enter, modify and delete course information and course modules; and provision of typical scenarios of use by the users to make sure that the supplier and customer have the same 'vision' of the system. The project demonstrated how expert evaluation of a current system can provide useful feedback into the requirements specification for the new system.

11.7.2 Re-design of intranet site (formative)

A study was carried out to evaluate and redesign an intranet site for a police service in the UK [24]. Human factors consultants performed the study working with a police officer who was project manager for intranet development, and a civilian co-ordinator with a knowledge of police procedures and human factors. Semi-structured interviews were performed with users and stakeholders covering: their needs and aspirations regarding the intranet; how well the current system meets those needs; and possible improvements that could be made. Interviewees were given access to the intranet site so they could demonstrate their comments. They included a constable, sergeant, inspector, senior officers and non-police administrative staff.

Following the user and stakeholder interviews, an expert review of the intranet pages was performed to establish the strengths and weaknesses of the current service. General recommendations for change were made following the expert evaluation. These were discussed with police representatives and different options for concept designs were proposed using storyboards and screen prototypes. These methods matched the requirement to create and discuss rapid prototypes within the design team. Having developed the design concept, several options for the graphic design for the site were produced as software prototypes to demonstrate both look and feel. A final design for the home page and secondary level content pages was then produced with web templates to allow the police service to install the new pages and maintain them in the future. The project showed how a combination of methods can produce an acceptable new system design within a relatively short time (three months).

11.7.3 Evaluation of a forest monitoring system (formative and summative)

The EC IST FOREMMS project (http://www.nr.no/foremms) developed an advanced forest monitoring and management system for Europe. Forest related data was stored in digital form collected from a network of sites across Europe. The system produced raster maps and datasets for scientists and public users showing the current state of forest eco-systems. In order to summarise the functions, qualities and data requirements for the systems, user interface simulations and prototypes have been developed. This was found to provide a good way to communicate the user and usability requirements within the design team, and to obtain confirmation from the users themselves. An example mock-up of a FOREMMS screen, developed in Powerpoint™, is shown in Figure 11.3.

When a prototype of the system was finished, a series of trials were conducted to measure the usability of the system [25]. A sample of potential users was recruited and asked to perform a series of typical tasks with the system. Measures were taken of task completion, task times and satisfaction. Recommendations were made to improve the system and a second round of tests were performed. This provided a direct comparison of usability results between the two trials. The main problem to address was maintaining equality of tasks between the two sets of trials as the capabilities between prototypes one and two changed and developed.

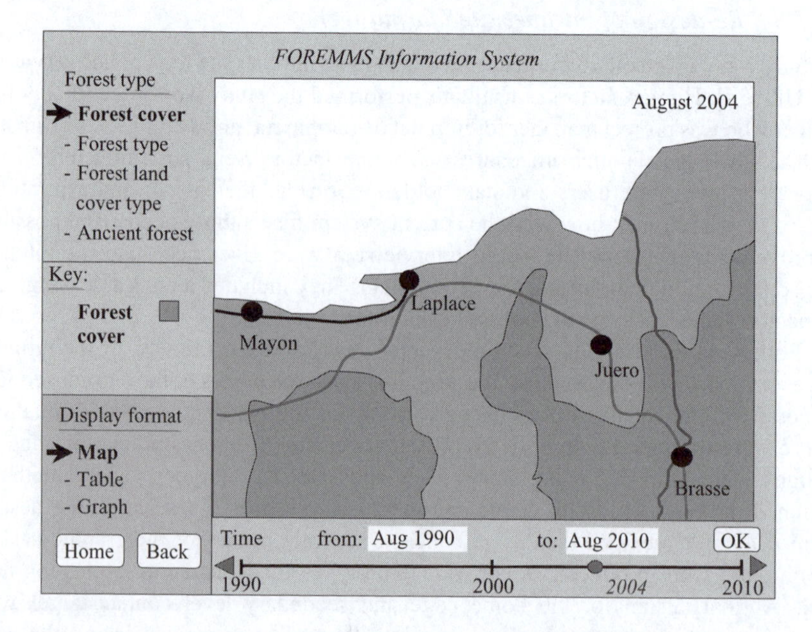

Figure 11.3 Example screen simulation for FOREMMS user interface

11.7.4 Application of the CIF to e-shopping (summative)

In 1997, the National Institute of Standards and Technology (NIST) developed a new industry-based reporting format for usability evaluation study, namely Common Industry Format (CIF) to help bridge the procurement decision for a software system between a consumer organisation and a supplier [26]. The main reason behind the initiative was the need to establish a standard criterion for a minimal indication of usability for companies or organisations in making large procurement decisions for software products. The CIF report format and metrics are consistent with the ISO 9241-11 [3] definition of usability. It was envisaged that a supplier would provide a CIF report to enable a corporate purchaser to take account of usability when making a purchase decision. A purchaser could compare CIF reports for alternative products (particularly if a common set of tasks had been used). The purchaser might specify in advance to a supplier the required values of the usability measures (for example, based on the values for an existing product).

This study was concerned with an online shopping website called Prezzybox [27]. The site offered a wide range of gift ideas for purchasers to select and have delivered, either to themselves, or to a friend or relative. The website was evaluated and reported using the CIF.

The main objective of the user test was to obtain user performance and satisfaction metrics for Prezzybox along with user comments to highlight any problems with the site. Testing was carried out with 12 users who were all experienced with the Internet and interested in buying from the Prezzybox site. Users were asked to make a real purchase from the site – this being their payment for taking part. Thus the evaluation was designed to test the success of users in completing an online purchase – a crucial success factor for any online shopping site.

The result of the evaluation was a CIF report that documented user performance and satisfaction with the site. It was found that 10 out of the 12 users made a purchase. If the consumer organisation could capture the two extra users, their sales would increase by 20 per cent. These performance results were also supported with user satisfaction ratings. The levels of satisfaction recorded were in general 'satisfactory' representing scope to improve the 'user experience' when using the site. Evaluator comments and recommendations for change to the site were also included, highlighting features that could be changed to help improve user success in making a purchase.

The Prezzybox site was refined based on the feedback and a second round of usability evaluation was carried out. Results showed a similar level of user performance in finding and purchasing from the site and an increase in user satisfaction. In general CIF was found to be good at reporting traditional usability metrics (effectiveness, efficiency, and satisfaction scores). However, the appropriateness of efficiency as a usability measure is not clear when the user may be browsing through a site. Longer shopping times may indicate pleasure in shopping online or a problem in finding goods. Shorter times may indicate an excellent site or a lack of interest leading to quick purchases.

11.7.5 User-centred design at IAI

The TRUMP project was set up by Serco Usability Services, working with IAI LAHAV to evaluate the benefits of applying user-centred methods on a typical project. The user-centred design techniques recommended by TRUMP [28] were selected to be simple to plan and apply, and easy to learn by development teams.

1. Stakeholder meeting and context of use workshop. The stakeholder meeting identifies and agrees on the role of usability, the usability goals, and how these relate to the business objectives and success criteria for the system. The context workshop collects detailed information about the intended users, their tasks, and the technical and environmental constraints. Both events each last for about half a day.
2. Scenarios of use. A half-day workshop to document examples of how users are expected to carry out key tasks in specified contexts, to provide an input to design and a basis for subsequent usability testing.
3. Evaluate an existing system. Evaluation of an earlier version or competitor system to identify usability problems and obtain measures of usability as an input to usability requirements.
4. Usability requirements. A half-day workshop to establish usability requirements for the user groups and tasks identified in the context of use analysis and in the scenarios.
5. Paper prototyping. Evaluation by users of quick low fidelity prototypes to clarify requirements and enable draft interaction designs and screen designs to be rapidly simulated and tested.
6. Style guide. Identify, document and adhere to industry, corporate or project conventions for screen and page design.
7. Evaluation of machine prototypes. Informal usability testing with three to five representative users carrying out key tasks to provide rapid feedback.
8. Usability testing. Formal usability testing with eight representatives of a user group carrying out key tasks to identify any remaining usability problems and evaluate whether usability objectives have been achieved.

IAI concluded that most of the techniques are very intuitive to understand, implement and facilitate. Practising these techniques in the early stages of design and development ensures fewer design mistakes later on. All participants and developers thought that most of the techniques were worthwhile and helped in developing a better and more usable system. The techniques were assessed as cost effective and inexpensive to apply.

11.7.6 Evaluation of home products (formative and summative)

The author participated in the European project FACE 'Familiarity Achieved through Common user interface Elements' [29]. The aim was to produce general design rules for consumer products. It involved collaboration with major companies manufacturing consumer electronics. The approach adopted was to simulate a range of alternative designs for different home product systems. These included a video recorder, a heating controller and a security system.

The process involved performing a user survey of 100 consumers, in each of four countries, to understand the attitudes of consumers towards existing devices and to gauge the importance of different usability characteristics. Interviews of 40 people were then performed to help study the way they use devices at present and to understand the problems consumers face with their own products at home.

From this work, a series of design rules for simpler usage was defined. These were implemented within a set of alternative software prototypes of new interface designs. The companies produced initial versions for human factors experts to comment on. This feedback was incorporated into revised simulations which were then tested with members of the public using assisted evaluations, performing tasks such as programming a video recorder or setting up a heating control programme. User attitudes were also measured using satisfaction questionnaires. The results enabled comparisons of prototype designs to be made and the best features of each design to be identified. These features fed into the refinement of the user interface design rules.

During the project, seven rounds of evaluation were conducted as part of an iterative design cycle. The project was able, by the use of the process described, to develop a set of design rules that were employed within consumer products. The exposure of the usability process to the industrial partners on the project enabled them to adopt similar procedures within future projects.

11.7.7 Evaluation of financial systems (summative)

Between 1993 and 1994, HUSAT performed a number of usability evaluations of a range of systems for money-market traders and dealers, for a major financial information organisation that recognised the need to improve the usability of its systems [30]. These evaluations were conducted at the client's site allowing them to employ representative users (traders and dealers) within the City of London. Again, due to the need for rapid feedback to design staff, these evaluations were conducted intensively – typically one to two days of preparation, one week of testing and three days to analyse the data and report the results. The aim was to obtain measures of performance and feedback from users, and to keep user sessions within a tight timescale. The approach adopted was a controlled user test where the users were asked to perform a series of tasks, and metric data and user satisfaction data were recorded using the SUMI satisfaction questionnaire. A typical problem encountered was availability of users (e.g. stock market traders) who often had to cancel their test sessions due to the demanding nature of their jobs. Following the successful application of this evaluation process to a number of financial products, it became company policy that all existing and future products would be evaluated formally with representative end users.

11.7.8 Evaluation of scientific search engine (summative)

A study was performed to compare a new search engine, designed to deliver information to science researchers, with two existing search engines as benchmark systems [31]. The aim was to assess how well the system matched competitor systems. It was decided to ask users to perform a controlled user test of the three search engines. This was carried out by recruiting a pool of scientific researchers from local universities. The study included 32 researchers from the disciplines of agriculture, biochemistry, biological science, chemical engineering, chemistry, computer science, life sciences and physics. Recruitment was carried out by advertising for participants via email through departmental distribution lists. Each volunteer was sent a document presenting the instructions to follow to test each of the search engines in turn. They self-recorded their results and responses onto the document and returned it to the experimenter. The order of usage of the search engines was randomised.

The basic test procedure for each user was to specify a query relevant to their own research, apply it to each search engine, and rate the relevance of the results. By rating the usefulness of the first 15 items returned, an interesting metric was calculated, i.e. mean relevance of items corresponding to their order in the list. Plotting these mean values for each search engine compared the relevance of items displayed high on the list of those retrieved. Having tried all three search engines, each user was asked to place them in order of preference, based on a range of design quality criteria, and to specify which search engine was preferred in general and whether it was likely to be used for their future research. These preference data corresponded well with the comparative relevance ratings.

The testing approach of providing users with a set of instructions to follow, and for them to test the search engines in their own time, proved a convenient way to organise remote user trials in a short time. Requesting users to perform a task to support their own work meant that the search engines were used for real tasks. This, together with a payment in the form of book tokens, increased their motivation to take part and provide feedback quickly. The approach proved successful and was applied to a broader scale evaluation employing users located around the world.

11.7.9 Evaluation of electronic programme guide (EPG)

A series of studies were carried out to evaluate a new electronic programme guide (EPG) for a digital television service [32]. The aim was to assess how usable a new digital TV system would be for typical TV viewers.

Context of use analysis highlighted the following characteristics of usage: viewing at a distance; limited space to display information; interaction via a handset; service integrated with TV programmes; system often used on a casual basis; without use of a manual; and in relaxed mode. A software prototype had been developed. The system included a range of features. A ‘Now and Next’ facility, displayed in a window, showed the name of the programme being watched on the current channel and what was coming next. The user could filter the large number of channels by type (e.g. sport, movies, children’s, documentaries) or select to see TV programmes by subject. Programme details were presented either in list format (as in a newspaper) or grid format. Recording, reminders to watch selected programmes and parental control facilities were also provided.

The prototype EPG was tested using assisted evaluation at the offices of the development organisation. This involved setting up three simulated lounge areas so that user sessions could be run simultaneously by different evaluators. Recording equipment was brought in from HUSAT's portable usability laboratory. Users were recruited according to specific characteristics through a recruiting agency. Users were required to perform a series of specified tasks in a co-operative fashion with the evaluator who observed their interactions and comments. Sixteen user sessions were carried out within a two-day period (over a weekend) when the system prototype was not being worked on by the development team.

The study demonstrated how user trials could be carried out within a simulated environment, off-site, over a short period of time. It also showed how human factors analysts from the consultancy organisation could work with a human factors specialist from the client organisation.

11.8 Conclusions

The achievement of usability within system design requires a combination of careful planning of the usability activities and a good understanding of the context of use for the system. Formative usability work normally requires work alongside the development team to help develop a usable system. This requires a flexible and pragmatic approach, concentration on the main usability issues, a good understanding of and empathy with the users and a sensitivity in reporting evaluation results to the development team.

Summative work requires clear specified usability goals and a rigorous approach in testing for their achievement. Usability testing results and recommendations for change must also be presented in a form that is palatable to the development team by emphasising the positive as well as the negative aspects of the system.

11.9 References

1 ISO/IEC DTR 9126-4: 'Software engineering – software product quality – Part 4: quality in use metrics.' Committee: ISO/IEC JTC 1/SC 7/WG 6, AZUMA, M., BEVAN, N., and SURYN, W. (Eds) (International Standards Organisation, Geneva, 2000)

2 SHACKEL, B.: 'Usability – context, framework, definition, design and evaluation', in SHACKEL, B., and RICHARDSON, S. J. (Eds); 'Human factors for informatics usability' (Cambridge University Press, Cambridge, 1991) pp. 21–37

3 BS EN ISO 9241-11: 'Ergonomic requirements for office work with visual display terminals (VDTs). Part 11 – Guidelines for specifying and measuring usability' (International Standards Organisation, Geneva, 1998) (Also available from the British Standards Institute, London)

4 BIAS, R. G., and MAYHEW, D. J. (Eds): 'Cost justifying usability' (Academic Press, Boston, 1994)

5 BS EN ISO 13407: 'Human-centred design processes for interactive systems' (International Standards Organisation, Geneva, 1999) (Also available from the British Standards Institute, London)
6 MAGUIRE, M. C.: 'Context of use within usability activities', *International Journal of Human-Computer Studies*, **55** (4), 453–484
7 THOMAS, C., and BEVAN, N.: 'Usability context analysis: a practical guide' (National Physical Laboratory, Teddington, Middlesex, 1995) (Now available from N. Bevan, Serco Usability Services, 22 Hand Court, London WC1V 6JF)
8 SHNEIDERMAN, B.: 'Designing the user interface: strategies for effective human-computer interaction' (Addison-Wesley, Reading, MA, 1992, 2nd edn.)
9 NIELSEN, J.: 'Usability engineering' (Academic Press, London, 1993)
10 NIELSEN, J.: 'Paper versus computer implementations as mock up scenarios for heuristic evaluation' Human-Computer Interaction – INTERACT'90 conference proceedings, 27–31 August, 1991, DIAPER, D., COCKTON, G., GILMORE, D., and SHACKEL, B. (Eds) (North-Holland, Amsterdam) pp. 315–320
11 RETTIG, M.: 'Prototyping for tiny fingers', *Communications of the ACM*, 1994, **37** (4), pp. 21–27
12 MAGUIRE, M.: 'Prototyping and evaluation guide' (ESRI (formerly HUSAT), Loughborough University, Loughborough, 1996)
13 PREECE, J., ROGERS, Y., SHARP, H., BENYON, D., HOLLAND S., and CAREY, T.: 'Human-computer interaction' (Addison-Wesley, Reading MA, 1994)
14 WILSON, J., and ROSENBERG, D.: 'Rapid prototyping for user interface design', in HELANDER, M. (Ed.): 'Handbook of human-computer interaction' (North-Holland, Amsterdam, 1988)
15 NIELSEN, J., and LANDAUER, T. K.: 'A mathematical model of finding of usability problems'. Proc. INTERCHI '93, Amsterdam, NL, 24–29 April, pp. 206–213
16 NIELSEN, J.: 'Finding usability problems through heuristic evaluation' Proc. ACM CHI'92, Monterey, CA, 3–7 May, pp. 373–380
17 LINDGAARD, G.: 'Usability testing and system evaluation – a guide for designing useful computer systems' (Chapman and Hall, London, 1994)
18 MONK, A., WRIGHT, P., HABER, J., and DAVENPORT, L.: 'Improving your human-computer interface: a practical technique' (Prentice Hall International, New York, USA, 1993)
19 RUBIN, J., and HUDSON, T. (Ed.): 'Handbook of usability testing: how to plan, design, and conduct effective tests' (John Wiley & Sons, Chichester, 1994)
20 DUMAS, J. S., and REDISH, J. C.: 'A practical guide to usability testing' (Intellectual Specialist Book Service Inc., UK, 2000)
21 KIRAKOWSKI, J.: 'The software usability measurement inventory: background and usage' in JORDAN, P. W., THOMAS, B., WEERDMEESTER, B. A., and McCLELLAND, I. L. (Eds): 'Usability evaluation in industry' (Taylor & Francis, London, 1996) pp. 169–178

22 QUESENBURY, W.: 'What does usability mean – five dimensions of user experience'. Presentation to the UK UPA, London, September 2002, www.WQusability.com

23 MAGUIRE, M. C., and HIRST, S. J.: 'Usability evaluation of the LINK project TIGS website and feedback on OVC specification' (ESRI (formerly HUSAT), Loughborough University, Loughborough, 2001)

24 MAGUIRE, M. C., and HIRST, S. J.: 'Metropolitan police service redesign of corporate intranet pages' (ESRI (formerly HUSAT), Loughborough University, Loughborough, 2001)

25 MAGUIRE, M. C., and SPADONI, F.: 'Deliverable D9: evaluation results, information societies technology (IST) programme project: FOREMMS – Forest Environmental Monitoring and Management System' (ESRI, Loughborough University, Loughborough, 2003)

26 National Committee for Information Technology Standards/NIST: Common Industry Format for Usability Test Reports, version 2.0. For NIST Industry Usability Reporting Project, American National Standards Institute, 27 July 2001

27 WONG, C.-Y., and MAGUIRE, M.: 'The effectiveness of the common industry format for reporting usability testing: a case study on an online shopping website', HCI International Conference, Crete, June, 2003

28 BEVAN, N., BOGOMOLNI, I., and RYAN, N.: 'Cost-effective user centred design', 2000, www.usability.serco.com/trump

29 BURMESTER, M. (Ed.): 'Guidelines and rules for design of user interfaces for electronic home products' ESPRIT 6994 (Fraunhofer IRB Verlag, Stuttgart, 1997) http://www.irb.fhgde/verlag

30 MAGUIRE, M. C., and CARTER, A.: 'Evaluation of Reuters RACE 3.0 Prototype' (ESRI (formerly HUSAT), Loughborough University, Loughborough, 1993)

31 MAGUIRE, M. C., and PHILLIPS, K.: 'User testing of search engines for science information', Project number HC/ES/220. HUSAT/ICE Ergonomics Limited[1], Loughborough University, 2001

32 MAGUIRE, M. C.: 'Report on the study of the management of domestic finances by family groups', NCR Knowledge Laboratory, HUSAT Research Institute[1], Loughborough University, 1999

[1] HUSAT and ICE have now merged to form ESRI – The Ergonomics and Safety Research Institute. Any correspondence regarding papers published by HUSAT or ICE should be addressed to: ESRI, Holywell Building, Holywell Way, Loughborough, Leicestershire, LE11 3UZ, UK.

Chapter 12

HF verification and validation (V&V)

Iain S. MacLeod

12.1 Introduction

This chapter will give guidance on how to assist the performance of system verification and validation (V&V) through the application of human factors (HF) to design during the various phases of a system's life cycle. Notably, a system will be considered as consisting of both humans and machines operating within social, cultural and operating environments. Figure 12.1 depicts this problem space and suggests that HF should be central to the consideration of systems. Therefore, the importance of empathy by the HF practitioner to the activities of the system engineer/maintainer and the user/operator should be emphasised. Moreover, it is argued that true expertise in HF is only acquired through a polymathical approach to work and an understanding of the approaches to the challenges of a system's life cycle by other disciplines such as systems engineering and integrated logistics support.

The earlier chapters of this book address many areas related to the application of HF methods to design and of issues related to human and system work. This chapter

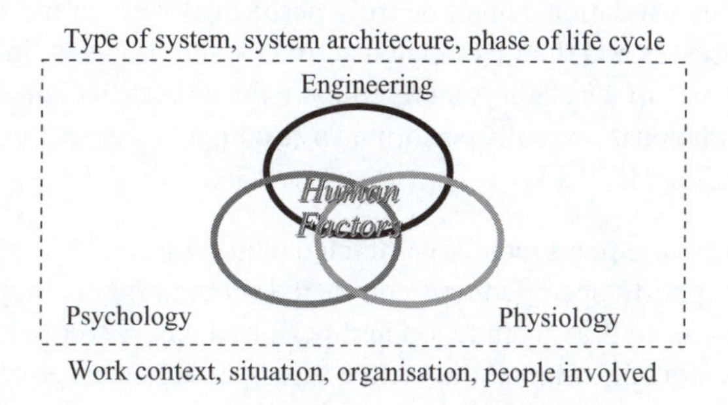

Figure 12.1 Influences on system consideration

will attempt to unite the previous considerations of HF into an overall perspective on the application of HF to systems. It will not enter into argument on specifics but rather will try and indicate the generics of good HF practice that support the V&V of HF work.

HF V&V is about checking that the required HF application within a system's life cycle processes are matched by a planned, knowledgeable, and traceable use of current HF methods, tools, and professional effort. HF is an area of professional endeavour with many diverse influences on its application. A wealth of publications address these issues but all are incomplete with relation to the matching of knowledge and professional competency to pertinent system life cycle processes. Nevertheless, there are recent examples of attempts to express competence requirements for work, however poorly matched to existing guidelines.

Two examples are:

- For industry: SEI CMM Model.
- For role: IEE Competency Guidelines for Safety-Related Systems Practitioners.

An understanding of the relationships within V&V is central to the discussions and arguments of this chapter. Two basic definitions will suffice as an introduction, namely:

- *HF validation*. The process of checking the correctness of planned and applied HF work to system life cycle processes, the attention of the work to system requirements, to the customer's duty of care responsibilities and to the achievement of system fitness for purpose (FfP). This process requires competence and adequate supervision within the HF team.
- *HF verification*. The process of confirming that the efforts and end products of applied HF have correctly addressed system safety issues, contracted performance requirements (usually based in a systems requirement specification), and strived for the achievement of system FfP: in particular, that the validity of the HF verification process has been substantiated throughout the system life cycle by a competent application of HF methods, an examination of applicable recorded arguments, decisions, and proofs, and maintained trace of evidence in support of the final system acceptance (or acceptance of upgrades during its period of service). Thus validation cannot be truly performed without the support of verification processes and their associated activities and products. In most cases the final acceptance of a system is a prerequisite for its certification for use – in this case HF validation is normally performed by a competent practitioner independent of the preceding HF V&V.

System FfP requirements include contracted requirements. However, FfP requirements are often poorly specified and contracted, containing as they do subjective issues of quality as well as of function and performance. Nevertheless, they are of high importance not only with relation to the safe operational usage of the system but also with relation to the duty of care responsibilities placed on the system owner by the UK Health and Safety Executive or other national equivalents.

12.1.1 Fitness for purpose (FfP)

The FfP requirements of a system can be said to include the following:

- Utility: can the system be effectively applied to its intended application?
- Ease of use: related to system supervision, management, direction, control, quality of life issues.
- Reliability: a system can be reliable but have poor utility but cannot have good utility without reliability.
- Safety: the system must be safe to operate and aid the human in recovery from system errors.
- Contracted requirements: the degree with which the system satisfies the contracted performance requirements.

There are many good publications/guidelines/standards giving information and advice on the application of HF to system work. They normally explain what should be done in a very prescriptive fashion, examples being:

- ISO PAS 18152, 2003 [1].
- Federal Aviation Administration, Human Factors Job Aid [2].
- Def Stan 00-25, Parts 12 and 13 [3].
- Do it by Design, An Introduction to Human Factors in Medical Devices [4].

A problem with the use of all the above publications is that any over-adherence to their contents can evoke complacency and mar the quality of the applied application of HF within its context.

12.2 Background

HF publications, such as the ones cited above, tend to lack arguments on *why* HF V&V should be done, *how* it should be done (outside the performance of methods), *what* iteration of activities should be made to progressively test results and assumptions, *where* these activities should occur, *what* the products of the activities should be, and in *what form* they should be presented considering the product user.

It is not within the scope of this chapter to detail all the above issues and arguments (see Chapters 1 and 2) but rather to show that it is necessary to consider all that is pertinent for the purposes of verification (that includes consideration on the validity of decisions). It is essential that sufficient robust and high-quality evidence is produced from the verification activities to support progressive and final validation activities to prepare for the acceptance of the system into service and continued acceptance of the system whilst in service. Thus verification evidence should have continuity and be without unexplained gaps.

HF applications to design imply the use of decision processes to determine the best compromises through trade offs possible for the optimisation of system design. Trade offs must be part of HF V&V to ensure that the best HF evidence is used to inform system life cycle decisions (this area is discussed in Section 12.2.2). This chapter considers a selection of HF areas of attention as appropriate and discusses related issues and implications of importance of the HF contribution to the V&V of systems. Practical approaches to the application of HF V&V are presented and examples given. The forms and timing of evidence required from HF V&V activities are discussed.

12.2.1 Limitations of HF discipline

It should be noted that at the time of publication there exists a far from perfect understanding and matching of physical HF to those human cognitive factors applicable to the performance of work. Indeed the meaningful matching of brain activities to system work activities is still in its infancy [5]. However, this should not thwart HF efforts to improve the safety and performance aspects of the design and performance of systems, if this is done with an awareness of the implications to systems quality of both the sufficiency and the limitations of our current knowledge. Indeed, HF has an important and strong life cycle contribution to make as it is the only professional systems-associated application with a sole focus on optimising and promoting the human contribution to systems performance and safety.

It is important to ask what system life cycle benefits necessitate the V&V of HF being performed. HF V&V is about maintaining focus and promoting the robustness and quality of HF application to systems. The root question must therefore be related to benefits that can be accrued through HF application. The benefits should include the achievement of a good system FfP. These benefits should also be related to savings in cost, both monetary and political. Cost benefits have been touched upon in Chapter 1, but for further discussions on this topic the reader should see References 6, 7 and 8. Various estimates have been made of the cost of late or inappropriate application of HF to system design with unplanned rectification costs increasing dramatically the later they are applied in the life cycle. Figure 12.2 illustrates the basic argument.

However, late HF application costs are not only related to monetary costs but to costs related to the influences of the diverse politics of stakeholders and the unwanted consequences to system utility and safe performance if the system is not FfP [9]. Poor system FfP is not only a result of inadequate integration of HF into the main stream of system activities, it is related to many diverse influences including a lack of adequate

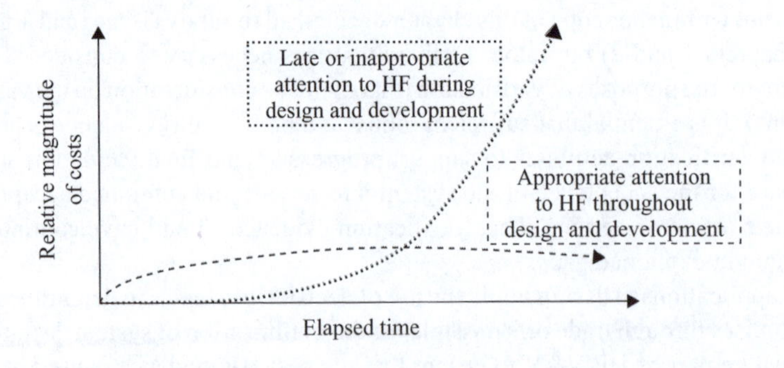

Figure 12.2 Example rectification costs post late/inappropriate attention to HF

consideration of system requirements and their specification. This latter shortfall arguably exists because system requirements engineering is poor at considering the human within its remit, and then mainly considers the human as a system constraint (constraints normally being without a performance attribute).

Further, there is often a poor use of the professional advice from subject matter experts (SMEs) (people with experience in the operation, management, maintenance, or control of similar systems or the system predecessor) of what system qualities to retain or strive for and what system attributes should be avoided. SMEs are a more problematical source of advice at the start of a system life cycle where the system is novel. However, the careful use of SME advice is an intrinsic part of the application of HF methods within system specification and test and evaluation (T&E) activities.

12.2.2 Consideration of trade offs within assist human factor integration

Human factors integration (HFI) is considered in Chapter 2. As a reminder its primary domains are system manning, personnel, Human Factors Engineering (HFE), training, and health and safety. Since we are considering trade offs from the orientation of the human aspects of systems work, we will start an example trade off consideration with the domains of manning and personnel.

Manning refers to the number of people applying effort to system performance whereas personnel refers to the particular skills and roles required of those manning the system. In a simple case, the manning numbers could be indirectly proportional to the level of personnel skills required. However, the technology adopted, system architectural complexity, and level of automation possible with the engineering components of the system have to be considered. To approach the effects of the engineered components on manning and personnel requirements, HFI can assist through considerations of the range of skills available to the customer, the mission requirements, the system functionality and performance requirements, and the task requirements of the system. Here, any changes brought about by technology should be carefully considered and understood. In turn, HFI consideration of FfP issues tie in the HFI to related issues such as systems utility, ease of use, and reliability. Also, safety engineering contributes to the trade off process by considering tasks that may be hazardous and confers with HFE on their amelioration either through engineering design, or levels of manning and personnel.

In turn trainers can contribute on issues of ease and effectiveness of training by considering the options available from the previous trade off processes. Finally, health and safety contributes to ensure that a reasonable level of care has been applied considering the health and safety people working with the system or those affected by the operation of the system.

From the above it can be seen that trade offs are complex and consider many aspects of HFI domains. Where possible the approaches to trade off should be planned. The accepted priority of HFI domains in trade off considerations will vary depending on the particular issue, its implications, and the phase of the system life cycle. However, the trade off process and its associated T&E process that assesses the quality

Table 12.1 Reasons for systems failure during design and development

Reasons for failure	Percentage indication
Incomplete requirements	13.1
Lack of user involvement	12.4
Inadequate resources	10.6
Unrealistic user expectations	9.9
Lack of management support	9.3
Requirements keep changing	8.7
Inadequate planning	8.1

Source: [10]

of HFI trade offs and other design decisions, must be conducted iteratively regardless of the adopted system development model. The results of trade off processes are always a compromise. It must be the duty of the HF V&V to ensure that the best compromise is strived for and that trade off processes and rules are made explicit.

12.2.3 Limitations in system manufacturing capabilities

The Standish Group [10] noted that in the UK over 53 per cent of the major civilian software intensive projects that they investigated had failed. Internationally, the figures have shown a worse picture indicating only a 16 per cent success rate in major digital electronics-based projects, accompanied by large cost and time overruns. Failure rates remain high. The same report indicated that 45 per cent of the functions of most systems were never used suggesting a lack of, or poor application of, HF V&V with regards to system FfP. Table 12.1 illustrates an assessment of some of the historical reasons behind system failures during their design and development. Note the top two reasons given for failure.

To understand how HF V&V is applicable to a system's life cycle we first have to consider what a system life cycle is. This subject is addressed in Chapter 2 but for completeness will also be addressed in the next section.

12.3 Systems, V&V and human factors

A system is a collection of parts where the product of the whole is greater than the sum of the products of the individual parts. A system life cycle is a consideration of a system from concept, through growth, to maturity, eventual demise, and then to system disposal or possible rebirth in another form. The specifics of what constitutes a system life cycle are much debated. However, in UK defence the current life cycle model is termed CADMID (Concept, Assessment, Development, Manufacture, In-Service and Disposal).

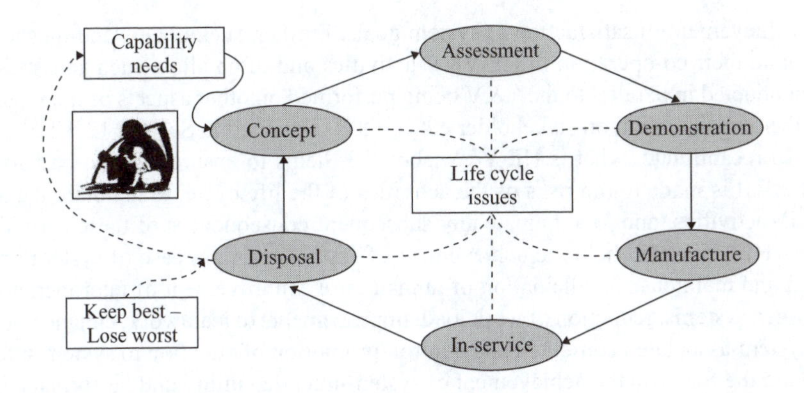

Figure 12.3 Depiction of a system life cycle

Different stages of system life cycle will require different levels and forms of HF application and practice. For example, at the concept or implementation-free stage of a system life cycle the HF application will mainly relate to high-level issues of system operation, manning, personnel, issues related to the operating environment, programme and operating risks, and information/knowledge requirements. In contrast, during system development the emphasis placed on HF work should be on a progressive test, evaluation and demonstration that the system meets the requirements as approached at that particular stage of development. Figure 12.3 illustrates a life cycle.

The forms and practice of V&V activities within the system life cycle must always consider the applicable forms and levels of the operating environment. Also of importance is the degree of FfP requirements, that exists at the time of the V&V activities with relation to operators, commanders and maintainers.

12.3.1 Issues with HF V&V

Consideration of the issues and implications arising from V&V may place a strong caveat on the robustness of the V&V results and flag the need for further checks later in the system life cycle. An understanding of the influences of the work and cultural organisation of the customer on V&V may also be important.

With relation to all of the above is a necessary degree of empathy with relation to the strictures affecting work and the quality of life issues pertinent to the situation. This empathy should not only relate to the end users of the system or their representatives – empathy is also important with relation to the managers and skilled professionals designing and developing the system because of their involvement in activities throughout the system life cycle. They, also, are stakeholders in the system. Indeed it is sensible that just as an HF practitioner should understand the system operating/maintenance tasks, they should also understand the problems, skills and tasks of the personnel involved in a system's design and development. It is argued that only through such an understanding can those personnel be advised and guided on the work needs of systems operating personnel and their relationships to the achievement of satisfaction of system goals. Further, such understanding should promote their co-operation in HF V&V activities and often allow such activities to be conducted in parallel to the V&V being performed on other aspects of the system. Further consideration on stakeholder concerns is contained in Section 12.4.1.

To recapitulate, what is HF V&V about? It helps to ensure that the best focus and effort is made to the risks of the activities of the life cycle, to maintain the aim of HF activities, and to anticipate any subsequent consequences of these activities. Good HF V&V will help the achievement of ease of training, ease of system operation and maintenance, alleviation of human error, improvement of interoperability between systems, reduction of work load, improvements to teamwork, enhancements of system-associated command and control, promotion of attention to system safety, and aid the focus on the achievement of system-intended utility and performance.

The above generics will include the matching of HF practices into the overall system life cycle activities, processes and management considering the multi-disciplinary work applicable to each phase of a system's life cycle. Table 12.2 summates some of the basic areas of HF V&V address for each phase.

All systems grow and mature from foundations based on a need for a system to fulfil the whole or part of a desired capability. Systems grow through stages of embryo, early growth, maturity, redundancy and an eventual death. These are truisms. However, the way a system develops and grows depends on several important influences.

Table 12.2 Summary of basic areas of address applicable to HF V&V activities

Life cycle stage	Some areas of HF application
Concept	Concept of operations or use, mission analysis, initial assessment of hazards, manning and personnel estimates, studies into aspects of system work, reliability and performance, user requirements document.
Assessment	Modelling, use of prototyping and synthetic environments, estimates of life cycle costs (life cycle main costs usually incurred by personnel, maintenance, training and rectification), evaluation, system requirements document.
Demonstration	Use of synthetic environments and prototypes as demonstrators. Decide on the physical model of the system.
Manufacture	Management and system specialisation plans including HF, safety and T&E. System physical build and evaluation. Decisions on trade offs to achieve best compromise and derived functionality. Evidence trace to requirements. System acceptance into service, use of prototypes and simulations for T&E during system development.
In-service	Maintenance, modification, and system updates. Use of simulations for training and T&E of system updates.
Disposal	Safety, cost, record good system properties and qualities and discard the bad, modelling of decommissioning process.

As a metaphor these influences may be compared to the nature/nurture debate on the source of human capabilities and personality.

The 'nature' of system-build mainly depends on traditional needs and established practices or traditions. This is with relation not only to the design and development of the adopted life cycle models but also with relation to the professions involved and the paradigms directing their education and professional practices. 'Nurture' is related to the changes and developments within societies and their uses of technology and encompasses:

- the influences that these changes bring to bear on an understanding of the benefits and drawbacks inherent in the particular technology;
- the risks associated with the traditional methods and their application to the adoption of the technology (see [11] for discussion on this area);
- the understanding of changes that the technology can bring to the nature of systems-associated work;
- the changes that can occur to organisational practices and culture within which the system resides.

12.3.2 Influences of technology

It is almost impossible to understand the relative importance of the influences on work of one technological change to another. However, technological developments do have major impacts on the nature of work and thus on the practice of HF and the quality of its application. Table 12.3 presents a rough guide on the speed and nature of change in technology over the last few decades.

Considering rapid changes in technology, automation (see Chapter 6) and associated changes to the nature of work, it is therefore all the more important to establish a careful routine of HF V&V and match this routine to system management plans and other system V&V activities.

From the above discussion it can be argued that an understanding of the generics of a system life cycle is not enough. What is also required is an understanding of the overall processes and skills as applied within that life cycle and the influence

Table 12.3 Example technology changes in the last few decades

Form of change	Approximate dates of application
Manual work assisted by analogue computing	1940s/1950s
Hybrid systems of analogue computing and electronics	1960s/1970s
Programmable electronic systems	1970s/1980s
Networked programmable electronics	1980s
Programmable communication nets	1980s
Real time handling and use of knowledge	Late 1990s/2000+

of new practices and technologies on the efficacy of these processes and skills. To understand the change influences of technology on work the technology itself must be understood. This understanding is aided by modelling, prototyping, demonstration (the former by using static and dynamic modelling techniques), and simulation of the system. The purposes and uses of simulation are addressed in Chapter 13.

It is important that all these activities are not done merely to occasionally demonstrate facets of the system to the customer. They are after all only analogues of reality. Understanding of most of the benefits and drawbacks of a system can only be realised through a careful programme of T&E and, eventually, the operation of the system in context and actual work situations.

In many cases the advances and foreseen gains in adopting technology have outstripped the developments of methods needed for their market application, management and use (the problems in failures of e-commerce being a recent example in 2002). There are many examples of technology being incorporated into design solely on arguments of cost saving and not on FfP. For example, the difficulties encountered by aircraft pilots in the operational use of flight management systems (FMSs) and mode confusion exacerbated by the adoption of the 'glass cockpit' of digital computer-driven displays in place of the previous analogue flight displays [12]. The question is how do we understand and approach the matching of technology to functionality and performance to achieve a required system capability.

12.4 HF V&V relationship to system-related human functions and responsibilities

A system function can be defined as a property of the system with an associated performance capability that is latent until evoked by an application of effort. A system is built to meet a defined capability within the bounds of its envisaged performance in a specified operating environment. A constraint defines the boundaries of system usage. A task is then a goal driven application of effort to a system function or functions, normally within defined constraints, to satisfy explicit system work goals. Moreover, system tasks can reside in the engineered system, the human components of the system, or both.

Discussion on human system-related functions, tasks and responsibilities must be made at the onset with reference with the human role within the system. Examples of roles are system operator, maintainer, engineer, controller, commander, manager or supervisor. Each has a different concern with the system, this concern being mediated not only by the type of system and its quality but also by their appreciation of the financial environment, the organisation they represent and the politics of that organisation. Table 12.4 presents some of the concerns of a sample of the stakeholders in a system.

12.4.1 Roles and stakeholders

Full consideration of human roles must also be accompanied by addressing the personnel and manning of the system with relation to the human effort required to

Table 12.4 Examples of typical concerns of system stakeholders

Stakeholder	Some typical concerns
Customers	Normally dictate the terms and type of contract, and the form of system development. Supervise the manufacturer over system life cycle phases. Accept the system into service on behalf of the end user.
Industry	Plans and manages work to meet contracted delivery dates. Works to specified functional and performance requirements. Relates work to standards and applicable professional practices. Influenced the profit expected by shareholders. Liaises with customer to govern their expectations.
Engineers	Work to specific interpretations of contracted requirements. Work to specialist plans under an overall management plan. Apply agreed practices of their profession.

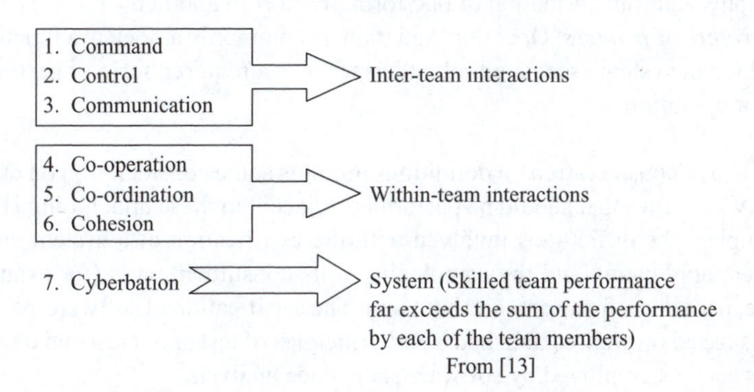

Figure 12.4 The seven 'C's of teamwork

operate, maintain, command, control, manage or supervise system operation. Within this area there should be HF V&V concerns about teamwork as well as on individual roles. Interactions between members of a team are just as important for HF consideration as the interactions between the individual and the engineered components of a system. Questions that should be raised concern issues of communications and work within and between teams. Figure 12.4 illustrates some of the considerations that might be placed on teamwork considering the HF V&V of the team's structure and disposition.

The relationship of system function and task requires further elaboration. The word 'function' has different meanings depending on the context of use and the profession of the user.

It is worth emphasising these differences as the word is used too glibly in discussions related to any defined system design life cycle [14]. This chapter will not attempt to put forward any one definition of the term. What is important to HF V&V is that the use of the term is accompanied by a meaning that is understood by other professions involved with the system.

There follows a consideration of some interpretations of the term 'function' by a few of the disciplines influencing system design (see [15]):

- *System engineering (SE)*. It can be argued that in SE a function represents something that can be implemented to help achieve a desired system behaviour and performance.
- *Software engineering*. In practice, a software engineer coding to a specification thinks of a function as the interrelationship of the coding activities that have to be performed to meet a part of their task requirement.
- *Engineering test*. Test engineers often consider functions to be subsets of a specified system function such that a discrete input evokes a discrete output that is only in part representative of the desired output from the system function being tested.
- *Mechanical engineering*. In mechanical engineering a function can be described as a physical transformation of one form of effort to another.
- *Operators or trainers*. Operators and trainers commonly associate a function with the human system associated role; they see it as a required form of performance, act or operation.

The above consideration on definitions prompts some consideration on other system V&V activities that should be performed adjacent to these undergoing HF V&V. As examples, the difficulties involved with the certification of a system vary with the system application and the complexity of its constituent parts (for example, its software, hardware, firmware, and humans). The certification of software-based electronics is based on well-established V&V principles of test and retest and developing methods such as employed by software static code analysis.

Hardware principles are also well established. With engineered components *verification* is the process of evaluating a system or component to determine whether the products of a given development phase satisfy the conditions imposed at the start of the phase [16]. In contrast, *validation* is based on the evaluation process that ensures that the developed product is compliant to the given product requirements. System acceptance principles are less well established but are usually based on specified system performance criteria and a regime of testing, evaluation and acceptance trials testing.

However, V&V methods for any new technologies will be difficult to apply because of the diverse technological developments that are liable to be involved in system build (as examples: neural nets, knowledge-based systems (KBS), developments in data base technologies, voice input command to systems, advances in computing technologies to name but a few). Existing approaches to the introduction of new technologies lack maturity and are too narrow in scope to meet most current system acceptance and certification requirements. Of note, the level of stricture on system acceptance will be less for non-safety critical systems.

Furthermore, the approaches to new technologies have a tendency to focus on issues purely related to the technology and ignore the real world issues such as those approached by HF V&V activities as applied to systems. However, real world issues must be considered by a system certification process if it is to be effective.

Current system engineering (SE) practices and methods are available to allow software build to meet the challenges of the existing certification processes. Software and human factors both exist outside the natural laws that bind the practice of most engineering disciplines. The approach to the acceptance of new technologies and the performance of HF V&V should consider examples from the best practices for incorporating software that exist with the current SE profession.

12.4.2 Forms of system development models

System testing requires a good knowledge of what is to be tested, how it is to be tested, and the expected system performance. Models of the system and of its constituent parts can depict such knowledge. Progressive testing of a system feeds the total V&V process, gives assurance that the system design is progressing as planned and assists in de-risking the design by highlighting inherent problems at as early a stage as possible.

A system model is traditionally rooted in specified system requirements. A requirements specification is argued to be just as essential for human components of a system as for any part of the engineered system. However, the associated functional specification of the human role should be constructed iteratively in the light of HF practitioners, and other system professionals, developing an understanding of problems and processes associated with the system work area. Indeed, especially if the system is novel, many required human functions will only be realised through an iterative and progressive programme of life cycle T&E.

Systems are developed throughout their life cycle using models as a template for the work. The traditional system construction model that has existed from before the creation of the Egyptian pyramids is the 'waterfall' model. This model has system developments occurring under a single procedural process that in its simplest form has phases related to system design, build, and test.

Therefore, the traditional engineering waterfall model applied to system design and development cannot represent the optimum when a more iterative approach is required to support the real world life cycle needs of most future complex systems. Figure 12.5 illustrates the waterfall model and alternatives to that model argued as more suitable for the support of the V&V associated with the introduction of new technologies. These models are the incremental and evolutionary models.

A mutation from these two developments is termed the spiral model. This model supports progressive test to the system throughout its design and development and indeed is applicable throughout the whole of the life cycle. This model is arguably the optimum to allow effective HF V&V as it allows trace and progressive checking on HF issues. However, it is a difficult model to contract. The spiral model is illustrated in Figure 12.6.

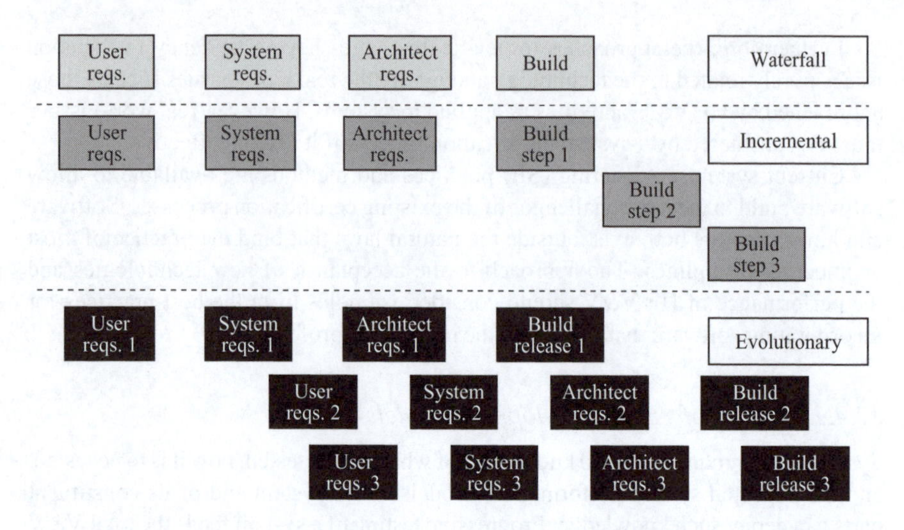

Figure 12.5 Waterfall, incremental and evolutionary models compared

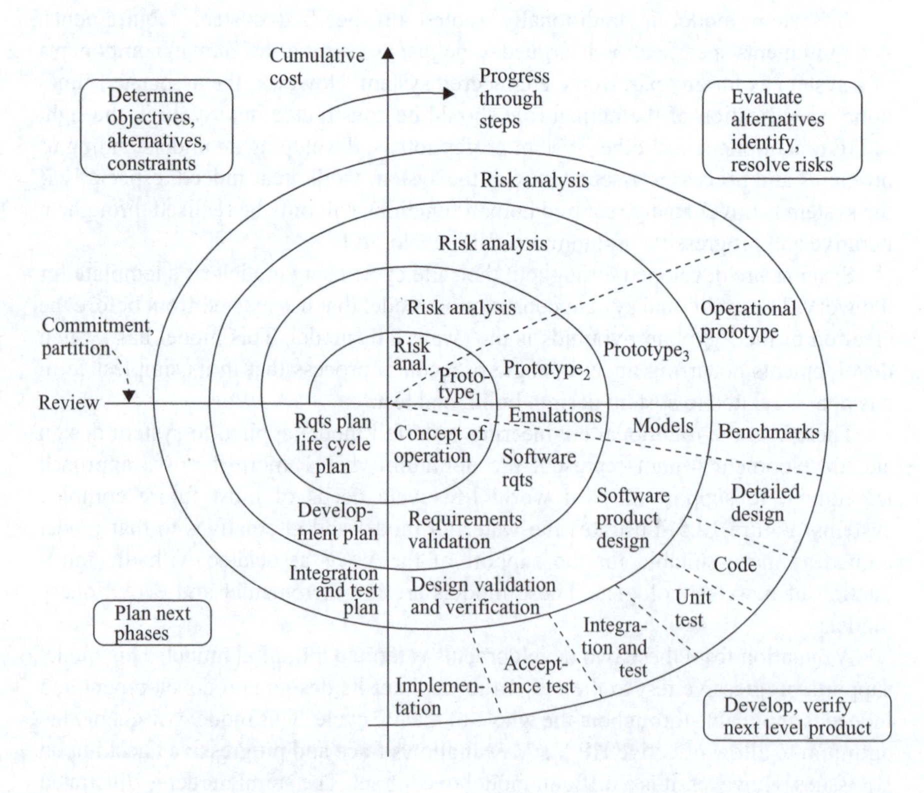

Figure 12.6 A spiral model for a system evolutionary acquisition. (Adapted from 17)

12.5 Pertinence of HF V&V activities within a system's life cycle

Increasingly the introduction of newer technologies and advances in computing techniques is increasing the potential power of systems to perform work. However, technology has a tendency to outstrip the methods and skills needed for its incorporation into systems, whether that is for new systems or for updates to systems already in service.

To reiterate, the introduction of new technology into a workplace changes the nature of work. This can be for the better or worse depending on the purposes for the introduction; also on the care with which it is introduced, and the V&V associated with its introduction.

Reliability of any new system's technology is not the only facet of system reliability that should be considered. Because of the changes to human work that are inevitable with the introduction of new technology it is imperative to assess the benefits and drawbacks of the levels of automation possible with that technology (see Chapter 6) and any inherent propensity for promoting forms of human error (Chapter 7) that may be associated with that automation.

In addition, the quality of design decisions within the system will have a large impact on the safe use, ease of use and utility of the system. All the above will impact on the quality of the system with relation to its FfP, part of which is the ease of training the system personnel in that system (Chapter 4). Quality of system design will also impact on the maintenance of trained skills and the development of expertise in role and, if required, on team working.

All the above show that efficient addressing of HF V&V is necessary to ensure that the best trade off and compromises are achieved throughout the system's life cycle with relation to the costs involved in its operation. These costs cover the quality imbued in the system to support the effectiveness of system operation and maintenance. Both operation and maintenance must be aided by the system through support to tasks of system command, control, management, and supervision. All these tasks require that system personnel are informed by the system in a useful and timely manner, and can likewise inform the system, to allow an ease of progression towards the satisfaction of system goals regardless of the system's particular tasks, actions or the assignment/mission it is required to perform.

Human Reliability Assessment (HRA) can be used (Chapter 8) as part of HF V&V and as a check on system safety both during development and in service. It is most appropriate to the study of human work involved with procedural and highly administered tasks.

HF V&V activities can check on all the above to ensure that the best practice, decision processes, and trade offs are applied. However, regardless of the professionalism of the HF endeavour and its associated V&V, it has to be appreciated that in the early parts of the system life cycle there will be facets of system design and operation that have been inadequately approached or missed through constraints placed by time, cost or lack of appropriate knowledge. This is where it is important that the operation of the system is carefully tested prior to its acceptance at the end of manufacture and that it is continually tested and reported throughout the in-service

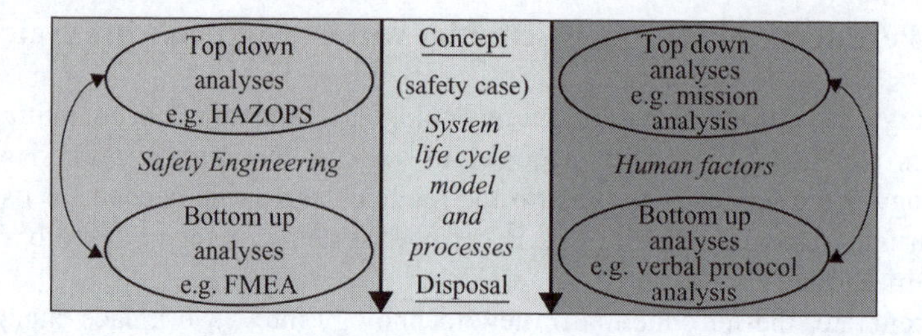

Figure 12.7 Illustration of the matching of safety engineering and HF models. (HAZOPS – Hazard and Operability Study. FMEA – Failure Modes and Effects Analysis)

phase. Therefore HF V&V is not only related to the progression of the initial phases of the system life cycle, it is appropriate to the assessment of system performance and updates, and the improvement and standardisation of its usage by its owners, throughout its in-service phase. Taking the general example of military systems, it is during the in-service phase that the human and engineering aspects of system usage come into the closest proximity through the continual V&V of the reliability of system performance, its effectiveness and its safe operation.

Furthermore, HF support to the safety case is important to the maintenance of the safety of system management, operation and maintenance. The safety case should be a continuous record of safety issues, implications, decisions and remedial actions throughout a system's life cycle. Indeed, there are parallels between the methods used by safety engineers and those used by HF practitioners [18]. These HF methods can all be used to contribute to building the safety case. Chapter 14 addresses issues of HF and safety (see also Figure 12.7). The following section will consider in detail some of the HF V&V activities that should be conducted throughout a system's life cycle.

12.6 Detail on HF V&V activities throughout a system's life cycle

There are many HF V&V activities that should be conducted and revisited throughout a system's life cycle. The primary drivers for such activities should be their relevance to the investigation of the effects of changes to the system or the usage of the system.

In most cases there are instigating activities and subsequent follow up activities depending on the phase of the life cycle and the changes to the system that have occurred. The use of HF V&V through the following activities will be illustrated:

- task analysis (see Chapter 5);
- use of modelling, prototyping, and simulation (see Chapter 13);
- T&E (Section 12.6.4).

12.6.1 HF V&V and task analysis (TA)

Task analysis (TA) is addressed in Chapter 5. This section will not present detail on TA methods but will outline the prerequisites and outcomes of such analysis with relation to HF V&V.

Analysis of any work task requires considerable background knowledge on the nature of the task and its scenario. Figure 12.8 indicates the influences on that creation and usage of such knowledge.

TA can be performed by several means and presented in many forms. It is an analysis of human-influenced system effort as directed towards the satisfaction of system goals through a planned use of system resident functionality. Arguably, a TA should include a consideration of the efforts of both human and machine. However, the human is considered as the decision maker and the provider of direction to the system efforts. TA can be performed on data and information collected by observation or data logging. It is frequently performed by using as a source system performance specifications and details of system equipments. It is usually used and checked for its robustness by one or several of the following means:

- review and update of the analysis by SMEs;
- checking of the analysis through task analytic simulation;
- by the use of synthetic environments;
- through system prototyping;
- by using system development rigs;
- through the use of simulators;
- by real world operation of the system;
- as a baseline for examining the effects of change to the system.

To perform any TA requires a detailed knowledge of the intended utilisation of the system (CONcept of OPerationS [CONOPS] in military terms) and a high-level analysis of the system processes required for system operation, this analysis

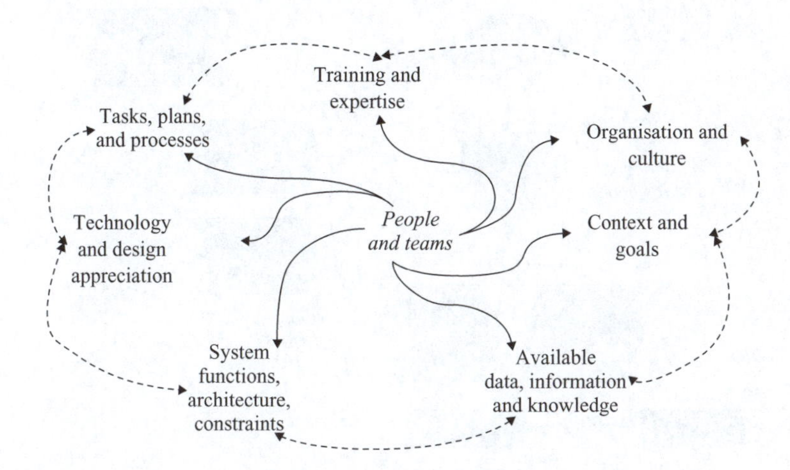

Figure 12.8 Influences on creation and usage of work knowledge

considering the relevance of subsystems to these processes (termed a mission analysis in military terms).

Furthermore, the required number and the roles of personnel should be investigated with relation to the requirements of the mission or system assignment. All the above has to be investigated under the performance constraints imposed by the system under consideration and its operating environment, work context and situation.

Therefore the first task of HF V&V for the above case is to check that the source data and information has been collected and argued as a sufficient baseline for the analysis. A further task is then to ensure that a suitable method for the form of TA has been chosen and that a competent person is associated with the direction of the work. Finally, TA should only be performed on that part of the mission or assignment necessary to support planned HF work with the system or in support of other system-related activities such as training needs analysis.

Made explicit within the scenario generation for the analysis should be all goals of the task work, plans governing the work, conditions necessary for the task work and expected results of that work. Figure 12.9 illustrates the main influences and considerations affecting task performance.

The second task of HF V&V is to check that all pertinent information related to the TA has been considered and sensibly included in the analysis. The analysis itself must be checked as being consistent in its method and its use of information. Proof must be available on analysis reviews and on changes made to the analysis as a consequence of the review. Where necessary, the analysis should be presented as a combination

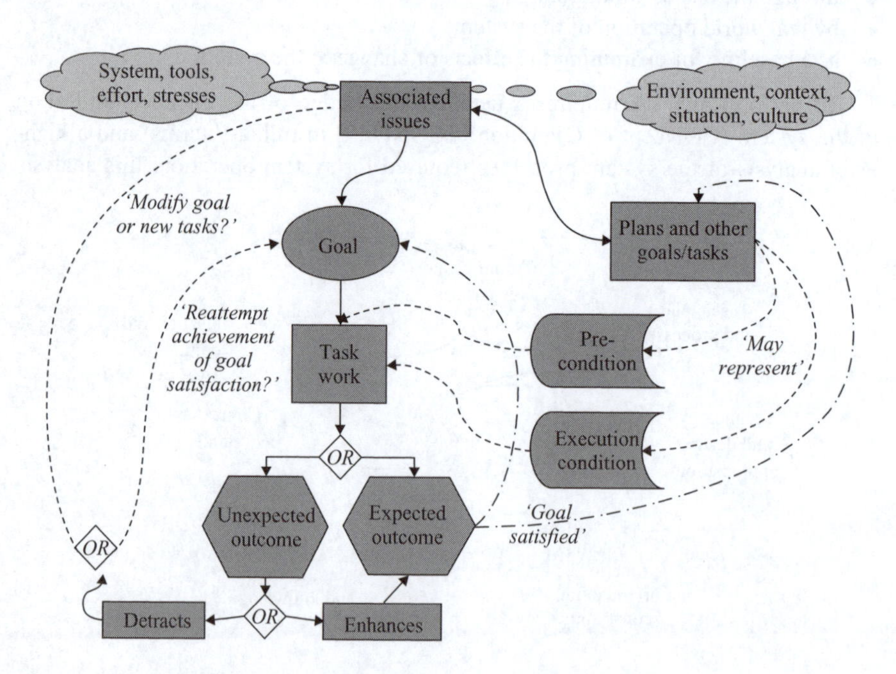

Figure 12.9 Main influences and considerations on task performance

of diagrams and accompanying text for explanation. Part of that explanation should concern the chosen depth of detail to which the analysis was performed.

A third task of HF V&V should be to examine the uses of the TA by its recipients and the suitability of the analysis product for that intended. In some cases the recipient may require certain tasks to be broken down to the expected operator activities associated with each task or to be expressed in a either a graphical or tabular format. For example, details of expected system operator activities may be especially important to the designer of a human-machine interface or to an assessment of the probabilities of human error.

The team work associated with task execution should also be checked. For example, the precondition for the start of a team member's task might be an instruction or information from another team member. As another example, an unwanted side effect from the performance of a task may inhibit the performance of another team member's task work.

Lastly, HF V&V should check that the TA is updated if this is indicated by new evidence obtained on task performance. The update of the analysis may not be necessary on the initial form of the analysis but on any subsequent modelling of that analysis (i.e. through a synthetic environment). The form of the update must be recorded along with its reasons to allow subsequent trace to the action. Further, the consequences of the update, its communication and subsequent actions must also be recorded. Note that the issues highlighted in Section 12.2 indicated that the matching of physical human factors to human cognitive factors is fraught with difficulties, making the iteration on TA and its updating all the more important.

All too often the only HF system related analyses performed are TAs. Thus without any precedent analysis, such as an analysis of the working environment or of the mission, TA often merely considers a facet of the system processes with little consideration of overall situation and context. Furthermore, without any subsequent activity analysis, the TA may provide poor information to the engineers of a system as it will have failed to inform them of the interpretation of the application of system functions to satisfy task subgoals.

TA verification is concerned with the authenticity of the hierarchy of analyses, and validation should be both a top down and bottom up check on the performed verification of the analyses. Thus, validation should be performed both with relation to the satisfaction of system requirements and to the confirmation that the tasks to be performed by the system are possible and adequate with relation to the form and level of effort to be applied and the time available for the task performance.

12.6.2 Use of modelling, prototyping and simulation

This area is addressed by Chapter 13 but discussion is made here for the sake of completeness. Modelling, prototyping and simulation are all methods of representing analogues of reality and can be used for many purposes within a system's life cycle. Examples of these purposes are:

- modelling the concept;
- demonstrating the system architecture and performance;

- demonstrating the human–machine interface;
- testing and evaluating system operation;
- Synthetic Environment Based Acquisition (SeBA) [19];
- real time studies on issues of work (i.e. its achievability, workload, error, timing);
- progressive HF testing of aspects of the system;
- stress testing of system software updates;
- training.

The following definitions are obtained from the UK MS&SE Web site [20] and are cited to aid explanation but are not intended to be prescriptive.

- *A model* is a representation of a system, entity, phenomenon or process. A model may be physical (e.g. a wooden mock-up) or virtual (computer-based). The 'something' may be physical (e.g. a vehicle) or abstract (e.g. a relationship between variables illustrated in a graph).
- *A simulation* is the exercising over time of models representing the attributes of one or more entities or processes. A simulation may be:
  - *'live'* exercising real people and/or equipments in the real world (e.g. a live trial or exercise);
  - *'virtual'* exercising real people and simulated people/equipments, possibly in a simulated world (e.g. a training simulator);
  - *'constructive'* exercising simulated people and/or equipments, usually in a simulated world.
- *A synthetic environment (SE)* is a computer-based representation of the real world, usually a current or future battle-space, within which any combination of 'players' may interact. The 'players' may be computer models, simulations, people or instrumented real equipments.

For detail in this area refer to Chapter 13. The appropriate use of any simulation methods will depend on the phase of the life cycle being considered and the information and knowledge available on system structure and performance. It is important when using these methods that the HF practitioner fully understands their benefits and limitations, considering the purposes of their application and the stage of the system life cycle at which they are to be applied. A check that should be made by HF V&V is that there is evidence that this understanding exists.

Using SE as an example to expand on the definitions used, SE can be considered to be a modelling tool, a prototyping tool, or a simulation depending on the application of the SE and the stage of the system life cycle at which it is used. For this reason SE will be used to discuss what HF V&V activities should be involved with its use. Overall:

> Verification and validation should be by review and assessment by scientific and military experts and independents, followed by further validation using exercises, trials, and historical events where possible and appropriate. [21]

The use of modelling, prototyping and simulation activities can be used to promote better understanding and realistic expectations from the customer during the progression of the system manufacturing processes. The MoD quote is a guideline on

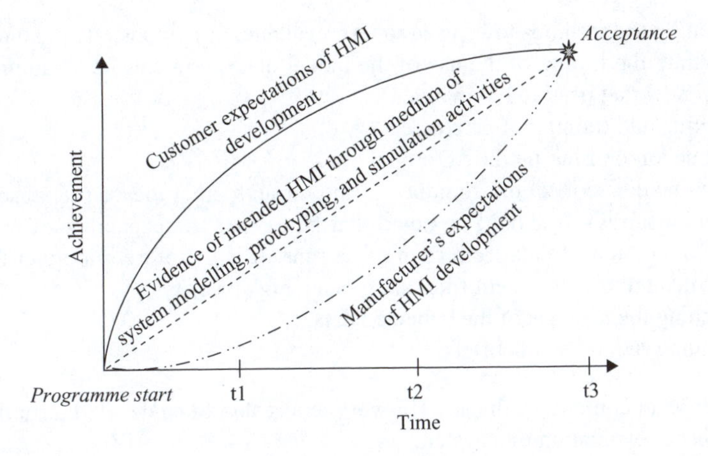

Figure 12.10 Need for liaison on customer expectations

the V&V that should be applied to the SE model throughout. However, when developing and manufacturing a system there is frequently a gulf between the customer's expectations, associated requirements of evidence of progress, and the evidence that can be presented by the manufacturer.

Modelling, prototyping and simulation can be used to decrease that gulf by providing the customer with visible evidence of the manufacturer's progress and intent. Figure 12.10 depicts that the gulf between customer expectations of evidence and the reality of the level of a completed design can be narrowed through proper liaison, progressive demonstrations of facets of the system and system evaluation.

12.6.3 Suggested actions and reporting

In addition, HF V&V should check that the actions listed below are considered, then completed satisfactorily, or argued as unnecessary. For example, it may be unnecessary to detail the purpose of the model to be used and its HF relevance if the model has been used previously for the same purpose in a similar system programme and the same phase of the system life cycle.

- Detailing the purpose of model use and its HF relevance.
- Detailing what the model is and its relevance to the issues being examined.
- Creating the system to be used including:
  - Form of SE (i.e. desktop or projected image);
  - hardware and software requirements;
  - number and form of federates (single location or distributed);
  - number and form of entities (players) (how many platforms or agents are to be considered?);
  - environments to be used (scenarios have to be generated);
  - purpose of the exercise (what is being investigated?);
  - form of architecture.

- Planning the methods to be used to test, evaluate, or demonstrate the model.
- Planning the timing of the use of the model or experiment, its location and the personnel that need to be involved.
- Briefing and training of personnel.
- Instruction on running the SE model.
- Planning data collection from the SE and through other means (i.e. observation, questionnaires). It should be noted that an SE has the capability of producing a great amount of data from each of its runs. It is therefore important that data capture and its subsequent filter are given a high priority.
- Planning the analysis of the collected data.
- Planning reports and debriefs.

The high-level contents of the any HF work report should be decided from the onset of any test or evaluation of a system.

All of the above actions and checks, or their equivalent, should be reviewed and confirmed by HF V&V.

12.6.4 Test and evaluation (T&E)

T&E will make use of modelling, prototyping and simulation tools. However, T&E can use various methods from static checks on software code to quality testing of threads of software, to rig tests conducted on both software and hardware, to tests on fully developed systems.

HF V&V in this case would be similar to the approaches used in Section 12.6.2 except that the planned use, application, combination of data and the analysis associated with HF methods would be the focus of the V&V activities.

Thus whilst investigation of system solutions or the demonstration of intended capabilities through analogue representations will require evidence in support, it is unlikely that the evidence need be as robust as that required for the T&E applied to real equipments and systems.

As previously discussed it is highly desirable that a traceable and progressive T&E programme is planned and conducted throughout a system life cycle. At certain stages of the life cycle the safety and performance requirements of a system have to be affirmed in order that the system can be certified as suitable for use. Typically, these stages are at the initial acceptance into service and after any major upgrades to the system during its in-service period. This affirmation or acceptance of a system is usually based on final operating trials on the system but these trials are partly based on evidence produced by the trace on previous T&E activities, that include HF.

All systems operate under an inviolate process governing several sub-processes. An example of an inviolate model of system process is that of the flight operation of an aircraft. This process has flight phases that involve the aircraft take off from the ground, its climb to a transit height, its performance of flight manoeuvres during flight, its descent to its destination, and the landing of the aircraft at that destination. If any of the phases are missed the aircraft cannot successfully perform its intended flight.

Considering the above, system effectiveness can be measured through a consideration of the system performance over the phases of a process.

12.6.5 Consideration on measures

Measures of effectiveness (MoEs) can be used to test if the aircraft has reliably performed its flight safely and within its designed or expected performance parameters. In other words, the check is whether the aircraft has achieved its basic flight goal. Associated with MoEs can be measures of performance (MoPs). Whilst MoE is concerned with the satisfactory performance of the high-level functionality of a system against a process, MoPs are concerned with the actual reliable performance of the aircraft in safely satisfying the goals of sub-processes.

The use of MoEs and MoPs in T&E requires that they are based on objectively stated conditions, the satisfaction of which indicates that the goal depicted by the MoE or MoP has been achieved. There are difficulties in constructing MoEs and MoPs that encompass all elements of a system including the humans [22]. In some cases there may need to be separate MoEs/MoPs for the system and for the system users. However, the system users' MoEs/MoPs should always complement those of the wider system.

Further, MoEs/MoPs can indicate if one or all of their associated conditions are satisfied, and usually when they have been satisfied, but often cannot indicate why that satisfaction has been achieved or not achieved as the case might be. Table 12.5 details the main rules that should be followed in constructing MoEs and MoPs.

It is in the area of substantiating the MoE/MoP results that HF can produce evidence supporting the results indicated by the MoEs/MoPs. The robustness of that evidence must be assured through the continued pertinence of HF V&V activities.

Table 12.5 Properties of MoEs (also applicable to MoPs)

MoEs must be	associated with system functions; related to high-level tactical goals; a measure of the satisfaction of goal performance; include a concept of limits; include a specific level of achievement; define the use of the MoE in assessment; define MoE weightings singly, collectively and by assessment topic; refined hierarchically as a top down process if required.
MoEs should be	unambiguous; evoked in context; have an explicit trigger for the measurement activity; define the type of measure and the method of measurement; define the form of recording the measure; agreed with all parties.
MoEs cannot be	merely an observation or simple 'measure'.

It is argued that the process model of any system has an associated Task Organisation Model (TOM). This model allows examination of the task performance conducted in support of the system processes. Thus, evidence can be produced by the TOM, which though mainly subjective, can nevertheless robustly answer questions related to why, how, what, where, and in what form work was performed to achieve the results.

The evidence is gleaned from task performance through the use of HF methods, system-based data loggers, professional observation both by SMEs and HF professionals, and post-work debriefs to name but a few sources. What has to be accepted is that depending on the level of expertise of the personnel participating in the system work, there will be many different task routes that can be taken to satisfy the work goals. This varied approach will occur except for very sequential tasks as might be regulated by procedures mandated by employers for certain work scenarios. It is important therefore to run both free play and scripted exercises where possible to determine the optimum work task routes for the particular test being performed. These routes can then be used as a template to investigate the reasons for test personnel choosing other routes.

Figure 12.11 indicates the association between the system process, MoEs, and MoPs. The process illustrated is the standard communication process of preparing the message, the transmission of the message, receipt of the message to the originator by its recipient, acknowledgement by the recipient that the content of the message is understood, and then a recipient's response post that acknowledgement to indicate

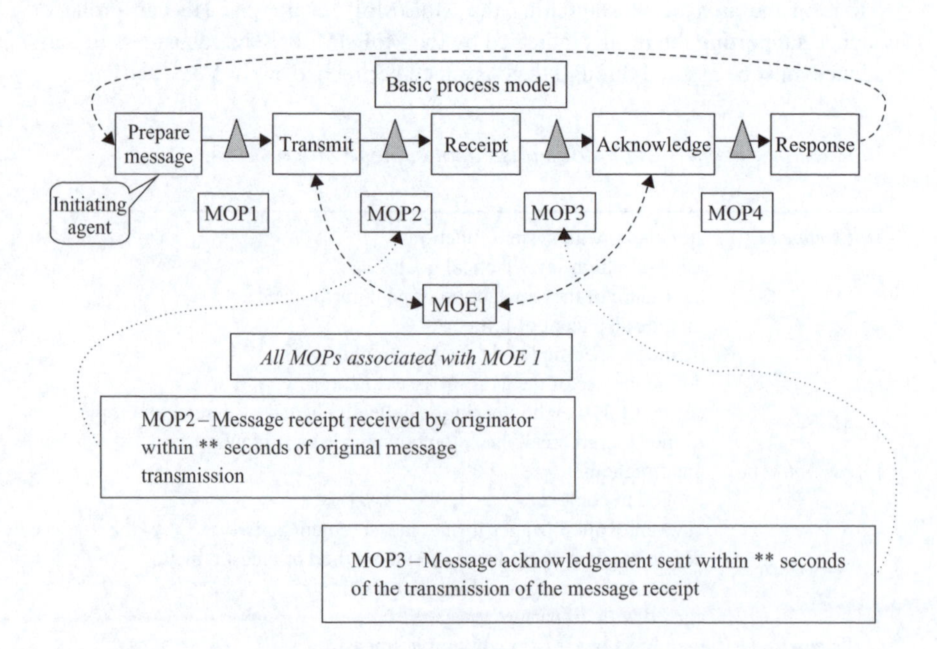

Figure 12.11 Association between system process, MoEs and MoPs

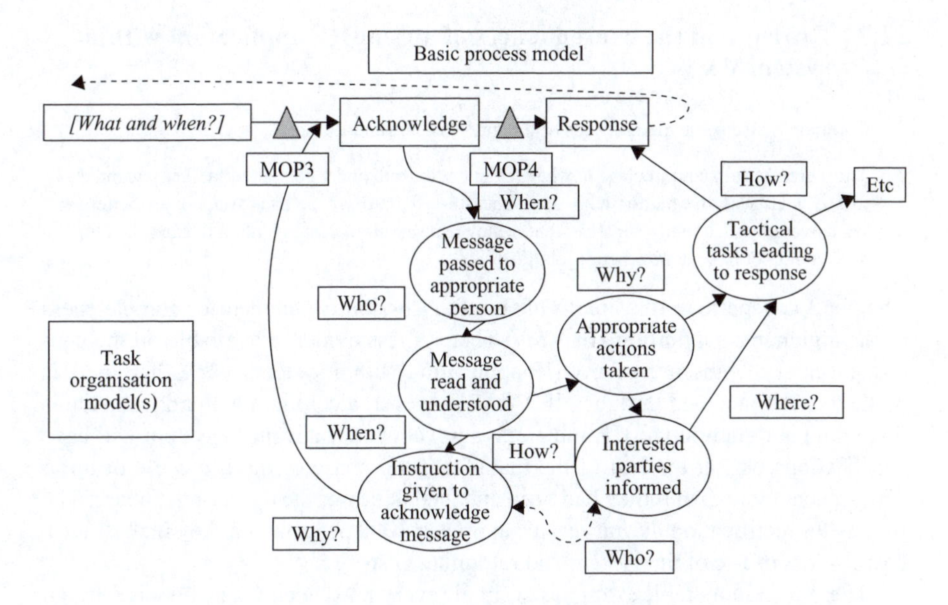

Figure 12.12 Associations between process model and task organisation model

whether there will be any further course of action on the message subject. For illustration, one simple MoE, MoE1, is taken as 'Full communication to be achieved on 75 per cent of occasions'. Figure 12.12 illustrates the association between part of the example process model and a representation of an associated TOM. This model is suitable for use with complex systems such as systems of systems. In many cases of simpler systems, a simpler model may suffice.

12.6.6 Ancillary issues affecting HF V&V

This chapter has given no guidance on how to effectively amalgamate subjective and objective data as might be collected under HF T&E. One method known as triangulation can be used [23, 24]. This method relies on assigning collected evidence as being top down on the area in question (i.e. rating scales covering work on several tasks over a period of time) or bottom up evidence (i.e. individual comments from SMEs).

The evidence is then sorted into categories (i.e. supporting excellent to poor) and amalgamated using a set of rules. For example, a simple rule might be three adverse SME comments on a task area will reduce a rating scale assessment down one category (i.e. say from good to average).

Additional problems exist with the practice of HF V&V when a system makes extensive use of Commercial off The Shelf (COTS) products. Many of these products will have been created to different standards of build and may also bring to the system more or fewer functions than required. For a consideration of the problems inherent in this area see MacLeod and Lane [25].

12.7 Evidence of the consequences of absent HF application within system V&V

Common Sense is the metaphysics of the masses. Bertrand Russell

The Three Mile Island nuclear power incident occurred, and I got called in. They wanted to see what was the matter with their operators. Could we perhaps train them better? I discovered, much to my surprise, that if they had set out to design a plant to cause errors, they could not have done a better job. [26]

Chapter 3 is generic to this area. This brief discussion is included for completeness of the arguments supporting HF V&V. There is less evidence available on the consequences of the absence of any HF application within systems V&V than there is on the benefits accrued through HF V&V. However, a common ballpark estimate is that with poor attention to HF and V&V a novel or new advanced system will have rectification costs at the end of the manufacturing phase of the life cycle of up to 300 per cent more than if they had been attended to nearer the onset of the phase. The figures for rectification during service approach 1000 per cent and they include such costs as loss of use of the system and retraining costs.

There are innumerable examples of simple everyday objects where poor HF design has made system use difficult or even impossible. Table 12.6 gives a mere two examples from the author's own experience; unfortunately in this case the identity of the actual projects cannot be disclosed for commercial reasons.

Table 12.6 Two examples of the consequences of absent or late HF application within system V&V brought about through HF V&V

Form	Problem	Ergonomic assistance
Aircraft mission system	System functionality added to throughout system development but the original Human Computer Interface (HCI) hardware was retained.	A great deal of ergonomic effort was placed on the system during the latter part of its development in an attempt to 'patch' problems with HCI. The effort resulted in only small successes.
New sensor system	Introduction of sensor equipment with operating controls and associated feedbacks depicted by VDU-based 'Buttons', and 'Windows' type dialogue boxes.	Sensor assessed by SMEs as totally unsuitable for the job. Equipment seen as: • undersensitive; • too slow in operation; • control feedback poor in form. Sensor operating system HCI needed a costly redesign to meet its operability requirements.

In contrast, a few examples are given on the benefits of HF V&V within different life cycle phases. Note the cited work was retrospective.

- Improvements by a major computer company to the efficiency of its security sign-up procedure resulted in a recovery of the costs of the improvement within the first half day of its operation [27].
- Measures to improve safety and productivity in the South African forestry industry, devised by ergonomists to protect against work injuries, produced estimated savings to the industry of US$ four million annually [7].

12.8 Summary and conclusions

The main areas of effort and considerations that should be the focus of a system's life cycle V&V are T&E, trade off processes, requirements and FfP trace and confirmation, and modelling (i.e. mock-ups, prototyping, SE, simulation). These efforts and considerations are all interrelated and rely on an understanding of system-related work gleaned through analysis and professional practice. This understanding in turn can only be obtained through working with and understanding the practices of other professions involved in work throughout the system life cycle. Such an understanding encompasses at least a good practical knowledge of:

- the system's purpose;
- its intended capabilities;
- the system working environments;
- appreciation of the culture and organisation of the end user;
- system functionality;
- the system architecture;
- the intended and achievable performance of the system;
- its design and use constraints;
- the system operating and maintenance tasks and goals;
- system-specific critical activities.

Of major importance is an understanding and empathy towards the needs and problems of the various stakeholders of the system, whether they are commanders, managers, supervisors, operators or maintainers. In addition, it is important to also have empathy with the work and problems of other work colleagues.

There are a considerable number of influences on the performance of HF activities throughout a system's life cycle, affecting the quality of the application of HF V&V. Not least is the fact that HF is a maturing science and profession that has yet to discover answers to many of the problems associated with the design of systems consisting of both engineered and human components. Compounding this incomplete knowledge are burgeoning advances with new technologies and their usage. These advances on their own change the nature of human work with systems, often in unexpected ways.

All professions have limitations and the realisation of these is necessary and honest. However, HF has an important and strong contribution to make as it is the only systems-associated application with a focus on optimising and promoting the human contribution to systems performance and safety. This HF contribution is especially important when it is realised that the human must still be the decision maker in the direction, command, control, management and supervision of a system towards the achievement of its goals.

Whilst there are no prescriptive solutions offered for the conduct of HF V&V, sufficient guidelines have been given for an interested party to start down the avenue of discovery and to obtain the HF expertise necessary to approach each system as a new challenge with its own inherent types of problems. The true challenge is to ably assist in imbuing the system with a good-quality FfP whilst also satisfying the contracted system performance requirements. Substantiation that the challenges to HF applied practice have been met is the production of a wealth of well-argued and associated high-quality evidence of HF activities and products that have contributed to the betterment of the system. Such evidence will always support effective HF V&V.

So, where possible plan all HF activities with consideration for the needs of other stakeholders, perform all activities, record everything, check quality of performance, check progress and results and re-plan as required to maintain the aim.

12.9 Questions

This section includes a sample of questions that may be addressed to assist consideration of HF V&V. The reader is advised that an attempt should be made to prioritise these questions with relation to their relevance to any intended HF V&V work. The questions can then be used as a form of high-level introduction to the completeness of the consideration of that work.

- Are the specified system performance requirements understood?
- Are the human-related requirements of the system specified?
- Are there any constraints from the planned work management on HF activities?
- Are the constraints on HF appropriate?
- Are the planned management work schedules and milestones understood?
- Has the HF work been planned in relation to the overall system work management plan?
- Has adequate HF expertise and effort been allocated to fulfil the requirements of the HF plan?
- Have the activities of the HF plan been associated with the work activities of other work plans such as those of test and evaluation (T&E) and safety engineering?
- Have the issues of the design, performance, analysis, and archive of tests and evaluations been adequately addressed by the plan?
- Does the plan consider an adequate suite of HF methods for test and evaluation and their application considering pre-test application, within-test application and post-test application?
- Has the HF planned programme indicated its planned deliverables, their timing, and their purpose?

- Have the HF planned programme activities considered the following in relation to the best consideration under the overall programme management plan:
  automation; tools and aids; the integration of HF domains with system domains; team working; trace of issues and decisions; trade offs; system error; training.
- Is there a progressive accumulation of evidence in support of system acceptance into service?
- How were SMEs selected and their biases and levels of expertise determined?
- Was evidence collected on the efficacy of the HF methods as applied?
- Was careful use made of both statistical and subjective evidence?
- Was there evidence available of the progressive address of HF-related issues?
- Was there progressive evidence of improvements resulting from the application of HF effort?
- Was there evidence of HF involvement in system safety case work?
- What is the role of HF in system acceptance?
- What HF issues remained after system in-service acceptance with post in-service updates?
- For remaining HF issues, is there a plan for addressing them in the future?
- What independent checks are planned for the validation of HF in system acceptance?

12.10 References

1 'A specification for the process assessment of human-system issues', ISO PAS 18152, 2003
2 Federal Aviation Administration, 'Human factors job aid' (Hughes Training Inc, Falls Church, VA, 1999)
3 Def Stan 00-25, 'Human factors for designers of equipment', MoD directorate of standardisation, Glasgow, Part 12 – Systems (1989) and Part 13 – HCI (1996)
4 SAWYER, D.: 'Do it by design: an introduction to human factors in medical devices' (US Food and Drug Administration, Center of Devices and Radiological Health, Rockville, MD, USA 1996)
5 HANCOCK, P. A., and SZALMA, J. L.: 'The future of neuroergonomics', *Theoretical Issues in Ergonomics Science*, 2003, **44** (1–2), pp. 238–249
6 BEVAN, N.: 'Trial usability maturity process: cost benefit analysis', ESPRIT Project 28015 TRUMP, D3.0. V1.1, September, 2000
7 HENDRICK, H. W.: 'Measuring the economic benefits of ergonomics', in SCOTT, P. A., BRIDGER, R. S., and CHATERIS, J. (Eds): 'Global ergonomics' (Elsevier Press, Amsterdam, 1998) pp. 851–854
8 HENDRICK, H. W.: 'The economics of ergonomics', in SCOTT, P. A., BRIDGER, R. S., and CHATERIS, J. (Eds): 'Global ergonomics' (Elsevier Press, Amsterdam, 1998) pp. 3–9
9 ANDERSON, M.: 'Human factors and major hazards: a regulator's perspective', in McCABE P. J. (Ed.): 'Contemporary ergonomics' (Taylor and Francis, London, 2003)

10 Standish Group (1995), CHAOS report, http://www.standishgroup.com/chaos.html
11 MOORE, G. A.: 'Crossing the chasm' (HarperCollins Publishers, New York, 1995)
12 SARTER, N. D., and WOODS, D.: 'How in the world did I ever get into that mode? Mode error and awareness in supervisory control', *Human Factors*, 1995, **37**, pp. 5–19
13 ALLUISI, E., SWEZEY, R. W., and SALAS, E. (Eds): 'Teams: their training and performance' (Ablex Publishing Corporation, Norwood, NJ, USA, 1992)
14 MACLEOD, I. S., and SCAIFE, R.: 'What is functionality to be allocated?' Proceedings of Allocation of Functions Conference (ALLFN '97), Galway, Ireland, 1–3 October 1997
15 MACLEOD, I. S.: 'A case for the consideration of system related cognitive function throughout design and development', *Systems Engineering*, 2000, **3** (3), pp. 113–127
16 UK Def Stan 00-55, 'Requirements for safety related software in defence equipments', 1991
17 BOEHM, H., and Hanson, W. J.: 'The spiral model as a tool for evolutionary acquisition', *Crosstalk: The Journal of Defence Software Engineering*, 2001, May, **14** (5), pp. 4–11
18 MACLEOD, I. S.: 'The need for congruence between the applications of human factors and safety engineering', in Proceedings of ESAS02, MoD Abbey Wood, Bristol, December, 2002
19 STONE, R. J.: 'Interactive 3D and its role in synthetic environment-based equipment acquisition (SeBA)', Contract Number MoD Dstl RD018-05103, Farnborough, April, 2002
20 MS&SE web site at http://www.ams.mod.uk/ams/content/docs/msseweb/
21 MoD Guidelines for the verification and validation of operational analysis modelling capabilities (http://www.ams.mod.uk/ams/content/docs/msseweb/)
22 SPROLES. N.: 'The difficult problem of establishing measures of effectiveness for command and control: a systems engineering perspective', *Systems Engineering: The Journal of the International Council on Systems Engineering*, 2001, **4** (2), pp. 145–155
23 MACLEOD, I. S., WELLS, L., and LANE, K.: 'The practice of triangulation', *Contemporary Ergonomics*, Conference Proceedings (Taylor & Francis, London, 2000)
24 RICHARDSON, T. E. (Ed.): 'Handbook of qualitative research methods for psychology and the social sciences' (BPS Books, British Psychological Society, Leicester, UK, 1996)
25 MACLEOD, I. S., and LANE, K.: 'COTS usability in advanced military systems', *Contemporary Ergonomics*, Conference Proceedings (Taylor & Francis, London, 2001)
26 NORMAN, D.: *mpulse Magazine*, 2001
27 NIELSON, J.: 'Usability engineering' (Morgan Kaufman, San Francisco, 1994)

Chapter 13
The application of simulators in systems evaluation

Ed Marshall

13.1 Background

Having designed a system, be it a piece of software, or the human–machine interface (HMI) for a complex system, there comes a point where the designers, or indeed the customers, will require an assessment to demonstrate that the product does actually perform safely and effectively. Simulators and other simulation techniques provide powerful tools for such assessments and ergonomists are often involved in the required investigation and evaluation exercises. This chapter will describe, using some examples, the use of simulators in the verification and validation of system performance.

13.1.1 Simulators and simulation

At the start it is perhaps useful to consider two definitions:

- *Simulator*: a machine designed to provide a realistic imitation of the controls and operation of a vehicle, aircraft, or other complex system, used for training purposes. For example, a flight simulator is a machine designed to resemble the cockpit of an aircraft, with computer-generated images that mimic the pilot's view, typically with mechanisms that move the entire structure in corresponding imitation of an aircraft's motion, used for training pilots (*New Oxford Dictionary of English*). Figure 13.1 shows a view of a modern flight simulator.
- *Simulation*: a research or teaching technique, used in industry, science, and in education, that reproduces actual events and processes under test conditions. Simulation may range from pencil and paper, or board-game, reproductions of situations, to complex computer-aided interactive systems (*Encyclopaedia Britannica* (2001)).

Figure 13.1 A modern aircraft flight deck simulator (reproduced by permission of Thales Simulation)

The distinction between these definitions is important and is adopted throughout this chapter. Two issues should be noted: first, a simulator is a complex, specific, and often elaborate, form of simulation. Second, the origins of simulators lie in approaches to teaching and training. Their application for the assessment of system performance tends to be a secondary concern.

The simulator builds on the more generic and simplified simulation techniques to provide the means to demonstrate and measure the performance of skills and the execution of complex tasks in an elaborate and realistic environment. However, although simulator studies are reported, because of concerns over confidentiality or simply because the processes are too technical to be described in the ergonomics literature, there tends to be little detail provided about the conduct of these studies. This chapter therefore addresses the practical issues when using a simulator to evaluate system performance.

13.1.2 History

It is probably fair to say that the problems associated with learning to fly aircraft have provided the driving force behind the development of simulators. One of the earliest simulators was the *Link* aircraft trainer, which was developed during World War II.

This device was used to provide initial training in basic flying skills. These primitive machines, which made use of pneumatics to move a very rudimentary cockpit in response to movements of the control stick, are still around today and the author saw one in use at an Air Training Cadets' Open Day only a couple of years ago.

The *Link* trainer demonstrated the two key objectives and benefits of the simulator: the student can learn the basic skills in a benign environment and the advanced user could learn and practise handling the difficult and complex manoeuvres similarly without risk. Also, back in the days of World War II, the 'Cambridge cockpit' studies used simulation to represent elements of the pilot's task to investigate aspects of performance, particularly the reasons for observed pilot errors. Although the initial impetus was military, by the 1970s simulator development had moved on to civil aviation and into power plant control, notably for the training of operators for the new nuclear reactors.

The progress in computing was harnessed to improve simulators and was increasingly focused on the generation of more realistic visual environments and more sophisticated mathematical models to replicate the behaviour of complicated processes such as nuclear reactor performance or the physical dynamics of jet aircraft. By the early 1980s simulators were being used routinely for training in the aviation and the power industries, but not as design aids or tools for ergonomics research.

For the nuclear industry, it was the accident at the Three Mile Island power plant in 1979 that had a dramatic effect on the use of simulators. In the wake of the accident, the main inquiry concluded that it had been caused by poor understanding of the plant by the operators. This lack of understanding had been exacerbated by inadequate attention to the ergonomics of the displayed information and inadequate simulator facilities for training. (See also Chapter 12 for a further reference.) The upshot was a demand for ergonomics research and better simulation. In the UK, this led to the demand by the Public Inquiry into the Sizewell 'B' Power Station, that a simulator must be available for use by trainee operators at least one year prior to the fuelling of the plant. The demands for increased safety and protection from human error meant that nuclear power plant simulators became available to ergonomists and psychologists interested in the performance of complex skills under extreme conditions.

13.1.3 The exploration of extreme situations

With this background, simulators have been used to look at how systems perform at the limits of design, during rare breakdowns, system failures and unexpected accident scenarios. The key advantages of the simulator in this respect are:

- The simulated system can be manipulated outside, or close to, its design limits.
- Rare events can be replicated and repeated at will.
- Fast changing system behaviour can be slowed down or acted out in a sequence of steps.

Obviously, it is these features that make simulators invaluable tools for training, but it is the same advantages that can be exploited for system evaluation and the investigation of human performance in extreme situations.

Because of the incidents at Three Mile Island, and then Chernobyl in 1986, the nuclear control room event and the performance of the operators when faced with a plant breakdown, has been the focus for a number of simulator studies. In this chapter, it is this environment that has provided the demonstration of the methods, the benefits and potential pitfalls when undertaking simulator-based studies of human performance.

13.2 What can the simulator do?

The simulator provides a powerful tool with which to explore human performance and to test this performance against design expectations and safety constraints. This is because the simulator provides a highly realistic and accurate representation of the system. A view of a nuclear power plant control room simulator is shown in Figure 13.2. The simulator can also deliver a range of scenarios that the experimenter can select and develop to meet the needs of the specific investigation. In addition, the simulator has the potential for delivering a range of accessible and detailed data on performance that is not feasible in the real world.

13.2.1 Fidelity

Fidelity is the degree to which any simulation, or simulator, represents the real world environment. However, fidelity is not a single dimension. For the experimenter, there are a number of ways in which fidelity must be considered when considering the validity of performance in a simulator. The experimenter must always be aware that it is the fidelity of performance by the participants that is under investigation, not the

Figure 13.2 The experimenter's view of a nuclear power plant control room simulator (reproduced by permission of British Energy)

fidelity of the simulator itself. The simulator is a tool that can deliver representative performance. In this respect it is useful to consider four aspects of fidelity:

- *Engineering validity.* The modern simulator is based on a series of mathematical models which represent the physical processes underlying the actual system. Manipulation of the inputs to these models generates outputs that mimic the behaviour of the system, be it an aircraft or a power plant or whatever. The use of models allows the experimenter to input accident or fault conditions so that the simulator then mimics the behaviour of the real system as if these events had really occurred. The simulator designers seek to ensure that this modelling is indeed accurate and that the simulated behaviour does actually replicate system performance. Designers will seek, therefore, to test the simulator models against observed real world behaviour where there are data available. However, it should be remembered that, when it comes to extreme conditions, the simulated representation is always, to some extent, hypothetical.
- *Physical fidelity.* For the instructor, or the experimenter, it is the actual physical reality of the simulator workspace that will be important. Thus, the simulator space must provide some degree of correspondence to the physical space in the real world. It is the flight deck in the case of an aircraft, whilst for the power plant it will be the control room that is copied. As well as the space, the arrangement of instruments, displays and controls will be duplicated. In addition, the whole instrumentation array and controls will be driven by the simulator models so that their physical appearance, their movement or colour change are duplicated in the simulator. For an aircraft this physical fidelity may include the view out of the cockpit window – this provides a considerable challenge for the model authors, though modern graphics systems permit increasingly faithful reproduction of the visual environment. For the nuclear plant control room, there are usually no windows, so this aspect of reality is not required.
- *Task fidelity.* The simulator seeks to combine the delivery of high degrees of engineering fidelity and physical fidelity. In the training world, such a device is termed a *full-scope* simulator. It is assumed that this combination results in a correspondingly high level of task fidelity. In other words, an operator placed in the simulator and faced with a situation, will carry out tasks at the same pace, in the same sequence and with the same workload as if meeting the same situation in the real workplace. Manipulation of controls will be matched by representative changes in instrumentation, displays and simulated views if appropriate. In vehicle simulators, the simulator will respond to control changes or scenarios with corresponding physical movements. All of which reinforces task fidelity that for the experimenter will be an important factor in generating realistic performance.
- *Face validity.* Face validity refers to the subjective reality as experienced by participants in simulator exercises and is a key issue in maximising the reality of operator performance. It is very dependent on the scenarios selected, the enactment of the scenario and co-operation between the experimenter and the operators. The participants always know they are in a simulator, yet they must behave as if in the real world. In the case of dealing with extreme scenarios, this entails a

degree of acting by participants and a *suspension of disbelief*. It is a difficult and sensitive matter to encourage and exploit this state of mind successfully, but the effort is essential if realistic performance by the participants is to be achieved. Realistic behaviour depends very much on the design of the selected scenarios. Although the scenarios are rare and unusual events, they must be credible. The experimenter should not be tempted to add too many ingredients to the situation or participants will withdraw from acting out realistic behaviour. In the training world, instructors are now using realistic scenarios, acted out in real time and with a full operating team in the simulator, in order to maintain face validity. Sustaining face validity also depends on the conduct and professionalism of the experimenters. They must foster an atmosphere of mutual trust and respect with the participants if effective and valid results are to be obtained.

13.2.2 Simulator scenarios

Simulators permit experimenters to run scenarios that represent events requiring intervention or interaction between the simulated system and the participants. When considering the selection and design of the scenario to be enacted in a simulator study, there are three main issues that should be ensured:

- The scenario must be relevant to the matters of concern.
- The scenario should engender performance that is realistic and representative.
- The scenario should generate the necessary data for recording and analysis.

With a complex system, it will be neither possible nor, indeed, desirable to examine in detail performance with all the situations or tasks that may be required. Generally, adequate information for the assessment can be obtained by focusing upon a limited sample of tasks or activities, provided that the selection is sufficiently representative to cover the issues of concern. The following three typical examples illustrate how scenarios may be selected:

- When investigating performance with a particular interface, whether this is a human–machine interface (controls and information displays) or a written procedure, the scenarios should be selected to exercise these interfaces. Adequate information in this case may be obtained using scenarios comprising a relatively small sample of tasks.
- If the evaluation is focused on the performance of a particular task in order to investigate different modes of operation or to compare different strategies, then one scenario may be sufficient.
- When an analyst wishes to investigate specific situations that have been identified as being particularly important (e.g. tasks that make the most onerous demands upon personnel, or situations identified as critical in developing a safety case), then the scenario must include these situations and may require longer duration to replicate a realistic work period in order to capture realistic performance.

Thus, in practical terms, the experimenter can select from three types of scenario: a sequence of several, relatively short situations, a single specific short and focused scenario, or the longer, more realistic representation of a work session or shift period. This choice may initially be based on the investigation's objectives, but, typically, scenarios will have to be tailored to meet the practical concerns of access to the simulator and the availability of participants.

13.2.3 Simulator data

By presenting something so close to a real-world environment, the simulator exercise provides the experimenter with the potential for generating a wide variety, and large amounts, of information. It is useful, in the first instance, to consider whether the data is objective or subjective. Objective data is data that is clearly defined, replicable, generally quantitative, and not likely to be affected by different experimenters. Subjective data is based on opinions, attitudes and feelings. These are generally obtained from the participants in the form of interviews, questionnaires and attitude scales. However, there is also subjective data derived from the experimenters' own opinions about a simulator exercise. These can also take the form of questionnaires or attitude scales in relation to their observations.

The data thus fall essentially into four types, all of which can be effective in making judgements about human performance. These range from hard, objective quantitative information through to qualitative and subjective judgements. These four types of data are:

- data from the simulated system itself;
- objectively observed aspects of human performance;
- subjective data from participants;
- subjective data from the experimenters.

13.2.4 Objective data from the simulator

A simulator can produce data relating to its own physical model through the progression of a scenario. For example, the simulator will generate alarm messages relating to a proposed incident. The number, content and rate of onset of alarm messages can be followed and recorded. This can be validated against real alarm reports recorded from actual plant incidents. Clearly, this can be a considerable list of items and the experimenter may have to sort the raw list in some way, such as by identifying particular target alarms and noting their time of activation and duration.

The simulator can also generate recordings of selected parameters, such as pressures and temperatures, and these can be used to provide an objective record of plant performance during a simulator trial. In order to aid the experimenter, raw data can be processed to provide specific support for particular evaluation requirements. For example, in assessing the performance of a pilot, the simulator can show the ideal glide slope for a landing with the actual track superimposed upon it.

The simulator can also provide, either by generated logs or by experimenter observation, a record of the interactions of the participants with the various plant interfaces. This permits accurate timings and the identification of responses by operators to particular events. For example, if the correct response to an incident is to initiate the starting of a pump within a required time, then this can be objectively determined from observations of the simulator.

13.2.5 Objective data on operator performance

Although the simulator can provide a record of events occurring through a trial, it is up to the experimenter to determine, on the basis of the pre-determined questions, the events that may be used to assess participant performance. If the designers have assumed a certain standard for operator performance, they may be interested in validating this assumption. The experimenter will, therefore, establish a criterion for performance which may, for instance, involve checking that taking account of each alarm is accompanied by a defined event that can clearly observed through the plant interface. In other words, the experimenter sets up an expected sequence for ideal performance which can then be measured against recordable events on the simulator.

In addition to process-related events, the experimenter may also record the time taken to complete actions and the occurrence of errors. These may be recorded directly as the trial progresses or from observation of video or audio recordings after the trial.

13.2.6 Subjective data from the participants

A rich source of data in a simulator exercise is information from the participants themselves. Indeed, it should be remembered that such subjective data may provide the only evidence of differences in performance between participants. For instance, when two trials with an interface show no variation in terms of the objective measures, it may only be in subsequent interviews that one operator may report that they performed well and that the interface provided effective support, whereas a second operator relates that they nearly made mistakes or found the interface cumbersome. In this way the subjective data amplifies and informs the more objective observations.

13.2.7 Subjective data from the experimenters

The experimenters will record their observations on a simulator trial session. This may comprise free notes or structured proformas set up to capture impressions about performance in a systematic way. Typical examples could be judgements on the quality of team behaviour in terms of communication, leadership or the quality of decision-making. This subjective data should be used to support the more objective recording of actual events, key actions and statements from the participants.

13.3 The participants

Participants are a vital component in a simulator exercise. In fact the availability of suitable volunteers to take part in an investigation can be the dominating factor in designing a simulator experiment. Simulators often represent complex environments and so the participants used for simulator trials may need the same complex technical skills as in the real world situation. Experienced nuclear plant operators or aircraft pilots are in high demand for their normal work and obtaining their release to take part in experiments can be very difficult. Classical experimental designs with balanced groups of participants are usually impractical in simulator studies. Ideally, an exercise will have access to sufficient numbers of these skilled personnel, but there are alternative ways for the experimenter to obtain valid data; these are described below.

13.3.1 Investigation without participants

Simulators can be used to establish whether an expected scenario will challenge the limits of human skill or will exceed accepted standards by observing the simulator interface without participating operators. For example, the simulator can be used to establish whether the task requirements require excessively short response times or whether the onset of information during an event exceeds the ability of operators to comprehend it. If, for instance, the ideal operator is required to be able to read and correctly interpret up to 15 alarm messages in one minute and the simulated trial generates 30 alarms per minute, then there is a mismatch between the interface and the design assumption with regard to operator performance.

13.3.2 Investigation using single participants or teams

Suppose a simulator exercise is proposed to justify a particular aspect of a safety case; for example, to show that a particular process fault can be detected, diagnosed and the correct remedial operations carried out within a predetermined time period. It may reasonably be deemed acceptable to demonstrate this using an appropriate scenario with a suitable trained and experienced crew scenario. If one crew succeeds and follows the correct procedures and decision processes then it can be argued that there is no pressing need to repeat the process with the rest of the staff complement.

Likewise, when validating procedures for dealing with particular plant accident scenarios, one experienced operating crew will probably provide an effective information source. There is probably little to be gained by repeating the exercise with other crews.

13.3.3 Investigation using trainees or students

Where there is a question of the use of novel interfaces to display specific types of information, to compare different media or different layouts, a more classical experimental design is required. However, it may well be considered that the simulator is necessary to provide a valid physical environment. Or it may be that the simulator is the easiest way of generating the various target stimuli at the correct pace and at the required distances. In such a case, the actual task of recognising a particular event or noting a particular indication can be quite simple. In such a case, volunteer participants may be gathered from trainees or student staff. In this way sufficient trials can be undertaken to enable valid statistical comparisons to be made.

13.4 Observation in the simulator

A simulator study involves close observation of human behaviour and there is always a risk that the process of observation can disrupt and distort the performance participants. The very presence of observers is clearly a threat to the fidelity of the environment. Methods for obtaining data must be practicable and the information must be available from the simulator but, first and foremost, the data gathering must be acceptable and non-obtrusive to the participants. The experimenter must always be aware that the process of observation itself may disrupt fluent performance of complex tasks and thus compromise the validity of the exercise.

13.4.1 Observer location

Some simulators provide an observation gallery from which instructors or experimenters can watch activities without disrupting or obstructing the participants in the simulator. This is helpful, but there can be problems of clearly viewing and examining detailed actions when the participant is between the observer and the console. If there is no observation gallery, or the experimenter needs to obtain close-up views, then observers will have to work in the same space as the participants. Clearly this is not ideal and the experimenter must take account of how this may distort observed performance. Over a long series of simulator sessions, the participants will become accustomed to the observers and will be able to ignore their presence. The experimenter must always keep out of the way. A good analogy and role model for the simulator observer is the cricket match umpire.

13.4.2 On-line observation

One of the most effective methods for recording data during a scenario is to use a pre-prepared proforma that has been tailored specifically for the particular investigation. For example, this could consist of a sequential checklist of expected events that the experimenter marks off with appropriate timings and notes as required. In this author's experience, the effort in preparing such a checklist can greatly assist in the subsequent analysis.

Computers can be used to record events. A simple approach is to code function keys on a keyboard to correspond with pre-defined events expected in the scenario. These key presses can then be identified on subsequent event logs as part of the timeline of the scenario. Ideally these should be synchronised and integrated with process events recorded by the simulator. However, the number of keys should be limited, otherwise the event recording may become too distracting or cumbersome.

When necessary, observers may interrupt the participants to ask specific questions or they may request completion of questionnaires at particular points to assess workload or attitudes to aspects of the scenario. This technique is valuable when seeking to investigate complex cognitive behaviour such as fault diagnosis or to measure features such as situational awareness.

13.4.3 The use of video and audio recordings

If fixed cameras are to be used, it is important to check that they cover the expected areas where key activities will take place and that they provide sufficient information to enable the experimenter to ascertain the tasks that are being undertaken. If a hand-held camera is to be used to track participants, then an additional person will be required to operate it. Microphones must be placed strategically, or, if the participants are moving in a large space such as a simulated control room, it may be preferable to provide portable radio microphones.

It should always be remembered that the analysis of video and audio recordings can be tedious and time-consuming. Digital video and modern computer-aided techniques for video analysis should be exploited wherever possible.

Video is valuable in the debriefing stage as it can be used to remind participants of the session and can prompt them to explain reasons for their actions. However, probably its main advantage is that it allows the analyst to slow down or repeat observation of complex or fast moving events. Video and audio records also permit the precise timing of sequences of events, which may not be possible during the scenario.

13.5 The simulator session

When offered the opportunity to investigate a system using a simulator there are thus a number of issues to be dealt with in preparing to carry out a study.

In a typical simulator session, participants take part by coping with scenarios imposed on them by the experimenter. As access to simulators is often limited, because they are very expensive and operating them requires considerable maintenance and expertise, sessions tend to be short and intensive. The experimenter must, therefore, make the most of the opportunity to capture the data required within the permitted timetable. Thus, if the session is to be successful, careful planning is required.

In a simulator session there are four phases: pre-briefing; familiarisation; the scenario or scenarios; then the final debriefing.

13.5.1 The preparation and pre-briefing

This may comprise a fairly brief introduction or may involve fairly lengthy preparation. In particular, extra time may be necessary if the participants have to be introduced to a novel system or an unfamiliar interface. It is useful during the pre-brief to give participants questionnaires or tests to gain a baseline indication of their knowledge. This applies specifically to questionnaire-based workload assessments or measures of subjective stress. Baseline measures can be used subsequently in comparison with measures taken during the experimental scenarios.

It is at the pre-briefing where the experimenter should ensure that all participants are fully aware of the purpose of the study, and that they understand the types of data being recorded and the reasons for these observations. If video and audio recording is to be used, the participants should not only be informed, but their consent to recording must be specifically obtained. They must also be assured that their responses will be treated confidentially and that any reports will not attribute comments to specific individuals. They should be informed that any recording will only be used by the experimenters and that they can request their destruction if they so wish.

In any event the subjects should be informed of how the recordings are to be made and used before being asked for their consent. Besides being polite to fully inform the participants, this also assists in gaining their trust. This can be influential in reducing the risk that any data is distorted because of the concerns about who will see the results.

The pre-brief also allows participants to get to know the experimenters and to establish a rapport between them, so that they become comfortable with the whole procedure. This process is a key to obtaining valid data.

13.5.2 Familiarisation

After the pre-brief, the simulator sessions commence. It is good practice to allow the participants time to become familiar with the simulator and the observation requirements. It is usual to present some introductory exercises, perhaps involving some sample scenarios or some relatively simple routine tasks. The familiarisation period can be extended to reduce the degree of expectation on the part of participants.

13.5.3 The scenario

Implementing the scenario or sequence of scenarios is the third stage. By the time that the first event targeted for observation commences, the participants should be used to their surroundings and no longer affected by the presence of the experimenters.

13.5.4 Debrief

After the tension of the scenario session, participants should take part in a debriefing session. They may well benefit from a short break before the debriefing begins.

There should be a structure to the debrief. It may comprise a structured discussion and will typically include completion of questionnaires and rating scales. The participant should be allowed to raise questions about the exercise though the answers may be qualified by requests not to mention the events to participants yet to take part in the study. It may be valuable to go over the events to gain subjective insights into the performance and to obtain details about the reasons for any actions or events that are of interest to the experimenter.

The debrief is crucial and should not be overlooked or rushed. It can be the source of more valuable insights and data than the scenario runs. It is often perceived as a benefit by the participants themselves as they have the opportunity to discuss their performance, and can put forward their own views on the exercise and its objectives.

13.6 Pitfalls

It can be seen that simulators present a tremendous opportunity to obtain valuable information about the performance of a system. However, there are a number of disadvantages to be considered and pitfalls to be avoided when undertaking simulator exercises.

13.6.1 The simulator itself

A simulator is a highly complex system that requires considerable maintenance to keep it running and expertise in its operation. Typically, ergonomists do not have this level of expertise and investigations can be delayed or even fail due to simulator breakdowns.

Solution: ensure that a competent simulator technician is on hand during all exercises.

Simulators are extremely expensive facilities and thus they are used as much as possible to justify their costs. This means that access for experimental purposes will be very limited. In exercises in which the author was involved that used a nuclear plant training simulator, the simulator was in so much demand that it was operated in a three shift 24 hour operation: trainees on the day shift, ergonomists and their exercises on the afternoon shift and computer system developers on the night shift! The problem with such a schedule was that the system developer's work often caused bugs, which meant that access was further delayed as they worked to restore operation.

Solution: ensure that the exercise is carefully planned to make best use of the available time.

13.6.2 Participants

Simulators represent the complete system, which means that operating the system necessitates skilled and experienced participants. Often, these are a scarce resource because they are required to operate the real world plant and today's emphasis on cost reduction means that operating crews in all industries are kept to a minimum. The experimenter's ability to obtain experienced subjects to act as participants will be very restricted.

Solution: carefully assess the need for participants and minimise the requirement. Make sure that the exercise is well planned so that there are no delays and any access to experienced staff is exploited to the full.

13.6.3 The data

Even a short simulator run can generate a considerable amount of raw data. This could be in the form of event logs, parameter plots, written notes, video or audio recordings. These can present insuperable problems for the experimenter attempting to extract the seminal data and then subjecting it to analysis. Often video tapes, although they provide rich data, are simply not analysed at all, but kept in case the findings derived from more directly amenable sources such as the experimenter's on-line notes need to be checked. Parameter logs and trends generated by the simulator itself are notoriously difficult to analyse, often just consisting of lists of numbers and alphanumeric codes.

In other words, although simulators are well equipped for their training function, they often do not have the data recording facilities to meet the needs of the experimenter. It can be cumbersome to set up the required scenarios and can be difficult to transform recorded data into a usable form.

Solution: identify the requirements for data beforehand. Make sure that the simulator generates the required information. Use computer-aided techniques wherever possible, in particular for identifying key events and for analysing video material. Use pre-prepared proformas for recording data on-line.

13.6.4 The participants' performance

Simulator studies, because they use well-motivated participants who will strive to do well and may be very competitive, can generate high levels of performance even with poor or sub-optimal interfaces. In addition, there is the expectancy in the simulator environment that a fault or emergency will inevitably occur, so that participants are highly aroused and prepared for action. This can again lead to falsely elevated results.

Solution: this is a difficult problem, particularly for comparative interface studies. If possible use long experimental sessions with extended familiarisation periods to diminish the expectation effect.

13.7 Conclusions

It is always a challenge to undertake a study based on a full-scope system simulator. Often the motivation for the study will be to address system performance under extreme conditions. For the experimenter, it is crucial to plan any simulator exercises carefully. Three general aspects are the key to a successful exercise:

- the fidelity of the simulator environment;
- the correct choice of scenarios;
- careful preparation for data collection.

Simulators are continuing to develop but, in the view of this author, there is still not sufficient attention given to data collection.

The trend to develop process control so that it is supervised, managed and controlled through computer displays has allowed the potential for real processes to be switched off-line to operate in *simulator mode*. This is seen as a valuable aid for live training, but also could be exploited in experimentation.

Nevertheless, there will often be restricted access to simulators and participants, which throws the onus back onto the experimenter to plan for efficient investigation methods. The message for the experimenter must always be: make the most of any opportunity to use a simulator, plan it carefully, select your data to match the goals of the exercise and treat your participants professionally. It is the relationship between the experimenter and participants that is at the heart of good and valid experimental results.

13.8 Exercises

For each of the following three examples, which are taken from actual system proposals, consider how you would use simulation to evaluate the proposed system.

(*a*) Would the provision of a stereoscopic (three dimensional) image assist surgeons carrying out keyhole surgery? Currently, they use micro-CCTV cameras to provide a single flat, two-dimensional screen image of the operational area.
(*b*) A novel, eco-friendly car proposes the application of computerised drive by wire controls using a single grip: twist the grip to accelerate, squeeze it to brake, with steering achieved by moving the grip up or down. Is such a control scheme feasible or desirable?
(*c*) Monitoring a nuclear power plant is effected mostly by visual information on screens, dials and indicators. Using simulated voice alarms has been proposed in order to reduce visual workload during emergencies. Is this likely to bring about a significant improvement in the handling of emergencies by plant operators?

You should include consideration of the following issues.

1. What are the key features of human performance involved?
2. What data would be required to assess the human performance features?
3. What degree of fidelity would be required of the simulator/simulation?
4. Give a specimen scenario to be used in the simulation exercise.
5. What level of expertise, skill or experience would you require of participants in any evaluation exercise?
6. How many participants would be required?
7. How would you record required data?
8. How would you analyse the data?

Further Reading

There is no single source for finding out more about simulator exercises and many simulator evaluation exercises, when they are reported, are not written up in any great detail. Some references that consider the use of simulators in a range of applications are listed below. For a general introduction and overview of evaluation and simulator application see:

WILSON, J. R., and CORLETT, E. N. (Eds): 'Evaluation of human work' (Taylor & Francis, London, 1990)

WISE, J. A., HOPKIN, V. D., and STAGER, P. (Eds): 'Verification and validation of complex systems: human factors issues' (NATO ASI Series, Springer-Verlag, Berlin, 1993)

For a description of ways to use simulators in task analysis see:

KIRWAN, B., and AINSWORTH, L. K.: 'A guide to task analysis' (Taylor & Francis, London, 1992)

For a description of ways to use simulators in human reliability assessment see:
KIRWAN, B.: 'A guide to human reliability assessment' (Taylor & Francis, London, 1994)

For a good overview of the issue of fidelity and its value in simulators see:
ROLFE, J. M., and CARO, P. W.: 'Determining the training effectiveness of flight simulators: some basic issues and practical developments', *Applied Ergonomics*, 1982, **13** (4), pp. 243–250

Chapter 14

Safety assessment and human factors

Carl Sandom

14.1 Introduction

This chapter examines the safety assessment of systems that include people. Specifically, the chapter is concerned with the analysis of safety in complex, information systems that characteristically support dynamic processes involving large numbers of hardware, software and human elements interacting in many different ways. The chapter assumes little or no knowledge of either safety assessment or human factors assessment techniques.

Information systems often require complex functionality to assist human operators with intricate tasks such as, for example, the conflict detection and resolution systems that assist air traffic controllers with critical decision-making tasks in modern air traffic management. As well as the complexity of modern technology and organisations, humans are also themselves inherently complex and the human factors relating to the physical and cognitive capabilities and limitations of system operators must also be addressed during the assessment of any complex, interactive system.

Information systems involving extensive human interactions are increasingly being integrated into complicated social and organisational environments where their correct design and operation are essential in order to preserve the safety of the general public and the operators. This chapter focuses on the safety assessment of information systems, typically operating in real time, within safety-related application domains where human error is often cited as a major contributing factor, or even the direct cause, of accidents or incidents.

14.2 Information systems and safety

The safety assessment of modern information systems is a growing concern; for example, command and control (C2) systems are now being developed with the potential for increasingly catastrophic consequences from a single accident. C2 systems that control everyday activities from power generation to air traffic management have the potential to contribute to – if not to cause – deaths on a large scale. Moreover, the contribution of modern Decision Support Systems (DSS), such as in modern battlefield digitisation applications, to hazardous events is often underestimated or worse still ignored.

Information systems typically provide organisational and environmental data to inform decision-making. Depending on its use, this data can lead to erroneous operator actions affecting safety, as was the case when the crew of the USS *Vincennes* incorrectly interpreted the data presented by their C2 system and a decision was taken to shoot down a commercial airliner killing 290 passengers [1]. Despite extreme examples like this, information systems are often not considered to be safety-related.

The view of information systems as not safety-related is often shaped by preliminary hazard analyses concluding that the system in question has no safety integrity requirements when applying guidance from risk-based safety standards such as UK MoD Defence Standard 00-56 [2] or IEC61508 [3]. This perception is often reinforced by anecdotal evidence of information systems that have previously tolerated relatively high *technical* failure rates without incident.

Another prevalent view is often that information systems are 'only advisory' and cannot directly cause an accident – the implication is that the human(s) in the system provide sufficient mitigation between the manifestation of a 'system' hazard and the credible accidents. While this perception may be correct, the assertion is often made without a rigorous analysis of the individual and organisational human factors involved and substantiated arguments being made for the validity of any mitigation claims. Also, this perspective overlooks the fact that credible but erroneous data can be presented by such systems and therefore no mitigation would be provided by operators if the data looks correct.

An approach to the safety assessment of information systems must be taken to identify risks within the system at the different levels, as complex interactions between different system levels can result in hazards caused by a combination of technical and human failures. Also, an analysis of the human factors is essential to understand the significance of the human mitigation that is typically significant within information systems. Without a holistic approach such as this, information systems tend not to be thought of as safety-related.

Before addressing specific issues relating to the assessment of human systems; it is necessary to understand some of the key concepts associated with the general process of system safety assessment. The following section will examine the concepts of safety, risk and hazards before looking at some of the issues relating to the specification of safety requirements for systems involving humans.

14.3 Safety assessment

The term 'safety' has many different connotations and it can be related to many different concepts such as occupational health and safety, road safety or even flight safety. It is important to make the distinction between these concepts and *functional safety* in order to appreciate what it is that safety-related system designers are trying to achieve. Storey [4] maintains that functional safety is often confused with system reliability; however, even the most reliable system may not necessarily be safe to operate. For example, the cause of the Airbus A320 Strasbourg accident was attributed to the fact that the pilot inadvertently selected a descent rate that was too fast – in this example, the aircraft behaved reliably but it crashed into a mountain with fatal consequences. System reliability is necessary but is not sufficient alone to ensure the functional safety of a system.

Functional safety is a difficult concept to define. The current drive towards enhancing system safety in the UK has its origins in the Health and Safety at Work Act 1974 [5] although this act is often incorrectly associated only with occupational safety. There are many different definitions of safety. For example, the Ministry of Defence (MoD) define safety as: 'The expectation that a system does not, under defined conditions, lead to a state in which human life is endangered' ([3], p. A-3). Alternatively, the British Standards Institution definition of safety is 'The freedom from unacceptable risks of personal harm' ([6], p. 4).

Although these definitions of safety and risk may be intuitively appealing, Ayton and Hardman [7] argue that a major theme emerging from the literature on risk perception is the emphasis on the inherent subjectivity of the concept. The subjectivity associated with risk can be illustrated by the way that an aircraft accident attracts much more media attention than the far greater number of road traffic accidents that occur in the UK each year.

It can be argued that safety should be defined in terms of acceptable loss or tolerability. The UK Health and Safety Executive [5] require risk to be quantified and it can be considered tolerable only if it has been reduced to the lowest practicable level commensurate with the cost of further reduction. This important concept is known as the ALARP (As Low As Reasonably Practicable) principle and it is illustrated in Figure 14.1.

Despite the difficulties in defining safety and risk, a common theme that links many definitions is that risk is a product of the *probability* of an accident occurring and the *severity* of the potential consequences [4, 7, 8, 9]. From the previous discussion, it should be clear that safety and risk are inextricably linked; indeed it may be argued that the task of producing a safety-related system can be seen as a process of risk management.

Lowrance's definition of safety [9], in terms of risk, has captured this sentiment succinctly and it will be adopted throughout the remainder of this chapter:

> We will define safety as a judgement of the acceptability of risk, and risk, in turn, as a measure of probability and severity of harm to human health. A thing is safe if its attendant risks are judged to be acceptable.
>
> ([9], p. 2)

Information systems developers must understand the issues and develop the skills needed to anticipate and prevent accidents before they occur. Functional safety must be a key component of the system development process and it must be designed into

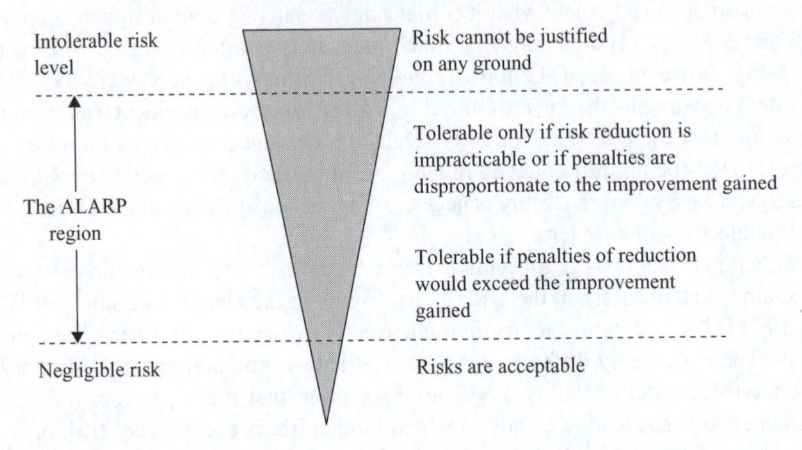

Figure 14.1 The ALARP principle for risk (adapted from [5])

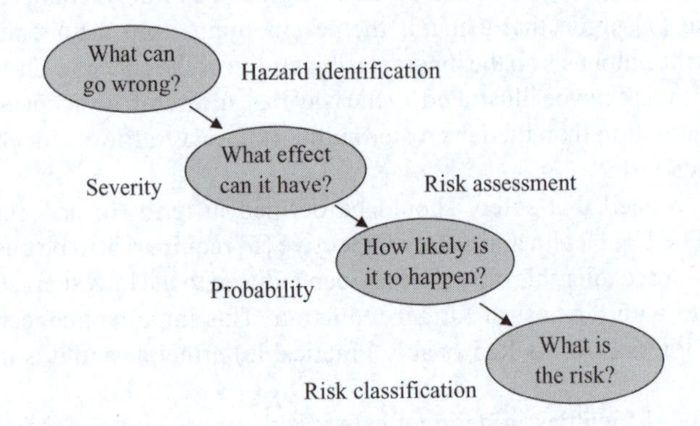

Figure 14.2 Safety assessment process

a system from the onset. System developers need different techniques for quantifying system risks and this must be preceded by the identification and analysis of system hazards. These processes are known collectively as system safety assessment.

Safety-related system designers must undertake safety assessments that integrate the concept of hazards with that of risk. To be comprehensive, a system safety assessment must address: hazard identification (what could go wrong?); hazard severity (how serious could it be?) and hazard probability (what are the chances of it happening?). This process will enable an assessment to be made of the system risk. This basic safety assessment process is depicted in Figure 14.2.

Safety-related systems designers must in some way identify the manner in which a system can cause harm in order to improve the safety of a system by preventing accidents before they occur. In simple terms, systems hazards lead to accidents;

therefore it is important to examine the fundamental question of what constitutes a hazard in a safety-related system.

Hazards have been defined in a number of ways. Defence Standard 00-56 ([2], p. A-2) defines a hazard as a: '*Physical* situation, often following some initiating event, that can lead to an accident'. This definition is unhelpful when considering where the hazards lie within a given system; however, it does imply the important point that a hazard will not always result in an accident. Leveson's definition of a hazard is useful in a systems context:

> A hazard is a state or set of conditions of a system that, together with other conditions of the environment of the system, will lead inevitably to an accident.
>
> ([8], p. 177)

Leveson's definition of a hazard is expressed in terms of both the environment and the boundary of the system. This is an important distinction from other definitions as system hazards can only be fully identified and analysed if a system is considered in the context of its operational environment. Leveson's definition as stated, however, implies that a hazard will inevitably lead to an accident. An alternative view of this hazard cause–effect relationship contends that a hazardous situation will not always lead to an accident and that a properly designed system can be returned to a safe state. From this discussion, a definition of a hazard is proposed which is useful in a *system* context:

> A hazard is a state or set of conditions of a system that, together with other conditions of the environment of the system, may lead to an accident.

Having examined some of the key concepts associated with safety assessment, it is useful to focus on the main reason for assessing any system which, at the start of any project, is to derive requirements and then later to ensure that these requirements have been fulfilled. As this chapter deals with system safety, the issue of safety requirements will now be examined.

14.4 Safety requirements

The terms 'safety requirements' and 'Safety Integrity Levels (SILs)' are often both used synonymously by systems designers. However, although the concept of SILs appears in almost all modern safety standards, there is no consensus about what they actually mean or how they should be used [10]. Nonetheless, the link between development processes and SIL is often used to determine what development activities implementation contractors are required to undertake.

Generally, the higher the SIL the more rigorous the development processes are required to be. However, a problem can arise if the system is deemed to be SIL 0 as this generally implies that the system does not affect safety and therefore no safety requirements are imposed on how it is developed. Information systems in particular are often deemed low or no SIL systems. This can be illustrated with a simple example. For a typical information system, let us assume that the following safety integrity

Table 14.1 SIL claim limits (adapted from [2])

SIL	Failure rate	Quantitative description (occurrences/operational hour)
4	Remote	1.0E-05 to 1.0E-06
3	Occasional	1.0E-04 to 1.0E-05
2	Probable	1.0E-03 to 1.0E-04
1	Frequent	1.0E-02 to 1.0E-03

requirement is specified, based upon an existing system's perceived acceptable failure rate:

The system will have a maximum failure rate of 100 failures/year

Given the typically high tolerable failure rates for information systems, this may not be unusual or unreasonable. If there is only one instance of the system, and it runs continuously, this equates to 8,760 operational hours per year. This gives:

$$\textit{Target system failure rate} = 100/8,760 = 0.01 \text{ failures/op hr}$$

To derive a SIL, reference needs to be made to the appropriate safety standard. If, for example, the SIL claims table of DS 00-56 [2], shown in Table 14.1, is used to determine the resultant SIL, then the system developer may conclude that the requirement is for a non-safety-related system as the target failure rate in this example is less than the SIL1 claim limit.

This simple example illustrates a common approach to determining if a system is safety-related or not based upon the view that safety integrity is *the only* safety requirement for a system. Integrity is not the whole picture, as previously discussed with reference to system reliability and the A320 crash. An alternative view is that it is the specified *functionality* and *performance* of an information system that determines both causes and mitigation of risks within the external environment. Leveson [11] presents compelling evidence, based on her review of major software-related accidents, that software (hence system) reliability has never been the cause of such disasters. Fowler and Tiemeyer [12] offer an excellent and detailed discussion on this perspective and this is extended to address human factors in Sandom and Fowler [13].

Traditional approaches concentrate on failures alone and the impact of the failure to provide the system functionality. However, it is equally important to analyse the protective aspects of the system in operation to identify what system functions mitigate external hazards rather than contribute to them. If the specified information system functionality were insufficient to achieve the necessary risk reduction then, no matter how reliable an information system was, it would not be safe.

Commonly used safety standards, such as DS00-56 [2], IEC61508 [3] and DO178B [14] adopt this failure-based view of safety without emphasising the importance of specifying safe functionality in the first place. Compliance with these standards, whilst giving some assurance of information system *integrity*, does little to ensure the correct functionality and performance of the information system and therefore that they are really safe. This has potentially serious implications for commonly used risk-classification schemes that appear to provide an easy way of converting the assessed severity of a hazard into a tolerable frequency of occurrence and thus into a safety integrity requirement. Setting realistic safety requirements must take into account system functionality, performance *and* integrity.

This has implications when assessing systems and, in particular, when assessing the human contribution to system safety. Any method of assessing the safety of human activities must address human functionality, performance *and* integrity. Before we return to this theme, it is essential to provide a common framework of systems terminology that can be used for the remainder of the chapter.

14.5 Systems and boundaries

It can be argued that if any component of the system directly or indirectly affects safety, then the system should be considered safety-related. However, before pursuing this line of argument, careful consideration must be given to what constitutes a system and where the system boundaries for information systems can typically be drawn.

Figure 14.3 shows a representation of a typical information system and its boundaries. In this representation the core system represents the various subsystems implemented in hardware, software or allocated to human operators. The service level

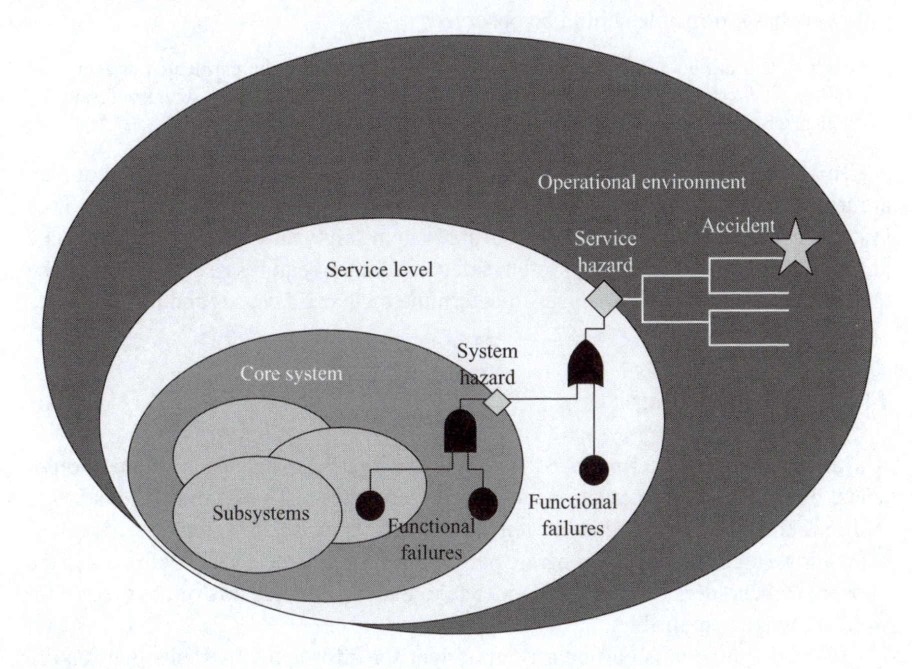

Figure 14.3 A system model

represents the service provided by the information system users or operators, encapsulating the operational procedures and other organisational factors. The service level is usually the responsibility of the operational authority. Finally, the operational environment represents the wider application domain within which the system resides and typically contains people, procedures and equipment issues. For example, an air traffic management system would be a core system used by air traffic control operators providing an ATC service to users in the operational environment that would provide aircraft control.

As discussed in detail in Chapter 1, human factors is a broad discipline concerned with the need to match technology with humans operating within a particular environment; this requires appropriate job and task design, suitable physical environments and workspaces and human–machine interfaces based upon ergonomic principles. Human factors analyses must examine the broad ergonomic, organisational and social aspects of an operational system in use within its operational environment. At the other end of the spectrum, human factors analyses must also examine how human–computer interfaces can foster the safe, efficient and quick transmission of information between the human and machine, in a form suitable for the task demands and human physical and cognitive capabilities.

Analyses of human factors issues in safety-related systems consistently reveal a complex set of problems relating to the people, procedures and equipment (or technology) interacting at each system level of a specific environment. These attributes are normally very tightly coupled and each of these attributes can interact with the other. To undertake a systematic analysis of all aspects of systems safety, the following basic principle should be observed:

> Each system attribute (people, procedures and equipment) must be considered at every system level (core, service, operational) to a depth of analysis commensurate with the human integrity requirements.

To achieve this goal, human integrity requirements must be determined and an argument must then be constructed using this framework to ensure that the whole systems context of use is considered during system safety analyses using appropriate analysis techniques. Generally, system safety assurance requires risk modelling to be undertaken using focused analyses to determine the hazard causes and mitigations.

14.6 Risk modelling

A systematic approach is required to define the interfaces between system components, boundaries and interactions to clearly define the risks and mitigations posed within each system level. This 'system of systems' approach to safety case development allows the risks of each separate part of the system to be clearly defined and the relevant stakeholders can provide evidence to underwrite the parts of the system for which they are responsible.

Such an approach is particularly important for information systems that require both technical and operational analyses to define the hazards and their mitigation.

This suggests how important it is to define the boundaries and targets for each of the system levels and use an integrated approach to support the practical development of a realistic safety case.

Figure 14.3 suggests that the scope of human factors analyses must address the whole system, service and operational environment. This vast scope presents a challenge for the systems engineer who needs to consider the safety-related aspects of the system and even then to focus the often limited resources available on the most critical system functions. The processes for determining human or technical safety requirements will necessarily be based upon different analysis techniques when dealing with human rather than technical subsystems.

The safety assurance of information systems must identify risks within a system at the different levels, as complex interactions between different system levels can result in hazards caused by a combination of technical and human failures. Accidents occur in the external environment as depicted in Figure 14.3. However, although a system developer may only be responsible for the safety of the core-system implementation, the effects of core-system hazards in the operational environment must be clearly understood. To undertake the overall process of modelling, risks and mitigation must be understood.

An accident is an unintended event that results in death or serious injury. With reference to Figure 14.3, accidents occur in the operational domain. As discussed, a hazard is a system state that may lead to an accident – whether it does or not, depends on the availability of mitigations to break the sequence of events that would otherwise lead to an accident (this is shown in Figure 14.4). Such mitigations are called consequential (since they relate to the consequences of a hazard, and can be

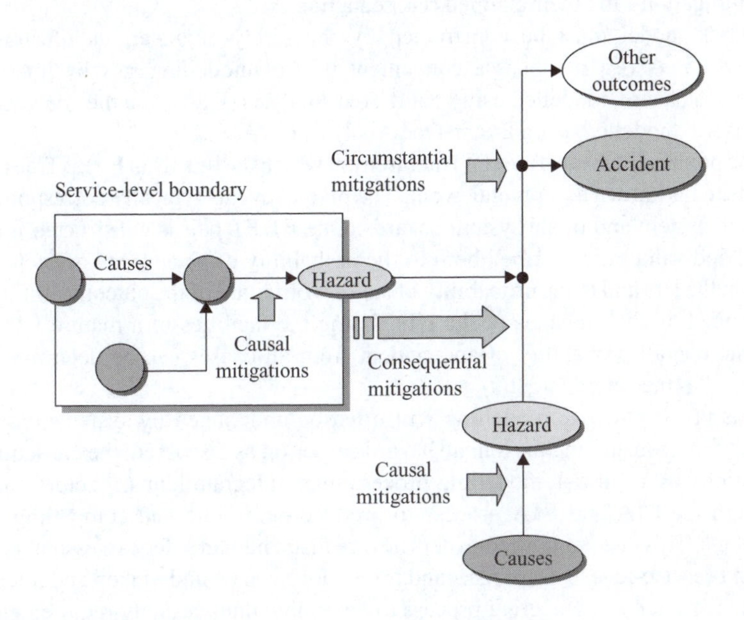

Figure 14.4 Accident sequence

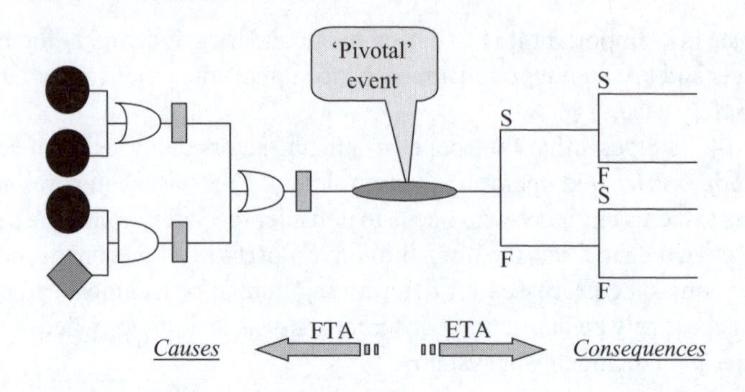

Figure 14.5 Risk model

either deliberately provided or circumstantial (i.e. purely a matter of chance). The likelihood of an accident is also dependent on the likelihood that the hazard would occur in the first place. This in turn is dependent on the frequency of occurrence of the underlying cause(s) of the hazard and on the availability of (causal) mitigations to break the sequence of events between the causes and the hazard itself.

A safety-related system may provide either causal or consequential mitigations. For a typical information system, human operators provide many of these mitigations. System operators often provide sufficient mitigation between the manifestation of a 'system' hazard and the credible accidents. However, as discussed, this is often without a systematic or rigorous analysis of the human factors involved to validate these mitigations and their claimed risk reduction.

A risk model must be constructed by the safety engineer, as illustrated in Figure 14.5. A risk model is a convenient way of modelling risk by linking the *causes* of a hazard, modelled using Fault Tree Analysis (FTA), and the *consequences* of a hazard, modelled using Event Tree Analysis (ETA).

The point in the Fault Tree (FT) hierarchy at which the link to an Event Tree (ET) is established is known as a pivotal event. The pivotal events typically correspond with the main system and or subsystem hazards. One FT/ET pair is constructed for each hazard and values are ascribed both to the probability of occurrence of each casual factor in the FTs and to the probability of success or failure of the outcome mitigations represented by the branches of the ETs. Using the facilities of a mature FTA/ETA tool, the overall probability of an accident from all causes can be determined and compared to the safety target(s).

This process of risk modelling will often be undertaken by systems engineers without fully considering the human hazard causes or, as discussed, the consequential mitigations. A valid risk modelling process must integrate human factors analyses into both the FTA and ETA process to produce defensible and compelling safety arguments. This chapter now considers how realistic human safety assessments of the core and service-level system risks and mitigations can be undertaken and integrated within the systems engineering process to determine human safety requirements for a typical information system.

14.7 Assessing the human contribution to safety

A pragmatic method of assessing the human contribution within the overall safety assessment process is required to focus human factors analysis on the safety-related aspects of the system using suitable human factors techniques. One approach for assessing the human contribution to system safety is outlined here integrating the use of appropriate human factors analysis techniques within the systems engineering lifecycle for the systematic determination and realisation of human factors safety requirements. As discussed previously, the key to safety assurance is to ensure that each causal factor (people, procedures, equipment) must be considered within and between each system level (Figure 14.1) to a depth of analysis commensurate with the integrity required of the human subsystem.

Although the analysis begins with the core system/subsystem level, the approach takes into account the human risks and mitigations at the service and operational levels. Although these techniques are described below in relation to the human subsystems, the method also provides for the analysis of human risks and mitigations at the service and operational levels.

Specific safety-related human factors activities comprise Critical Task Analysis (CTA) and Human Error Analysis (HEA) as depicted in Figure 14.6.

These safety-specific activities should be planned to ensure that there is no overlap with wider, system-level human factors activities while taking maximum advantage of system hazard and risk assessment analyses for the other subsystems. The CTA and HEA activities are tightly coupled and are based upon, and integrated with, the FTA and ETA risk modelling analyses described previously and depicted in Figure 14.5.

Two iterations of each CTA and HEA should be undertaken during the typical systems development lifecycle and, as the analyses become more focused, the results will inform each other as shown in Figure 14.6. These activities are complementary

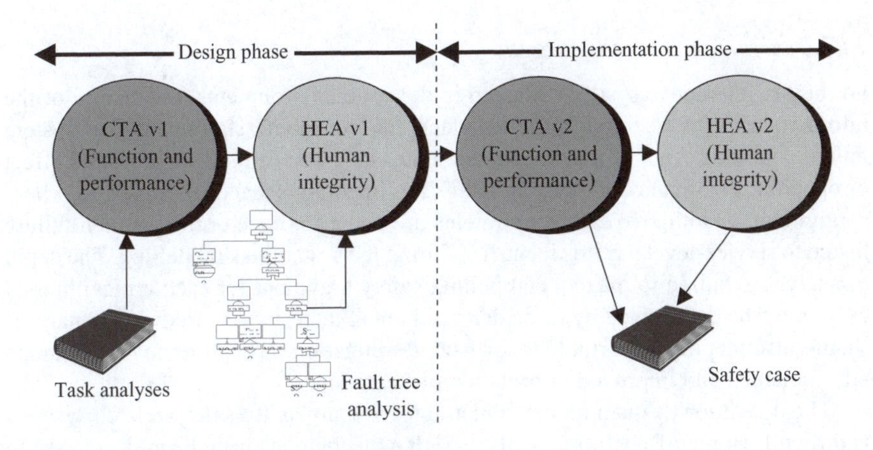

Figure 14.6 Human safety assessment

as CTA and HEA are bottom-up and top-down analysis techniques respectively (from a hazard to human event perspective). This combination of top-down and bottom-up analyses significantly increases the probability of identifying inconsistencies in the individual techniques and thus enhances safety assurance.

As discussed previously, the specification of system safety requirements must address functionality and the associated performance and integrity levels. Integrating human factors analyses into the general safety assessment process facilitates the specification of safety-related human functionality and its associated performance and integrity levels and this process is now described in some detail.

14.7.1 Human safety functions

A CTA can be undertaken to identify and analyse the potential for human performance errors in critical operational tasks. CTA concentrates on the human factors aspects of the Human Machine Interface (HMI). This analysis is a bottom-up technique used broadly to analyse the relationships between system hazards; operational tasks (identified by task analysis; see Chapter 5 for a detailed discussion) and the HMI design. The analysis works in a bottom-up fashion from operational tasks, related to basic error events, to identified service-level hazards.

A CTA can initially focus on the identification and analysis of the relationships between system hazards and safety-related operational tasks. This analysis will enable any hazard and task analyses to be checked for consistency, providing confidence in subsequent safety assurance claims. Any deficiencies – such as hazards with no related operational tasks or operational tasks (deemed as safety-related by subject matter experts) with no relationship to identified hazards – can be highlighted. The CTA will also identify opportunities for hazard mitigation through removal of human error potential and improved information presentation by comparing the task analysis with HMI design guidelines from appropriate sectors (for example Federal Aviation Authority ATM HMI design guidelines [15]). Undertaking a CTA can therefore allow the system developers to identify the human safety functions.

14.7.2 Human integrity targets

For highly interactive systems situated in dynamic environments, the quality of the information acquired through the interface can contribute significantly to system failure, and the design of the human–computer interface can have a profound effect on operator performance and system safety. It is imperative that qualitative and, where appropriate, quantitative safety arguments are made for each critical human failure linked to service-level hazards identified during the system risk modelling. The depth of analysis required to make a compelling safety argument for each critical human event must be determined by these derived human integrity requirements. Analyses should also identify opportunities for hazard mitigation through removal of human error potential and improved information presentation.

The derivation of quantitative human integrity targets for safety-related systems is difficult. Human Reliability Analysis (HRA) techniques have been developed to address this issue (see Chapter 8 for a detailed discussion). However, HRA techniques have mainly been applied successfully within process control environments, such as the nuclear industry for example, where the operating and environmental conditions are relatively more easily quantifiable. HRA techniques are much more difficult, and expensive, to apply meaningfully to human systems with highly dynamic environmental contexts.

A pragmatic method of addressing this issue is to undertake a Human Error Analysis (HEA) focused specifically on the basic human events identified in the system fault trees. HEA analysis is a top-down technique used broadly to model the relationship between service-level hazards and critical human failures, and the mitigating aspects of the system design. For systems which typically have a high degree of operator interaction, many of the FTA basic events will be identified as human interactions. An example fragment of an FTA is shown in Figure 14.7 with the basic human event OP NOT DET.

Once each fault tree is modelled, predictive, quantitative failure data can be input at the bottom from availability and reliability data for all hardware and software based events. By subtracting these values from the associated hazard target, quantitative Human Integrity Targets (HITs) can then be calculated for each critical human event. It should be understood that these basic human events originate from both the system and service levels taking the operational context into account.

The HEA would then focus on developing specific safety arguments for each basic human event to provide evidence that the HITs can be achieved. For critical areas, where the HEA reveals that the HITs are unrealistic, mitigations can be re-assessed and recommendations developed for further action. In this way, no predictions are being made about the human error rates; rather, the HITs are derived from the remaining integrity requirements once the hardware and software failure data is input and a qualitative analysis is undertaken to ascertain if the remaining human integrity requirements are realistic.

14.8 Summary

The technical, social and human complexity involved in the development of modern interactive systems presents a number of problems that are exacerbated when the failure of an interactive system has potentially lethal consequences. Safety-related systems are used in complex social contexts and the integrity of their design and operation is essential in order to ensure the safety of the public and the environment.

The prevalent view is often that information systems are not safety-related even when used within safety-related environments; this perception is often reinforced by anecdotal evidence of information systems that have previously tolerated relatively large *technical* failure rates without incident. The view is often that information systems are 'only advisory' and cannot directly result in an accident – the implication is that the human(s) in the system provide sufficient mitigation between the manifestation of a 'system' hazard and credible accidents. While this perception may be correct, the assertion is often made without a rigorous analysis of the human factors involved and arguments being made for the validity of the mitigation claim.

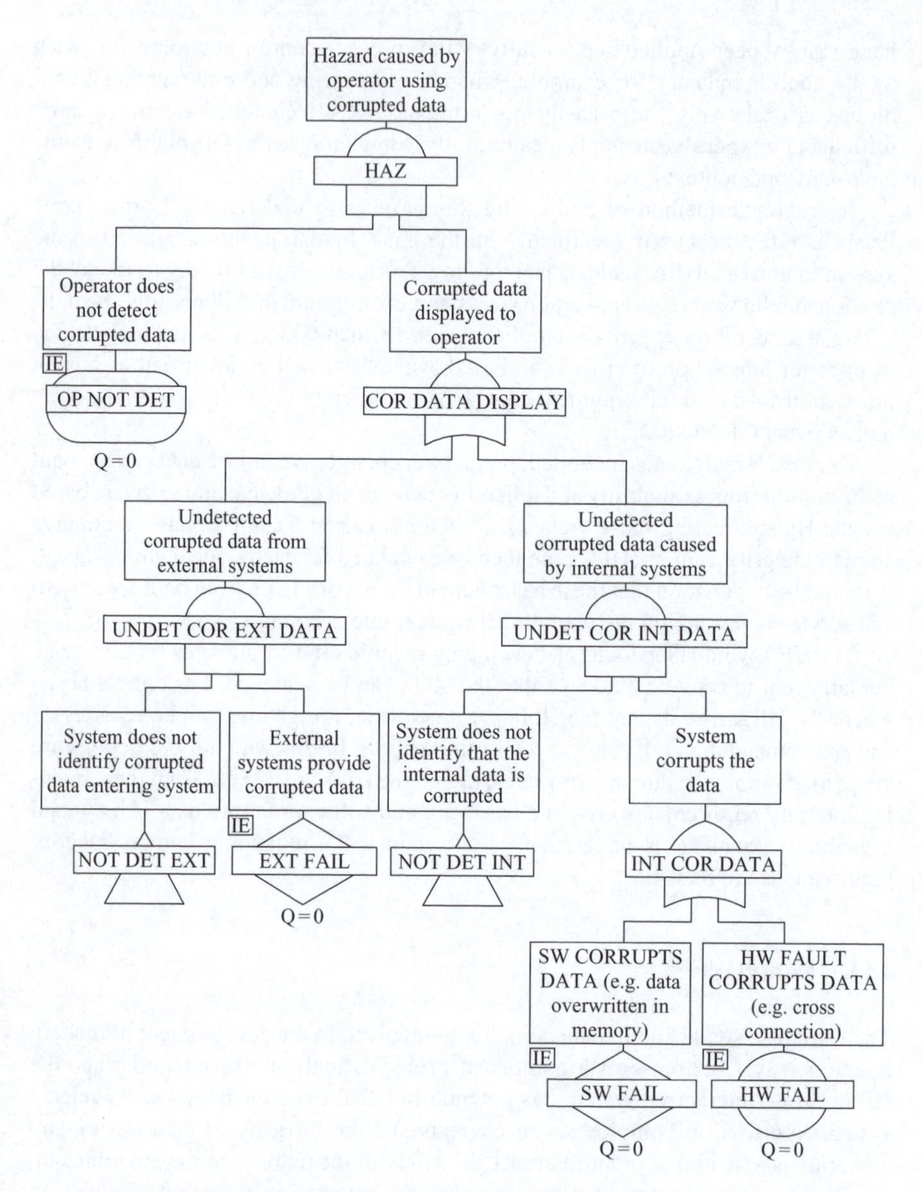

Figure 14.7 Fault tree example

If human factors risks are not considered, a system will not achieve the required level of integrity. If human factors mitigations are not considered, the technical system components may be over-engineered at additional cost to achieve a target level of safety. This chapter has shown how the use of appropriate human factors techniques and methods can be integrated with typical systems engineering techniques for the assessment of human safety requirements in safety-related information systems.

14.9 Further questions

1. Explain how an 'advisory' system might be safety-related.
2. Define the term 'safety'.
3. Explain the ALARP principle.
4. Describe a basic safety assessment process.
5. With reference to Figure 14.3, draw a simple air traffic control system showing its boundaries and where the operators and users are located.
6. Draw a typical accident sequence and describe the different mitigations, particularly those provided by people.
7. Explain how system risk can be modelled and describe how human risks and mitigations relate to this process.
8. Describe a human safety assessment process.
9. Explain the inputs, outputs and process of a critical task analysis.
10. Explain the inputs, outputs and process of a human error analysis.

14.10 References

1 GREATOREX, G. L., and BUCK, B. C.: 'Human factors and systems design', *GEC Review*, 1995, **10** (3), pp. 176–185

2 DS 00-56: Safety management requirements for defence systems, Part 1: requirements, UK MOD Defence Standard, December 1996

3 International Electrotechnical Commission, IEC 61508, Functional Safety of Electrical/Electronic/Programmable Electronic Safety Related Systems, 65A/254/FDIS, IEC, 1999

4 STOREY, N.: 'Safety-critical computer systems' (Addison-Wesley, London, 1996)

5 HSE: The Health and Safety at Work Act 1974, HMSO

6 BS4778: British Standards Institution 4778, 1995

7 AYTON, P., and HARDMAN, D. K.: 'Understanding and communicating risk: a psychological view', in REDMILL, F., and ANDERSON, T. (Eds): 'Safety-critical systems: the convergence of high tech and human factors'. Proc 4th Seventh Safety-Critical Systems Symposium (Leeds, 1996)

8 LEVESON, N. G.: 'Safeware: system safety and computers' (Addison-Wesley, London, 1995)

9 LOWRANCE, W. W.: 'Of acceptable risk: science and the determination of safety' (William Kaufman Inc., Los Altos, CA, 1976)

10 HAMILTON, V., and REES, C.: 'Safety integrity levels: an industrial viewpoint', in REDMILL, F., and ANDERSON, T., (Eds): 'Towards system safety', Proceeding of the Seventh Safety-Critical Systems Symposium (Springer, London, 1999)

11 LEVESON, N. G.: 'The role of software in recent aerospace accidents'. Proceedings of the 19th International System Safety Conference, Huntsville, Alabama, USA, September 2001

12 FOWLER, D., TIEMEYER, B., and EATON, A.: 'Safety assurance of air traffic management and similarly complex systems'. Proceedings of the 19th International System Safety Conference, Huntsville, USA, 2001
13 SANDOM, C., and FOWLER, D.: 'Hitting the target: realising safety in human subsystems'. Proceedings of the 21st International Systems Safety Conference, Ottawa, Canada, 2003
14 DO-178B/ED-12B: 'Software considerations in airborne systems and equipment certification', 1992
15 FAA: 'Human factors design guide update (DOT/FAA/CT-96/01): a revision to Chapter 8 – computer human interface guidelines', National Technical Information Service, Springfield, VA, USA, April 2001

Further Reading

The author is not aware of any texts addressing the detailed integration of human factors techniques into the systems development lifecycle, specifically to address safety and risk management. There are, however, some classic texts on both safety and human factors issues that address the broad principles and a selection of these is given here.

LEVESON, N. G.: 'Safeware: system safety and computers' (Addison-Wesley, Boston, MA, 1995)

A classic text on system and software safety that addresses some human factors issues.

PERROW, C.: 'Normal accidents' (Harvard University Press, Cambridge, MA, 1984)

An easy and provocative read with an excellent discussion of why complex systems involving humans will inevitably lead to accidents.

REASON, J.: 'Human error' (Cambridge University Press, Cambridge, 1990)

The classic text on human error, although the book does not address how to apply the theory to systems engineering.

REASON, J.: 'Managing the risks of organisational accidents' (Ashgate, London, 1997)

Seminal coverage of organisational issues relating to safety and the management of organisational safety risk.

STOREY, N.: 'Safety-critical computer systems' (Addison-Wesley, Boston, MA, 1996)

Another classic text on safety engineering, although human factors are not addressed as an integral part of the systems engineering process.

Index